Vertebrate Zoology

Vertebrate Zoology

Editor:

Dr. Nisha Dahiya

ArclerPress

www.arclerpress.com

Vertebrate Zoology

Editor: Dr. Nisha Dahiya

© 2017 Arcler Press LLC
708 3rd Avenue, 6th Floor
New York
NY 10017
United States of America
www.arclerpress.com

ISBN: 978-1-68094-515-7

Library of Congress Control Number: 2016951372

About the Editor

Dr. Nisha Dahiya

Dr. Nisha Dahiya is serving as a Research Scientist at the Indian Council of Medical Research (ICMR), New Delhi, India. She completed her PhD in 2014 from University of Camerino, Italy, under the Malaria and Human Development Program, which was supported by World Health Organization, Global Malaria Program. Her main interests are drug discovery for infectious diseases and surveillance and molecular epidemiology of rising drug resistant pathogens. Currently, she is assisting in the coordination of national network for surveillance of anti microbial resistance through out the country.

Contents

List of Contributors

Adam Woolfe
Medical Research Council Rosalind Franklin Centre for Genomics Research, Hinxton, Cambridge, United Kingdom

Martin Goodson
Medical Research Council Rosalind Franklin Centre for Genomics Research, Hinxton, Cambridge, United Kingdom

Debbie K. Goode
Medical Research Council Rosalind Franklin Centre for Genomics Research, Hinxton, Cambridge, United Kingdom

Phil Snell
Medical Research Council Rosalind Franklin Centre for Genomics Research, Hinxton, Cambridge, United Kingdom

Gayle K. McEwen
Medical Research Council Rosalind Franklin Centre for Genomics Research, Hinxton, Cambridge, United Kingdom

Tanya Vavouri
Medical Research Council Rosalind Franklin Centre for Genomics Research, Hinxton, Cambridge, United Kingdom

Sarah F. Smith
Medical Research Council Rosalind Franklin Centre for Genomics Research, Hinxton, Cambridge, United Kingdom

Phil North
Medical Research Council Rosalind Franklin Centre for Genomics Research, Hinxton, Cambridge, United Kingdom

Heather Callaway
Medical Research Council Rosalind Franklin Centre for Genomics Research, Hinxton, Cambridge, United Kingdom

Krys Kelly
Medical Research Council Rosalind Franklin Centre for Genomics Research, Hinxton, Cambridge, United Kingdom

Klaudia Walter
Medical Research Council Biostatistics Unit, Institute of Public Health, Addenbrookes Hospital, Cambridge, United Kingdom

Irina Abnizova
Medical Research Council Biostatistics Unit, Institute of Public Health, Addenbrookes Hospital, Cambridge, United Kingdom

Walter Gilks
Medical Research Council Biostatistics Unit, Institute of Public Health, Addenbrookes Hospital, Cambridge, United Kingdom

Yvonne J. K. Edwards
Medical Research Council Rosalind Franklin Centre for Genomics Research, Hinxton, Cambridge, United Kingdom

Julie E. Cooke
Medical Research Council Rosalind Franklin Centre for Genomics Research, Hinxton, Cambridge, United Kingdom

Greg Elgar
Medical Research Council Rosalind Franklin Centre for Genomics Research, Hinxton, Cambridge, United Kingdom

Zhan Yu
Department of Biology, McGill University, Stewart Biology Building, 1205 Docteur Penfield Ave., Montreal, QC, H3A 1B1 Canada

David Morais
Department of Biology, McGill University, Stewart Biology Building, 1205 Docteur Penfield Ave., Montreal, QC, H3A 1B1 Canada

Mahine Ivanga
Department of Biology, McGill University, Stewart Biology Building, 1205 Docteur Penfield Ave., Montreal, QC, H3A 1B1 Canada

Paul M Harrison
Department of Biology, McGill University, Stewart Biology Building, 1205 Docteur Penfield Ave., Montreal, QC, H3A 1B1 Canada

Nicholas I. Mundy
Department of Zoology, University of Cambridge, Cambridge, United Kingdom

Rahul Bale
Department of Mechanical Engineering, Northwestern University, Evanston, Illinois, United States of America

Izaak D. Neveln
Department of Biomedical Engineering, Northwestern University, Evanston, Illinois, United States of America

Amneet Pal Singh Bhalla
Department of Mechanical Engineering, Northwestern University, Evanston, Illinois, United States of America

Malcolm A. MacIver
Department of Mechanical Engineering, Northwestern University, Evanston, Illinois, United States of America
Department of Biomedical Engineering, Northwestern University, Evanston, Illinois, United States of America
Department of Neurobiology, Northwestern University, Evanston, Illinois, United States of America

Neelesh A. Patankar
Department of Mechanical Engineering, Northwestern University, Evanston, Illinois, United States of America

Walter Jetz
Department of Ecology and Evolutionary Biology, Yale University, New Haven, Connecticut, United States of America

Paul V. A. Fine
Department of Integrative Biology, University of California, Berkeley, California, United States of America

Tatsuya Hirasawa
Evolutionary Morphology Laboratory, RIKEN, 2-2-3 Minatojima-minami, Chuo-ku, Kobe, Hyogo 650-0047, Japan

Shigeru Kuratani
Evolutionary Morphology Laboratory, RIKEN, 2-2-3 Minatojima-minami, Chuo-ku, Kobe, Hyogo 650-0047, Japan

Laure Cattin
Department of Ecology and Evolution, Biophore Building, University of Lausanne, 1015 Lausanne, Switzerland

Johan Schuerch
Department of Ecology and Evolution, Biophore Building, University of Lausanne, 1015 Lausanne, Switzerland

Nicolas Salamin
Department of Ecology and Evolution, Biophore Building, University of Lausanne, 1015 Lausanne, Switzerland
Swiss Institute of Bioinformatics, Quartier Sorge, 1015 Lausanne, Switzerland

Sylvain Dubey
Department of Ecology and Evolution, Biophore Building, University of Lausanne, 1015 Lausanne, Switzerland

Yuguo Yu
Centre for Computational Systems Biology, State Key Laboratory of Medical Neurobiology, Institutes of Brain Science, Department of Physiology and Biophysics, School of Life Sciences, Fudan University, Shanghai 200433, People's Republic of China
Department of Neurobiology, Yale school of Medicine, Charité Universitätsmedizin Berlin, 333 Cedar Street, New Haven CT 06510, Berlin 10117, Germany

Jan Karbowski
Institute of Applied Mathematics and Mechanics, University of Warsaw, Warsaw 02-097, Poland.
Institute of Biocybernetics and Biomedical Engineering, Polish Academy of Sciences, Warsaw, Poland

Robert NS Sachdev
Department of Neurobiology, Yale school of Medicine, Charité Universitätsmedizin Berlin, 333 Cedar Street, New Haven CT 06510, Berlin 10117, Germany

Jianfeng Feng
Centre for Computational Systems Biology, State Key Laboratory of Medical Neurobiology, Institutes of Brain Science, Department of Physiology and Biophysics, School of Life Sciences, Fudan University, Shanghai 200433, People's Republic of China

Adriana M Montaño
Department of Biosystems Science, School of Advanced Sciences, the Graduate University for Advanced Studies (Sokendai), Shonan Village, Hayama, 240-0193, Japan
Department of Pediatrics, School of Medicine, Saint Louis University, 1100 S. Grand Blvd, Doisy Research Center, Saint Louis, MO, 63104, USA

Fumi Tsujino
Department of Biosystems Science, School of Advanced Sciences, the Graduate University for Advanced Studies (Sokendai), Shonan Village, Hayama, 240-0193, Japan

Naoyuki Takahata
Department of Biosystems Science, School of Advanced Sciences, the Graduate University for Advanced Studies (Sokendai), Shonan Village, Hayama, 240-0193, Japan

Yoko Satta
Department of Biosystems Science, School of Advanced Sciences, the Graduate University for Advanced Studies (Sokendai), Shonan Village, Hayama, 240-0193, Japan

Takayuki Onai
Kuratani Evolutionary Morphology Laboratory, RIKEN Center for Developmental Biology, 2-2-3 Minatojima-Minamimachi, Chuo-ku, Kobe 650-0047, Japan.

Toshihiro Aramaki
Pattern Formation Group, Graduate School of Frontier Biosciences, Osaka University, 1-3 Yamadaoka, Suita, Osaka 565-0871, Japan.

Hidehiko Inomata
Laboratory for Axial Pattern Dynamics, RIKEN Center for Developmental Biology, 2-2-3 Minatojima-Minamimachi, Chuo-ku, Kobe 650-0047, Japan

Tamami Hirai
Kuratani Evolutionary Morphology Laboratory, RIKEN Center for Developmental Biology, 2-2-3 Minatojima-Minamimachi, Chuo-ku, Kobe 650-0047, Japan

Shigeru Kuratani
Kuratani Evolutionary Morphology Laboratory, RIKEN Center for Developmental Biology, 2-2-3 Minatojima-Minamimachi, Chuo-ku, Kobe 650-0047, Japan

Izabela Makałowska
The Huck Institutes of the Life Sciences, the Pennsylvania State University, University Park, PA 16802, USA,

Chiao-Feng Lin
Institute of Molecular Evolutionary Genetics and Department of Biology, the Pennsylvania State University, University Park, PA 16802, USA
Institute of Bioinformatics, University of Muenster, 48149 Muenster, Germany

Krisitina Hernandez
Institute of Molecular Evolutionary Genetics and Department of Biology, the Pennsylvania State University, University Park, PA 16802, USA

Linda Z Holland
Marine Biology Research Division, Scripps Institution of Oceanography, University of California at San Diego, La Jolla, CA 92093-0202, USA.

João E Carvalho
Laboratoire de Biologie du Développement de Villefranche-sur-Mer (UMR 7009 – CNRS/UPMC), Observatoire Océanologique de Villefranche-sur-Mer, 181 Chemin du Lazaret, B.P. 28, 06230 Villefranche-sur-Mer, France

Hector Escriva
CNRS, UMR 7232, BIOM, Université Pierre et Marie Curie Paris 06, Observatoire Océanologique, 66650 Banyuls-sur-Mer, France

Vincent Laudet
Institut de Génomique Fonctionnelle de Lyon (CNRS UMR5242, UCBL, ENS, INRA 1288), Ecole Normale Supérieure de Lyon, 46 allée d'Italie, 69364 Lyon Cedex 07, France

Michael Schubert
Laboratoire de Biologie du Développement de Villefranche-sur-Mer (UMR 7009 – CNRS/UPMC), Observatoire Océanologique de Villefranche-sur-Mer, 181 Chemin du Lazaret, B.P. 28, 06230 Villefranche-sur-Mer, France

Sebastian M Shimeld
Department of Zoology, University of Oxford, The Tinbergen Building, South Parks Road, Oxford OX1 3PS, UK

Jr-Kai Yu
Institute of Cellular and Organismic Biology, Academia Sinica, Taipei 11529, Taiwan

Anthony Graham
MRC Centre for Developmental Neurobiology, King's College London

Thomas Butts
MRC Centre for Developmental Neurobiology, King's College London

Andrew Lumsden
MRC Centre for Developmental Neurobiology, King's College London

Clemens Kiecker
MRC Centre for Developmental Neurobiology, King's College London

Bei Huang
State Key Laboratory of Freshwater Ecology and Biotechnology, Institute of Hydrobiology, Chinese Academy of Sciences, Wuhan, Hubei Province, 430072, China
Graduate School of Chinese Academy of Sciences, Beijing 10049, China

Zhi T Qi
State Key Laboratory of Freshwater Ecology and Biotechnology, Institute of Hydrobiology, Chinese Academy of Sciences, Wuhan, Hubei Province, 430072, China

Zhen Xu
State Key Laboratory of Freshwater Ecology and Biotechnology, Institute of Hydrobiology, Chinese Academy of Sciences, Wuhan, Hubei Province, 430072, China

Pin Nie
State Key Laboratory of Freshwater Ecology and Biotechnology, Institute of Hydrobiology, Chinese Academy of Sciences, Wuhan, Hubei Province, 430072, China

Neil C. Chi
Department of Biochemistry and Biophysics and Programs in Developmental Biology, Genetics, and Human Genetics, University of California San Francisco, San Francisco, California, United States of America
Cardiovascular Research Institute, University of California San Francisco, San Francisco, California, United States of America
Department of Medicine, University of California San Francisco, San Francisco, California, United States of America

Robin M. Shaw
Cardiovascular Research Institute, University of California San Francisco, San Francisco, California, United States of America
Department of Medicine, University of California San Francisco, San Francisco, California, United States of America

Department of Physiology, University of California San Francisco, San Francisco, California, United States of America
Howard Hughes Medical Institute, University of California San Francisco, San Francisco, California, United States of America

Benno Jungblut
Department of Biochemistry and Biophysics and Programs in Developmental Biology, Genetics, and Human Genetics, University of California San Francisco, San Francisco, California, United States of America

Jan Huisken
Department of Biochemistry and Biophysics and Programs in Developmental Biology, Genetics, and Human Genetics, University of California San Francisco, San Francisco, California, United States of America

Tania Ferrer
Department of Pediatrics, University of Utah, Salt Lake City, Utah, United States of America
Nora Eccles Harrison Cardiovascular Research and Training Institute, University of Utah, Salt Lake City, Utah, United States of America

Rima Arnaout
Department of Biochemistry and Biophysics and Programs in Developmental Biology, Genetics, and Human Genetics, University of California San Francisco, San Francisco, California, United States of America
Harvard Medical Schools, Boston, Massachusetts, United States of America

Ian Scott
Program in Developmental and Stem Cell Biology, The Hospital for Sick Children, and Department of Molecular Genetics, University of Toronto, Toronto, Ontario, Canada

Dimitris Beis
Developmental Biology, Biomedical Research Foundation, Academy of Athens, Athens, Greece

Tong Xiao
Department of Physiology, University of California San Francisco, San Francisco, California, United States of America
Programs in Neuroscience, Genetics, Human Genetics, and Developmental Biology, University of California San Francisco, San Francisco, California, United States of America

Herwig Baier
Department of Physiology, University of California San Francisco, San Francisco, California, United States of America,
Programs in Neuroscience, Genetics, Human Genetics, and Developmental Biology, University of California San Francisco, San Francisco, California, United States of America

Martin Tristani-Firouzi
Department of Pediatrics, University of Utah, Salt Lake City, Utah, United States of America
Harvard Medical Schools, Boston, Massachusetts, United States of America,

Didier Y. R. Stainier
Department of Biochemistry and Biophysics and Programs in Developmental Biology, Genetics, and Human Genetics, University of California San Francisco, San Francisco, California, United States of America,
Cardiovascular Research Institute, University of California San Francisco, San Francisco, California, United States of America,

Preface

Vertebrates belong to the phylum Chordata of kingdom Animalia and constitute subphylum vertebrata. Subphylum vertebrata includes organisms that have a spinal column and brain enclosed in cranium (as indicated by their Clade Craniata). In vertebrates, during early development, three different segmental systems are established: the somites, the rhombomeres and the pharyngeal arches. In each segmental system, an iteration of parts along the long axis is established. The formation of the somites emphasizes the segmental organization of trunk. Somites are the unique characteristic of developing vertebrate embryo, which are derived from mesenchymal cells in the presomitic mesoderm. The segmental boundary of somites in vertebrates is determined by Notch signaling. Rhombomeres indicate the segmentation of the hindbrain and the pharyngeal arches designate segmentation of the pharynx. In a study, the whole-genome comparison between humans and the pufferfish, revealed 1400 highly conserved non-coding sequences, most of which were found to be located in and around developmental regulator genes. A medium throughput functional assay using 25 sequence elements around four unrelated developmental regulators (SOX21, PAX6, HLXB9, and SHH) indicated that 23 show significant enhancer activity in one or more tissue. These highly conserved non-coding sequences are likely to form part of the genomic assembly that uniquely defines vertebrate development.

Vertebrates possess a circulatory system; the unique features of such a system are specialized cardiac conduction system and cardiac valves, which ensure unidirectional blood flow. Using a transgenic zebra fish line Tg(cmlc2:gCaMP)s878 four distinct stages of cardiac conduction development were identified. Other factors such hemodynamic flow and contraction, regulate the fast conduction network of this specialized conduction system. It was also found that a homeobox transcription factor gene tcf2 regulates the articulo-ventricular conduction. In vertebrates, two major skeletal systems, the endoskeleton and the exoskeleton have been identified where exoskeletons arise from the neural crest and mesoderm. In these organisms, both the systems cannot be differentiated by their differences in embryonic histogenesis but by their relative positions. The brain size in vertebrates increases among mammals and birds. It is postulated by some researchers that temperature is a critical factor in brain enlargement with ectothermic animals thriving in tropical climates having brain sizes larger than those of ectothermic animals living in cold climates. In a similar study, it was estimated that the species from higher latitudes tended to be younger, and colour-polymorphic species were older than monomorphic species. In addition, those organisms, which have persisted for a long time, may be more likely to deal with future modifications of their environment. Terrestrial vertebrates exhibit latitudinal gradient in diversity where the warm and productive biomes such as tropical forests possess greater number of vertebrate species. A model that integrates area and productivity over temperature and geological time can best explain species richness in 32 bioregions globally.

Editor
Dr. Nisha Dahiya

INTRODUCTION

Although not the most numerous group of animals in either number of individuals or in number of taxa, the vertebrates are an endlessly fascinating group of animals. Vertebrates are interesting partly because we are intimately familiar with them, but also because we are them. Our species, *Homo sapiens*, is included within the Vertebrata.

Vertebrates are also interesting because they are easy to find. Many other groups of organisms cannot be seen with the naked eye or reside in places inhospitable to humans, like the sea floor. We coexist with many other vertebrates and are often quite aware of them in our environment. We also have many biological, social, cultural and historical ties with vertebrates. We have domesticated dogs, cats, and birds, and have used horses as a means of transporation. We have used carrier pigeons as a means of communicating over long distances. We also use many products from bovids, like milk and leather. In some religions, cows are sacred, objects of reverence. Vertebrates have a long history on this earth -- more than 500 million years, from the late Cambrian up to today. These first vertebrates lacked jaws, like the living hagfish and lampreys. Jawed vertebrates appeared 100 million years later, in the Silurian. The vertebrates show affinities with other chordates but share some characters that make them unique. Some have argued that many of the characters that describe vertebrates have been derived from the same set of cells, the neural crest cells. These cells appear early in development, and only vertebrates have them. From neural crest cells are derived the skull and jaw bones.

While vertebrates have a long and interesting fossil record, humans are unfortunately helping to add to the list of organisms that no longer exist. These extinctions reduce the diversity of life on the planet and destroy our ability to understand all aspects of biology. We hope to show you the diversity of vertebrates that has existed in the past and the diversity that exists today. We also hope to show how, taken together, vertebrates extinct and extant can help us understand a great deal about the forces that have shaped the evolution of our world and ourselves in the past and help us predict and understand the future.

Vertebrates comprise all species of animals within the subphylum Vertebrata (chordates with backbones). Vertebrates represent the overwhelming majority of the phylum Chordata, with currently about 64,000 species described. Vertebrates include the jawless fish and the jawed vertebrates, which include the cartilaginous fish (sharks and rays) and the bony fish.

A bony fish clade known as the lobe-finned fishes is included with tetrapods, which are further divided into amphibians, reptiles, mammals, and birds. Extant vertebrates range in size from the frog species *Paedophryne amauensis*, at as little as 7.7 mm (0.30 in), to the blue whale, at up to 33 m (108 ft). Vertebrates make up less than five percent of all described animal species; the rest are invertebrates, which lack vertebral columns.

The vertebrates traditionally include the hagfish, which do not have proper vertebrae, though their closest living relatives, the lampreys, do. Hagfish do, however, possess a cranium. For this reason, the vertebrate subphylum is sometimes referred to as "Craniata" when discussing morphology.

Molecular analysis since 1992 has suggested that the hagfish are most closely related to lampreys, and so also are vertebrates in a monophyletic sense. Others consider them a sister group of vertebrates in the common taxon of craniata.

Vertebrate Basics

Vertebrates are the most advanced organisms on Earth. The traits that make all of the animals in this section special are their spinal cords, vertebrae, and notochords. It's all about having a series of nerves along your back (dorsal side). If you are an organism, you can't just have the nerves sitting there. You need to give those nerves support and protection. That need brings us to the backbones and a rod of cartilage called the notochord.

Not So Many Species

Fifty thousand species might seem like a lot. Compared to the invertebrates, there are not that many species of vertebrates. You might be asking why. One reason is that vertebrates are usually larger than invertebrates. They need more space. Another reason is that, even though they are more advanced, there are many limitations on the environments that are available to them.

Think about it this way. If you are smart mammal, would you rather live near the ocean or in the frozen tundra of the arctic? Many land animals can make that decision and move to more desirable areas for living. Those nicer areas can only support so many species of animals.

They have Got The Brains

Vertebrates are smart. Some of them are very smart. We mean you. Most vertebrates have very advanced nervous systems. While a goldfish might not compare to your intelligence, when you compare a goldfish to a sea anemone, a goldfish is like Einstein. Octopi are probably the smartest invertebrates and may equal or be smarter than some vertebrates. Octopi are the exception in the invertebrate category.

More cool traits about vertebrates are that they have muscles and skeletons. While the materials may vary, muscles allow vertebrates to move around very efficiently and perform complex moves. That ability to move and the intelligence to go with it gives vertebrates such as reptiles and birds an advantage in the natural world.

Classes of Vertebrates

The subphylum Vertebrata includes all of the familiar large animals and some rare and unusual ones as well. The 7 living classes of vertebrates are distinguished mostly on the basis of their skeletal system, general environmental adaptation, and reproductive system.

subphylum:	Vertebrata						
class:	Agnatha	Chondrichthyes	Osteichthyes	Amphibia	Reptilia	Aves	Mammalia

The subphylum Vertebrata includes all of the familiar large animals and some rare and unusual ones as well. The 7 living classes of vertebrates are distinguished mostly on the basis of their skeletal system, general environmental adaptation, and reproductive system.

Three of the vertebrate classes are fish. The most primitive of these is Agnatha. It consists of jawless fish that do not have scales. These are the lampreys and hagfish. Fish that have skeletons consisting of hard rubber-like cartilage rather than bone are members of the class ***Chondrichthyes. These are the sharks and rays. All of the bony fish are members of the class Osteichthyes.*** Tuna, bass, salmon, and trout are examples of *Osteichthyes*.

Animals in the class Amphibia spend part of their lives under water and part on land. Frogs, toads, and salamanders are amphibians. Many of these species must keep their skin moist by periodically returning to wet areas. All of them must return to water in order to reproduce because their eggs would dry out otherwise. They start life with gills, like fish, and later develop lungs to breathe air.

The class Reptilia includes turtles, snakes, lizards, alligators, and other large reptiles. All of them have lungs to breathe on land and skin that does not need to be kept wet. They produce an amniotes egg which usually has a calcium carbonate rich, leather hard shell that protects the embryo from drying out. This is an advantage over fish and amphibians because the amniotes egg can be laid on land where it is usually safer from predators than it would be in lakes, rivers, and oceans.

Tortoise, snake, and lizard (class *Reptilia*)

Amniote egg

The class Aves includes all the birds. They also produce amniote eggs but usually give them greater protection from predators by laying them high off of the ground or in other relatively inaccessible locations. In the case of both reptiles and birds, the eggs are fertilized within the reproductive tract of females. There are other striking similarities between reptiles and birds in their anatomies and reproductive systems. This is not surprising because birds are descendents of theropod dinosaurs (two-legged mostly carnivorous dinosaurs).

Birds (class Aves)

Dogs, cats, bears, humans and most other large animals today are members of the vertebrate class Mammalia. All mammals conceive their young within the reproductive tract of the mother and, after birth, nourish them with milk produced by their mammary glands. Mammals are heterodonts with strong jaws. That is to say, they have a variety of specialized teeth (incisors, canines, premolars, and molars). This allows them to chew their food into small pieces before swallowing it. Subsequently, they can eat any size plant or animal. Many reptiles must swallow their prey whole, which limits them to hunting smaller game.

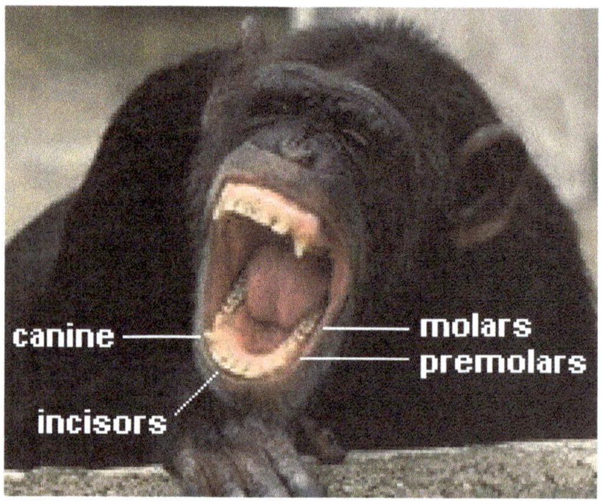

Mammalian heterodontism

Like birds, mammals are endothermic or warm blooded. They are able to maintain a relatively constant body temperature regardless of external environmental conditions mainly by using internal physiological mechanisms. In other words, they are homeothermic, or stable in core body temperature, as a result of endothermy. All of the living species of insects, fish, reptiles, and amphibians are ectothermic, or cold blooded. They keep their body temperature in a normal range mainly by avoiding exposure to environmental temperature extremes. For instance, reptiles usually remain in shaded areas on hot days to prevent fatal overheating. On cold nights, their lowered body temperature can cause them to become sluggish and inactive. In contrast, endothermic animals are able to remain active at night and often in the winter when the air temperatures are especially cold. They can also move about in the heat of very warm days. This ability most likely provided an advantage for the early small mammals in surviving alongside dinosaurs and other large reptiles, which apparently were mostly ectothermic. The downside of endothermy is the need to consume far more calories relative to body size in order to maintain a constant core body temperature. Small mammals, such as moles with their rapid metabolism rates, must eat insects or other high calorie foods every half hour or so in order to stay alive. By comparison, cold blooded rattlesnakes usually eat only once every 3-6 weeks and have been known to go without food for as long as two years.

Aiding in mammal body temperature control is their insulating hair and sweat glands. Sweating helps to dissipate heat by evaporative cooling. Compared to most other land mammals, humans are relatively hairless, but they have far more sweat glands. Mammals have four chambered hearts (like birds), complex nervous systems, and large brains relative to the size of their bodies. This broad range of useful features has made mammals highly adaptive and successful. They first appeared about 200,000,000 years ago, early in the age of dinosaurs, and replaced reptiles as the dominant class of land animals after 65,000,000 years ago. As the rapidly changing environment at that time led to the mass extinction of most large reptiles, it left vast evolutionary possibilities which mammals took advantage of by rapidly diversifying through adaptive radiation.

Important to mammalian success is their reproductive system. Their bodies took the amniote egg revolution of reptiles and birds one step further In effect, the uterus functions as the protective eggshell. Young mammals spend a long period of their early development within their mother's uterus. After birth, they are provided with protein and fat rich milk to eat and are usually protected until maturity. Pregnancy and milk production require mothers to significantly increase their own calorie consumption in order to provide nutrients for their infants. A nursing human female normally uses about 30% of her body's energy just to produce milk. Despite their success, mammals still only make up about .4% of known animal species. It is humbling to realize that all chordates together are only just over 3.7% of known animal species. By comparison, well over 1/2 of all animal species are insects.

How are they classified?

Vertebrates are classified by the chordate subphylum vertebrata. Invertebrates are any other animal that is classified outside of that class.

Are there a lot of vertebrate species?

There are currently around 65,000 known species of vertebrate animals. This sounds like a lot, but vertebrates are only around 3% of all the animals on Earth. Most of the animal species are invertebrates.

What are some vertebrate animals?

Fish - Fish are animals that live in the water. They have gills that allow them to breathe under water. Different species of fish may live in fresh water or salt water. Some examples of fish include the brook trout, the great white shark, lionfish, and the swordfish. Birds - Birds are animals that have feathers, wings, and lay eggs. Many, but not all, birds can fly. Some examples of bird species include the bald eagle, the cardinal, the flamingo, ostriches, and the red-tailed hawk. Mammals - Mammals are warm-blooded animals that nurse their young with milk and have fur or hair. Some examples of mammals include humans, dolphins, giraffes, horses, and spotted hyenas. Amphibians - Amphibians are cold-blooded animals. They start out their lives living in the water with gills just like fish. Later they develop lungs and can move to dry land. Amphibians include frogs, toads, newts, and salamanders. Reptiles - Reptiles are cold-blooded animals which lay eggs. Their skin is covered with hard and dry scales. Reptile species include alligators, crocodiles, snakes, lizards, and turtles.

Cold-blooded and Warm-blooded

Vertebrate animals can be either warm-blooded or cold-blooded. A cold-blooded animal cannot maintain a constant body temperature. The temperature of their body is determined by the outside surroundings. Cold-blooded animals will move around during the day between the shade and the sun to warm up or cool down. Cold-blooded animals are ectothermic, which means outside heat. Reptiles, amphibians, and fish are all cold-blooded. Warm-blooded animals are able to regulate their internal temperature. They can sweat or pant to cool off and have fur and feathers to help keep them warm. Warm-blooded animals are called endothermic, meaning “heat inside”. Only birds and mammals are warm-blooded.

Big and Small

The smallest vertebrate is thought to be a tiny frog called the Paedophryne amauensis. It only grows to about 0.3 inches long. The largest is the blue whale, which can grow to over 100 feet long and 400,000 pounds.

Fun Facts about Vertebrates

The only mammals that lay eggs are monotremes such as the platypus and spiny anteater. There are reptiles that live on every continent except Antarctica. Most fish have skeletons made of bone, they are called bony fish. Other fish have skeletons made of cartilage. These include sharks and rays. Frogs can breathe through their skin. The shortest childhood of any mammal is the hooded seal. They are considered adults when they are just four days old. Vertebrates tend to be much more intelligent than invertebrates.

CHAPTER 1

VERTEBRATE DEVELOPMENT

Adam Woolfe, Martin Goodson, Debbie K. Goode, Phil Snell, Gayle K. McEwen, Tanya Vavouri, Sarah F. Smith, Phil North, Heather Callaway, Krys Kelly, Klaudia Walter, Irina Abnizova, Walter Gilks, Yvonne J. K. Edwards, Julie E. Cooke, Greg Elgar

INTRODUCTION

Identification and characterisation of *cis*-regulatory regions within the non-coding DNA of vertebrate genomes remain a challenge for the post-genomic era. The idea that animal development is controlled by *cis*-regulatory DNA elements (such as enhancers and silencers) is well established and has been elegantly described in invertebrates such as *Drosophila* and the sea urchin [1,2,3,4]. These elements are thought to comprise clustered target sites for large numbers of transcription factors and collectively form the genomic instructions for developmental gene regulatory networks (GRNs). However, relatively little is known about GRNs in vertebrates. Any approach to elucidate such networks necessitates the discovery of all constituent *cis*-regulatory elements and their genomic locations. Unfortunately, initial *in silico* identification of such sequences is difficult, as current knowledge of their syntax or grammar is limited. By contrast, computational approaches for protein-coding exon prediction are well established, based on their characteristic sequence features, evolutionary conservation across distant species, and the availability of cDNAs and expressed sequence tags (ESTs), which greatly facilitate their annotation.

The completion of a number of vertebrate genome sequences [5,6,7,8,9], as well as the concurrent development of genomic alignment, visualisation, and analytical bioinformatics tools (for an overview see [10]), has made large genomic comparisons not only possible but an increasingly popular approach for the discovery of putative *cis*-regulatory elements. Comparing DNA sequences from different organisms provides a means of identifying common signatures that may have functional significance. Alignment algorithms optimise these comparisons so that slowly evolving regions can be anchored together and highlighted against a background of more rapidly evolving DNA that is free of any functional constraints.

One of the key decisions inherent in comparative genomics is the choice of organisms for which the comparison will be made. A number of successful pairwise and multiple-species sequence comparisons have already been carried out to identify novel enhancer elements in mammalian genomes [11,12,13,14,15,16,17,18,19,20]. Unfortunately, owing to differences in mutation rates across the genome, many slower-evolving regions show a significant degree of non-coding sequence conservation that reflects the short evolutionary distance between mammals and the slow rate of neutral

Citation: Woolfe A, Goodson M, Goode DK, Snell P, McEwen GK, Vavouri T, et al. (2005) Vertebrate Developments. PLoS Biol 3(1): e7. doi:10.1371/journal.pbio.0030007

divergence in vertebrates [20]. Consequently, although approximately 40% of the human and mouse genomes is alignable, only approximately 5% is estimated to be under selection, making it difficult to identify functionally relevant sequences [8]. One approach has recently been described [21] that identifies only those sequences that are identical over at least 200 bp between human and mouse genomes (termed ultra-conserved elements) and examines their distribution in the genome. Around half of the 481 elements identified showed no evidence of transcription and are therefore likely to be regulatory. Another highly successful approach to increasing the resolving power of comparative analyses is to use multi-species alignments combining both closely related and highly divergent organisms [14,22,23,24]. By using large evolutionary distances, even the slowest-evolving neutral DNA has reached equilibrium, thereby significantly improving the signal to noise ratio in genomic alignments. Although non-coding sequences generally lack sequence conservation between highly divergent species [22], there are a number of striking examples where comparison between human and pufferfish *(Fugu rubripes)* gene regions has readily identified highly conserved non-coding sequences that have been shown to have some function in vivo [25,26,27,28,29,30,31,32,33,34]. Humans and *Fugu* last shared a common ancestor around 450 million years ago [35], predating the emergence of the majority of all extant vertebrates, implying that any non-coding sequences conserved between these two species are likely to be fundamental to vertebrate life. The *Fugu* genome has the added advantage of being highly compact, reducing intronic and intergenic distances almost 10-fold [7,36]. Without exception, all reported examples of non-coding conservation between these two species have been associated with genes that play critical roles in development. This suggests that some aspects of developmental regulation are common to all vertebrates and that whole-genome comparisons may be particularly powerful in identifying regulatory networks of this kind. As a first step towards identifying such networks in humans, we have used comparative genomics to identify and begin to characterise non-coding sequences that are highly conserved between human and *Fugu*. A general strategy for testing whether non-coding regulatory sequences of this type are functionally relevant involves assaying their ability to up-regulate reporter gene expression in vivo. "Enhancer" assays using transgenic animals, in particular mouse, are both slow and laborious, but have provided some exciting data on the function of non-coding DNA around developmental genes (e.g., [31]). An alternative approach uses transient expression in zebrafish *(Danio rerio)* embryos [37,38,39], which are particularly suited to this form of analysis. Whilst transient expression is highly mosaic, the availability of large numbers of fertilised eggs, ease of micro-injection, and transparency of the developing embryos means that hundreds of individuals may be screened at a time. This provides a rare opportunity for increasing the throughput of this kind of functional assay. We have adopted a medium-throughput strategy to test DNA sequences for enhancer activity in zebrafish embryos, whereby results may be cross-referenced and compared through a generalised scheme. We present functional data for 25 highly conserved non-coding sequences that are located around four unrelated developmental regulators, SOX21, PAX6, HLXB9, and SHH in order to demonstrate the suitability of this screen to a wide range of genes and expression patterns.

RESULTS

Identification of Highly Conserved Non-Coding Sequences in Vertebrate Genomes

To locate conserved non-coding sequences, we masked the majority of the coding and tRNA content of the *Fugu* genome assembly [7] and compared the remaining regions using MegaBLAST [40] with the human genome sequence contained in Ensembl release v18.34.1 [41]. From this analysis we identified 19,744 sequences with similarity between the two genomes. By only including alignments of at least 100 bp in length, the number of sequences was reduced to 4,400. We then removed telomere-like sequences and transposons, and excluded any known protein-coding sequence or non-coding RNA species that may have been missed (see Materials and Methods). Sixty-five unique human sequences had matches to two independent locations in the *Fugu* genome. This is due to additional gene or genome duplications in the teleost lineage [42], where regulatory elements have been retained together with both copies of the fish gene [43]. To avoid redundancy in the human set, the longest matching sequence was retained and the duplicate removed. Finally, from the 1,373 sequences that remained, we determined whether any formed part of untranslated regions (UTRs) of mRNA molecules. Eighty sequences (approximately 6%) are situated in the 5′ or 3′ UTRs of known mRNA molecules. In addition, a similar number match one or more EST sequences, although most of these appear to be unspliced genomic contamination within EST libraries or incompletely spliced pre-mRNA. We did not remove these potentially transcribed sequences as, unlike vertebrate UTRs in general, they demonstrate a remarkable degree of conservation, and it is not clear whether they might be functional at the genomic or the transcript level. The remainder had no match against any expressed sequence in any database. This core set of 1,373 highly conserved non-coding elements (CNEs) forms the basis of this study.

The set of CNEs comprise a total of 273 kb of sequence, with a maximum length of 736 bp (average = 199 bp) and identity ranging from 74% to 98% (average = 84.3%). This is considerably higher than the level of identity seen between coding regions in these two organisms. Unsurprisingly, virtually all of the CNEs are conserved in rodent and chicken genomes, as well as a majority in the zebrafish genome. Of the 1,373 CNEs, 1,365 are conserved in the mouse, 1,316 in rat, and 1,310 in chicken, aligning to the human sequence with average identities of 97% for mouse and rat and 96% for chicken; 1,093 are also found to be conserved in the zebrafish genome, aligning with an average identity of 87.6% to the *Fugu*sequence. The zebrafish, chicken, mouse, and rat genomes are at different stages of completeness, and therefore missing sequence information may account for the missing CNEs (as well as the lower percent identity in zebrafish), although it may also reflect regulatory differences between the lineages.

Although CNEs are found throughout the human genome in all chromosomes except 21 and Y, their distribution is not uniform; in fact, they appear highly clustered. To examine their distribution in more detail, we plotted the position of each CNE on its respective chromosome in the human genome (Figure 1A). We then calculated the percentage of CNEs that were located in close proximity to another. We found that

90% of CNEs are less than 1 Mb apart, 85% of CNEs have a neighbouring CNE within 370 kb, and 75% are located within 158 kb of another CNE. The probability that over 85% of CNEs would be within 370 kb of another in a random distribution is less than 10^{-76} (Figure 1B). By carefully examining the distribution of CNEs across the genome, we generated a total of 165 clusters, including 19 singletons. Over 85% of the CNEs (1,172/1,373) are located in clusters containing five or more CNEs. The 20 largest clusters each contain 20 or more CNEs, comprising 43% (594/1,373) of the total number of elements.

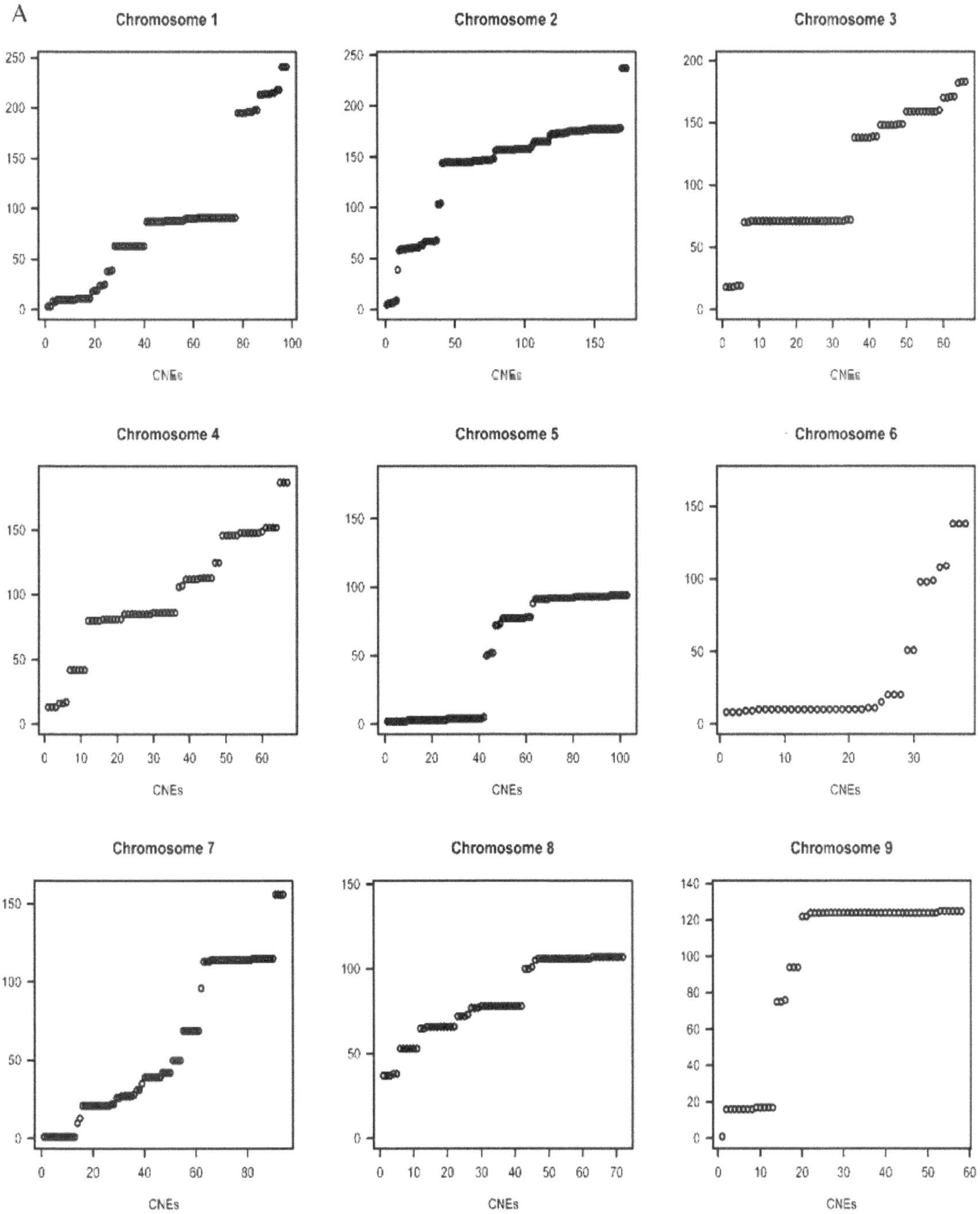

B

Percentage of CNEs	50%	55%	60%	65%	70%	75%	80%	85%	90%	95%
Distances between consecutive CNEs (kb)	49	61	77	96	124	158	236	370	1051	9024
χ^2 test p-value	$<10^{-14}$	$<10^{-18}$	$<10^{-24}$	$<10^{-30}$	$<10^{-41}$	$<10^{-50}$	$<10^{-63}$	$<10^{-76}$	$<10^{-54}$	<0.001

Figure 1. Distribution of CNEs along the Human Genome. (A) Each CNE is plotted relative to its position along each of human Chromosomes 1 to 9 (data for other chromosomes not shown). The y-axis represents length along the chromosome (in megabases). (B) Distribution of the fraction of CNEs that are within certain distances of each other; e.g., 85% of the distances between CNEs are less than or equal to 370 kb. χ^2 tests were carried out by comparing observed cluster sizes with those generated randomly for each chromosome (see Materials and Methods).

We then looked at the type of genes that are associated with CNEs in the human genome. For each CNE, we extracted the closest gene from Ensembl and submitted the resulting set of genes to GOstat [44] in order to identify the most statistically over-represented Gene Ontology (GO) terms [45]. Critically, 12 of the most over-represented 13 terms ($p < 0.001$) relate to transcriptional regulation and development.

We examined each cluster in turn to see how many were situated close to genes implicated in transcriptional regulation or development (we have termed these *trans-dev* genes). Over 93% of the clusters (154/165) have a *trans-dev* gene located within 500 kb of one or more of its CNEs (Figure 2; Materials and Methods). Of the remaining 11 clusters, five are closest to genes with zinc finger domains as identified by InterPro [46], one is in a gene desert, one maps to the AUTS2 gene region [47], and four are located adjacent to uncharacterised genes.

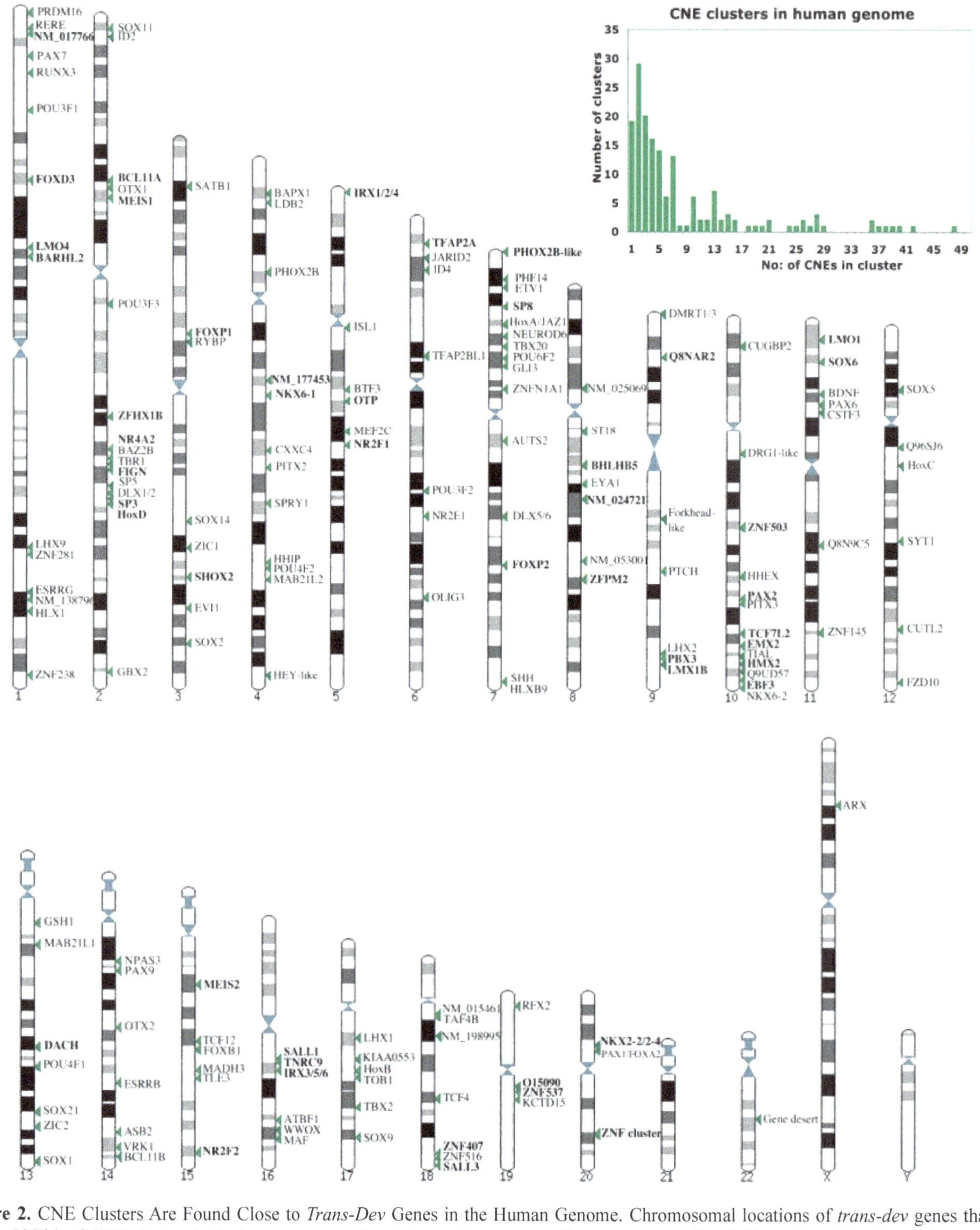

Figure 2. CNE Clusters Are Found Close to *Trans-Dev* Genes in the Human Genome. Chromosomal locations of *trans-dev* genes that are within 500 kb of CNE clusters in the human genome (each cluster is represented by a green arrowhead). Genes in bold script are located next to clusters of ten or more CNEs. Gene names are taken from Ensembl v23.34e.1. Graph inset shows distribution of CNE cluster sizes in the human genome.

Whilst most of the clusters can be associated with one *trans-dev* gene, there are 15 clusters in which CNEs are located close to two or more *trans-dev* genes. In nine of these cases, the CNEs associate with a group of paralogous genes, including the HOX, IRX, Nkx2–2/2–4, and DLX clusters, although there are three instances where a pair of unrelated *trans-dev* genes are located next to a CNE cluster (SHH and HLXB9, PBX3 and LMX1B, and PAX1 and FOXA2). Finally, there are three clusters that associate with two or more zinc finger genes.

Trans-dev genes associated with CNE clusters tend to be located in regions of low gene density. We counted the number of genes situated within 500 kb upstream and downstream of a *trans-dev* gene, and compared this with the average for all human genes. Whereas the average for all human genes is 17, it is just six for the *trans-dev* genes. This is similar to the "gene desert" phenomenon described around the DACH gene [31]. Interestingly, the CNEs themselves are generally located large distances from their nearest gene. The average distance between a CNE and the 5′ end of the closest human gene is 182 kb (median = 120 kb), with 93 CNEs more than 500 kb, and 12 CNEs more than 1Mb, from any known gene.

A number of the *trans-dev* genes that we identified have previously been shown to have highly conserved *cis*-regulatory elements associated with them, including the Hox clusters [24,33], PAX6 [48], PAX9 [32], SOX9 [28], OTX2 [34], SHH [30], DLX genes [29], and DACH [31]. Five CNEs do not appear to cluster with any known genes in either the human or *Fugu*genomes and are located in a large gene desert on human Chromosome 22. Given that gene annotation and genomic sequencing of parts of the human genome are not yet fully complete, the discovery of CNEs here may well point to the existence in this region of an important transcriptional or developmental regulation gene with which they are associated. Indeed we find the largest number of CNEs (48) clustered around a relatively uncharacterised gene with zinc finger domains, ZNF503 on human Chromosome 10, the rat orthologue of which was recently characterised as a probable transcriptional regulator in brain development [49].

All CNEs were compared with each other to look for local similarities. Forty-three elements show significant similarity to at least one other CNE, and in each case are situated close to genes with clear paralogous relationships, e.g., the HOX and IRX clusters. The remainder of the sequences appear to be unique in the human genome.

In order to identify additional conserved sequences around specific genes for further functional assay, localised multiple-alignment comparisons were performed using the multiple LAGAN (MLAGAN) alignment tool kit [50]. This tool kit provides the opportunity to introduce genomic sequence from additional species, in this case mouse and rat, which significantly enhances the signal to noise ratio. For a random subset of 25 of the *trans-dev* genes associated with CNE clusters, stringent whole-genome alignment located 408 CNEs, whereas MLAGAN identified over twice as many conserved regions (871) of at least 100 bp in length. The whole-genome analysis was more stringent in that we used a minimum exact word match of 20 bp, whereas MLAGAN uses short inexact words to create anchors from which a more sensitive (Needleman–Wunsch) alignment is carried out. It is important to note that similar alignments on genes that are not implicated in developmental regulation do not identify conserved non-coding sequence (e.g., [22,51]).

The alignment of a known transcription factor, SOX21, identifies a large number of conserved non-coding sequence elements in addition to the CNEs found in the whole-genome analysis. We have called these "regionally defined CNEs" (rCNEs) (Figure 3A). In mammalian genomes, the distance between the first and last element around SOX21 is over 450 kb. As is the case for a number of the larger CNEs throughout the genome, some of the CNEs around the SOX21 gene are more highly conserved than the gene's coding exon. For example, in multiple alignments of mouse, rat, human, and *Fugu* sequence, one CNE (SOX21_19) has 90% identity over 558 bp whilst another (SOX21_1) contains a 112-bp region of 100% identity (Figure 3B), demonstrating an extraordinary level of conservation for genomes separated by 900 million years of divergent evolution.

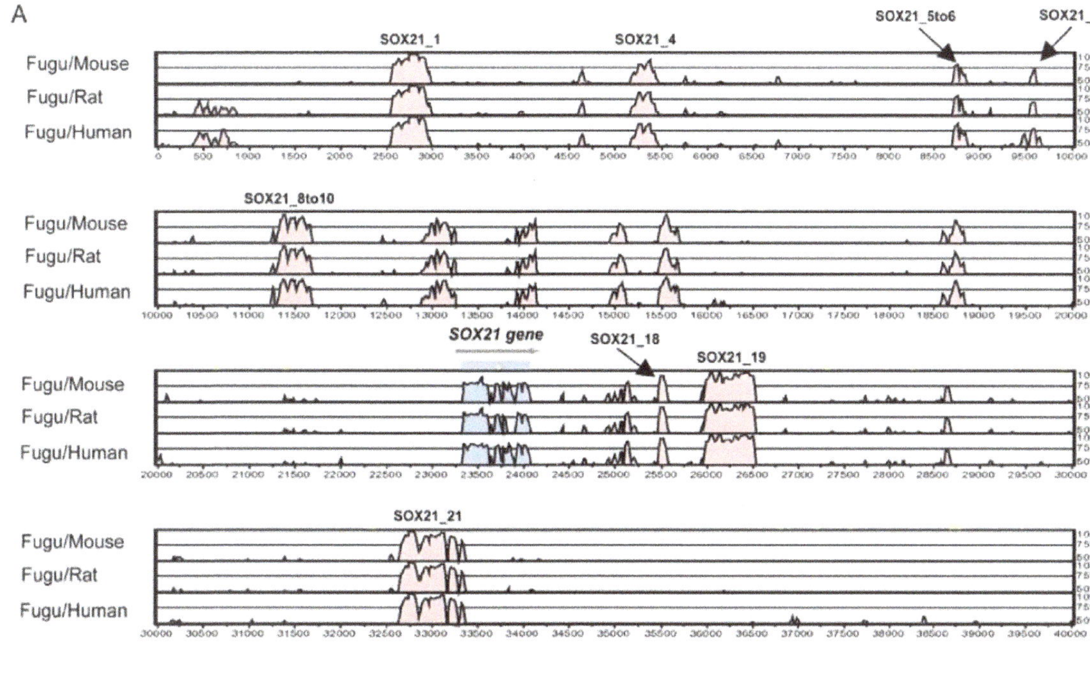

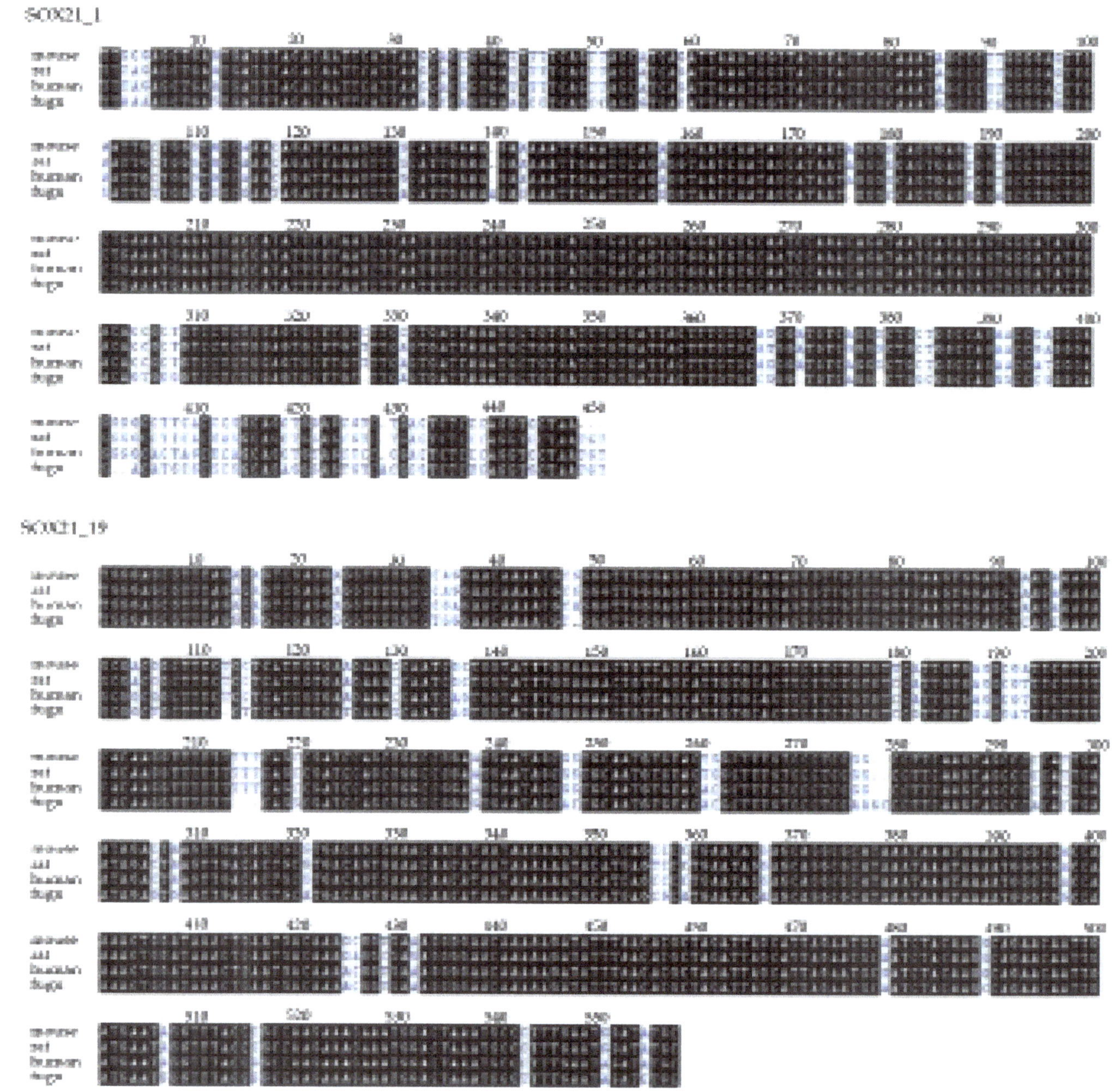

Figure 3. Comparative Sequence Analysis of the SOX21 Gene. SOX21 genomic regions for mouse, human, and rat were extracted from Ensembl to include all flanking DNA up to the nearest neighbouring genes (ABCC4 and NM_180989 in the human genome and their orthologues in the rodent genomes). The region covering *Fugu* SOX21 (138–178 kb of *Fugu* Scaffold_293 [M000293]) was extracted from the*Fugu* Genome Server at http://fugu.rfcgr.mrc.ac.uk/fugu-bin/clonesearch. (A) MLAGAN alignment of the SOX21 gene using *Fugu* DNA as the base sequence compared with mouse, rat, and human genomic DNA. Coloured peaks represent regions of sequence conservation above 60% over at least 40 bp. The SOX21 coding region (SOX21 is a single exon gene) is annotated, and sequence identity is shaded in blue. Non-coding regions of sequence identity are shaded in pink. The eight elements that have been functionally assayed are labelled. Six of these are identified in the global analysis as seven CNEs (SOX21_8–10 covers two CNEs). SOX21_7 and SOX21_18 are rCNEs. (B) Multiple DNA sequence alignments of CNE SOX21_1 and CNE SOX21_19 between mouse, rat, human, and *Fugu*.

Finally we searched invertebrate sequence databases, including the whole-genome sequences of *Ciona intestinalis, Drosophila melanogaster,* and *Caenorhabitis elegans,* to see whether we could identify any of these highly conserved vertebrate sequences within the invertebrate lineage. Although many of the genes identified in our analysis have clear homologues within these genomes, we found no significant matches to any CNEs. More sensitive alignment using MLAGAN also failed to identify any conserved non-coding sequence similarity between vertebrates and non-vertebrates (including C. elegans, D. melanogaster and A. gambiae), whilst in each case the coding sequences were identified. This is surprising, given that the degree of identity between CNEs in vertebrates is higher than that of the coding regions for these genes. Thus, it is unlikely that the same set of sequences that appear to regulate important vertebrate *trans-dev* genes are found in invertebrates.

Functional Assay

We have assayed the ability of conserved non-coding sequences identified both from the whole-genome MegaBLAST analysis (CNEs) and from regional MLAGAN alignments (rCNEs) to up-regulate green fluorescent protein (GFP) reporter expression in zebrafish embryos (seeMaterials and Methods). We chose four cluster regions that contain different types of developmental genes: SOX21, PAX6, HLXB9, and SHH. Elements are co-injected with a minimal promoter–GFP reporter construct into early zebrafish embryos. This co-injection strategy [37,38] is an efficient, yet simple and rapid method for identifying enhancer activity; indeed enhancer activity of elements is more striking when tested in a co-injection assay than when ligated directly to a promoter–reporter construct [37].

A total of 25 conserved non-coding regions were selected (Figures 3, 4), of which ten were CNEs and 15 were rCNEs (Table 1). GFP expression was analysed in live embryos on the second day of development and recorded both schematically and in tabular form. A mean of 188 embryos were screened for each element, compared with a mean of just over 200 embryos per control (Table 1).

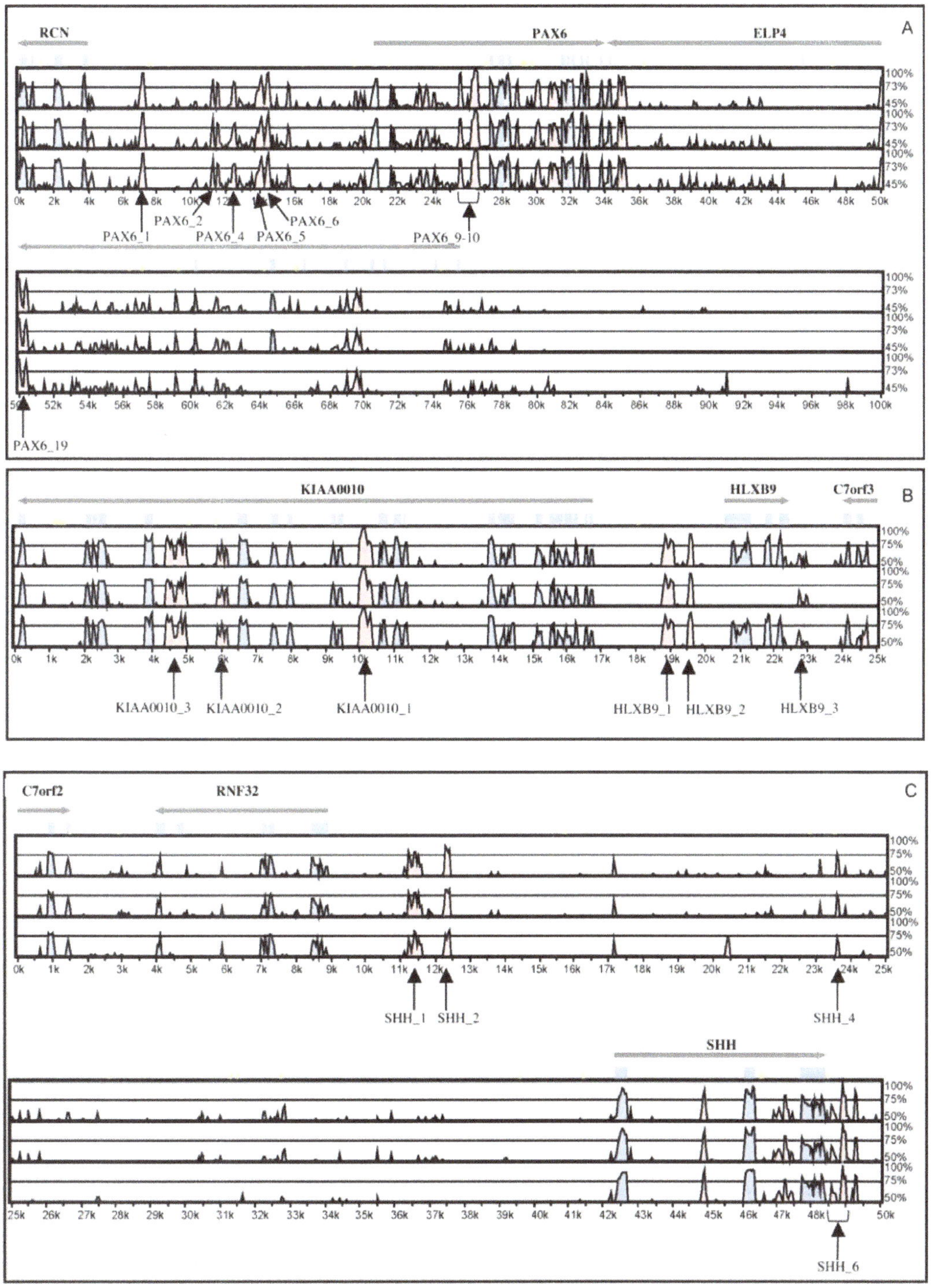

Figure 4. MLAGAN Alignments of Regions Encompassing the PAX6, HLXB9, and SHH Genes PAX6 (A), HLXB9 (B), and SHH (C). In each panel, human (top), mouse (middle), and rat (bottom) genomic DNA from Ensembl is aligned with *Fugu* genomic DNA from orthologous regions. Alignment parameters are the same as in Figure 2. Seventeen elements that have been functionally assayed from these regions have been labelled. The following were identified as CNEs: PAX6_6, PAX6_9–10, KIAA0010_1, and KIAA0010_3.

Table 1. Elements Used in Functional Assay

Element Type	Element Name	Percentage of Embryos Expressing GFP	Length (bp) of Element[a]	Length of PCR (bp)	EMBL Accession Number
SOX21	SOX21_1	10.4	388	771	CR846634
	SOX21_4	24.1	366	457	CR846632
	SOX21_5-6	14.6	146	740	CR846631
	SOX21_7	17.9	97	444	CR847493
	SOX21_8-10	5.3	133 + 191 (340)	973	CR846637 + CR846638
	SOX21_18	0	92	306	CR847492
	SOX21_19	12.2	553	897	CR846629
	SOX21_21	3.7	499	1008	CR846633
PAX6	PAX6_1	12.5	281	672	CR847482
	PAX6_2	6.7	93	259	CR847484
	PAX6_4	7.4	48 + 88 (175)	501	CR847485
	PAX6_5	0	26 + 100 (140)	334	CR847486
	PAX6_6	20.8	169	750	CR846537
	PAX6_9-10	22.1	184 + 400 (1064)	1601	CR847487 + CR846536
	PAX6_19	34	113	699	CR847483
HLXB9	HLXB9_1	30.1	313	748	CR847478
	HLXB9_2	14.4	147	371	CR847479
	HLXB9_3	26	65	327	CR847480
KIAA0010	KIAA0010_1	23.5	349	603	CR847029
	KIAA0010_2	18.4	228	803	CR847481
	KIAA0010_3	13.7	251 + 294 (682)	838	CR847031 + CR847030
SHH	SHH_1	24.4	370	672	CR847488
	SHH_2	17.7	142	227	CR847489
	SHH_4	27.7	63	174	CR847490
	SHH_6	44.2	412	596	CR847491
Control	No element	0.5 (1/199)	N/A	n/a	
	SOX21 non-coding_1	0.9 (1/112)	N/A	416	
	PAX6 non-coding_1	0 (0/156)	N/A	505	
	PAX6 non-coding_2	0.3 (1/296)	N/A	1265	
	SOX21 coding_1	0 (0/214)	N/A	467	
	PAX6 coding_1	1 (2/194)	N/A	314	
	SHH coding_1	0.4 (1/247)	N/A	221	

[a] In some cases, two conserved regions that are very close together have been included in one PCR. In this case, the length of each element is given with the region they span in parenthesis.
N/A, not applicable.

Controls in which no element was injected (GFP reporter construct injected alone), in which non-conserved, non-coding genomic DNA from the PAX6 or SOX21 regions was co-injected with the GFP reporter, or in which conserved, coding DNA from PAX6, SOX21, or SHH exons was co-injected with the GFP reporter produce essentially no up-regulation of GFP expression (Table 1). When conserved non-coding sequences were injected, up-regulation of GFP expression was observed with all but two of the elements tested, with between 4% and 44% of embryos screened being positive (Table 1). Furthermore, GFP expression was generally observed in consistent patterns, specific to the element injected (Figure 5).

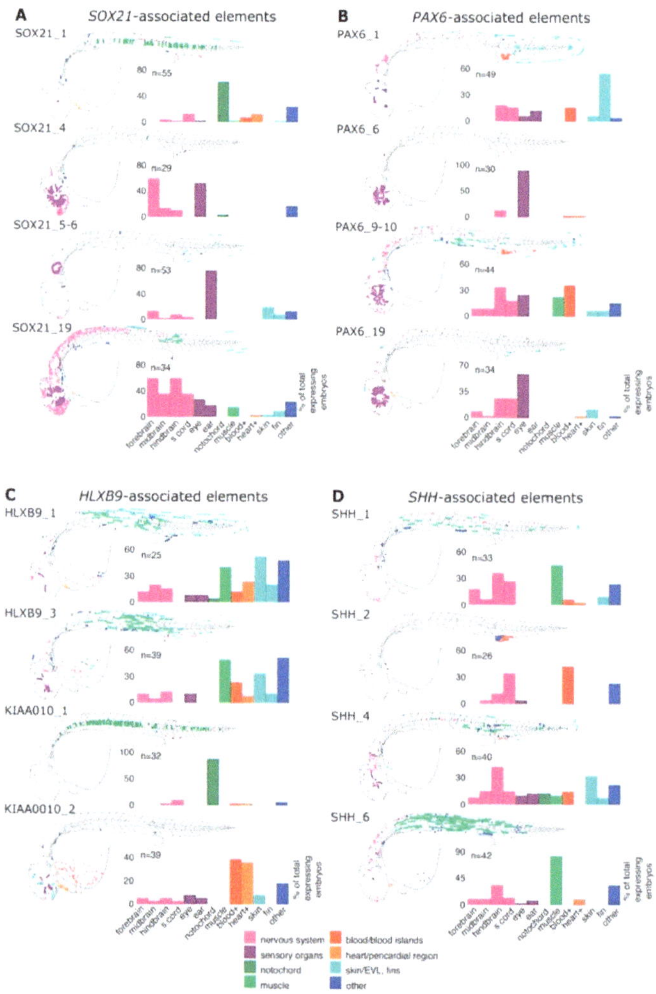

Figure 5. Composite Overviews of GFP Expression Patterns Induced by Different Elements Tested in the Functional Assay. Cumulative GFP expression data, from *SOX21*-associated elements (A), *PAX6*-associated elements (B), *HLXB9*-associated elements (C), and *SHH*-associated elements (D). Cumulative data pooled from multiple embryos per element on day 2 of development (approximately 26–33 hpf) are dis-

played schematically overlayed on camera lucida drawings of a 31-hpf zebrafish embryo. Categories of cell type are colour-coded: key is at bottom of figure. Bar graphs encompass the same dataset as the schematics and use the same colour code for tissue types. Bar graphs display the percentage of GFP-expressing embryos that show expression in each tissue category for a given element. The total number of expressing embryos analysed per element is displayed in the top left corner of each graph. Legend for the bar graph columns accompanies the bottom graph in each panel; "blood+" refers to circulating blood cells plus blood island region, "heart+" refers to heart and pericardial region (Please note: Some cells categorised as heart/pericardial region may be circulating blood cells), and "skin" refers to cells of the epidermis or EVL. s. cord, spinal cord.

In order to build up a comprehensive picture of the GFP expression pattern induced by each of the elements, the expression profiles from multiple embryos positive for a given element were overlaid onto a schematic diagram, so providing a composite overview for each element (Figure 5). This also provided a convenient format for data storage and comparison between elements.

SOX21-Associated Elements

Of the eight SOX21-associated elements tested in our functional assay, seven enhance GFP expression (Table 1). Three of these enhancing elements direct reporter gene expression most prominently to the central nervous system (CNS) (SOX21_4 and SOX21_19 [Figures 5A, 6A, and 6B] and SOX21_7).

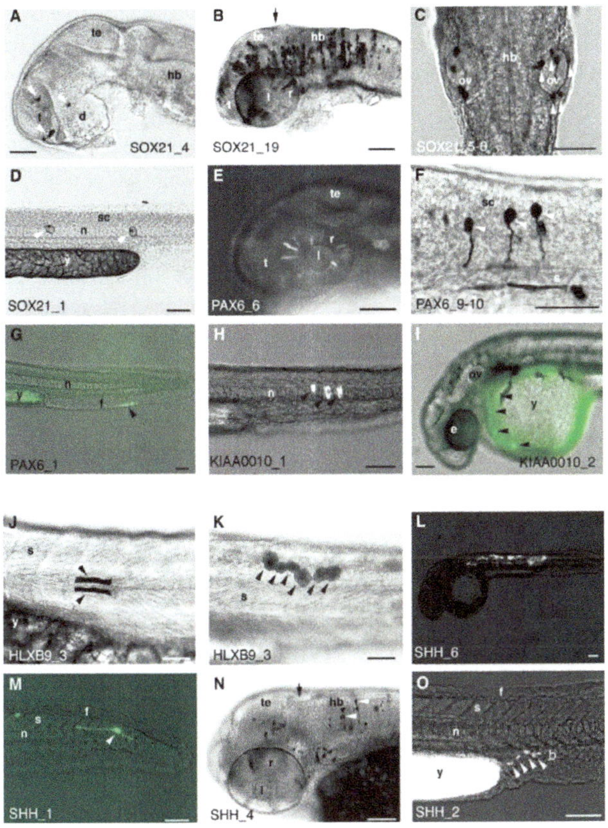

Figure 6. Different Elements Enhance GFP Expression in Specific Tissue and Cell Types. GFP expression is shown in fixed tissue following wholemount anti-GFP immunostaining, bright-field views (A–D, F, J, K, and N), or in live embryos as GFP fluorescence, merged bright-field and fluorescent views (E, G–I, L, M, and O). Lateral views, anterior to the left, dorsal to the top (A, B, and D–O) or dorsal view, anterior to the top (C). Embryos approximately 28–33 hpf (A, D–I, L, and O), approximately 48 hpf (B, C, J, K, and N), or approximately 26 hpf (M). The identity of the element co-injected with the GFP reporter construct is shown at the bottom of each panel. Black arrows indicate the approximate position of the midbrain–hindbrain boundary; black and white arrowheads indicate GFP-expressing cells.

SOX21_19 strongly directs remarkably widespread GFP expression throughout the brain and rostral spinal cord (88% of expressing embryos show GFP-positive cells in the CNS; Figures 5A and 6B). SOX21, a member of the SRY-related HMG-box (SOX) gene family of DNA-binding proteins, acts as a transcriptional repressor during early development [52], and is expressed in a complex, dynamic pattern in the developing vertebrate CNS [53,54,55].

Scale bars approximately 100 µm (A–E, G–I, and L–O) and 50 µm (F, J, and K).

b, blood island; d, diencephalon; e, eye; f, fin fold; hb, hindbrain; l, lens; n, notochord; ov, otic vesicle; r, retina; s, somite; sc, spinal cord; t, telencephalon; te, tectum; y, yolk.

(A) SOX21_4. Head region (eyes removed): neurons in the telencephalon and diencephalon are GFP-positive (arrowheads).

(B) SOX21_19. Head region: numerous GFP-expressing neurons are visible in the forebrain, midbrain, and hindbrain. Retinal expression is also apparent.

(C) SOX21_5–6. Hindbrain region: white arrowheads indicate GFP expression by several cells in the epithelium of the right developing ear (ov). GFP-expressing cells in left deveoping ear are in slightly different focal plane.

(D) SOX21_1. Trunk region: two individual notochord cells express GFP (arrowheads).

(E) PAX6_6. Head region of live embryo: GFP is expressed in several retinal cells.

(F) PAX6_9–10. Anterior trunk region (at the level of somites 1–3): three spinal cord neurons with ventrally projecting axons express GFP (arrowheads).

(G) PAX6_1. Tail region of live embryo: arrowhead indicates GFP expression in the developing median fin fold.

(H) KIAA0010_1. Trunk region, three notochord cells express GFP (arrowheads).

(I) KIAA0010_2. Anterior end of embryo: arrowheads point to circulating blood cells expressing GFP.

(J) HLXB9_3. Trunk region: GFP-expressing muscle fibres in somite 5 (arrowheads) lie immediately dorsal and ventral to the horizontal myoseptum.

(K) HLXB9_3. Trunk region (at the level of somites 13–15): arrowheads mark GFP expression in six cells forming the epidermis or EVL.

(L) SHH_6. Whole live embryo: numerous GFP-expressing muscle fibres can be seen in the trunk.

(M) SHH_1. Tail region of live embryo: GFP is expressed in a single bipolar neuron near the caudal end of the spinal cord (arrowhead marks cell body).

(N) SHH_4. Head region (dorsolateral view): cells labelled with anti-GFP include midbrain and hindbrain neurons and cells in the retina (slightly out of focal plane). Arrowheads indicate cell bodies of hindbrain neurons, from which axons can be seen projecting ventrally.

(O) SHH_2. Trunk region of live embryo: GFP-positive cells in the region of the blood islands (caudal to the urogenital opening; arrowheads) show a slightly elongated morphology, suggesting they may be blood vessel precursors rather than blood cells.

Three elements strongly enhance GFP expression in the sense organs: SOX21_4 and SOX21_19 direct GFP expression to the developing eye (in 52% and 27% of expressing embryos, respectively; Figures 5A and 6B), and SOX21_5–6 strongly enhances reporter expression in the developing ear (75% of expressing embryos; Figures 5A and 6C). These observations draw parallels with prominent regions of endogenous SOX21 expression in the sense organs: i.e., the nasal epithelium, the lens and retina of the eye, and the sensory epithelia of the developing inner ear [55]. SOX21_1 strongly enhances expression in the notochord (62% of expressing embryos; Figures 5A and 6D), a domain not normally associated with SOX21 expression.

PAX6-Associated Elements

Six out of seven PAX6-associated elements tested in our functional assay enhance GFP expression (Table 1). Four of these six functional elements direct GFP expression most frequently to the developing eye (PAX6_6, 90% of expressing embryos; PAX6_19, 59% of expressing embryos [Figures 5B and 6E]; PAX6_2, 92% of expressing embryos; and PAX6_4, 100% of expressing embryos). A fifth element, PAX6_9–10, also directs reporter gene expression to the eye in a significant proportion (25%) of expressing embryos (Figure 5B) as well as to neurons most frequently in the hindbrain and spinal cord (Figures 5B and 6F).

Significantly, PAX6 is a paired-box-containing transcription factor, expressed in and playing essential roles in the developing eye; it is also expressed in the forebrain, hindbrain, and spinal cord (data from the Zebrafish Information Network; http://zfin.org). PAX6 is associated with the loss-of-function disorder aniridia. Some aniridia cases show chromosomal rearrangements downstream of an intact PAX6 gene, indicating that *cis*-acting elements can influence PAX6 gene expression in the eye at a significant distance from the coding region [56]. Indeed, PAX6 expression is known to be influenced by *cis*-acting elements in upstream, intronic, and downstream positions. For example, 5′ elements drive expression in the lens, pancreas, and parts of the neural tube [27], intronic elements drive expression in the retina, forebrain, and hindbrain [27,57], and several 3′ regions direct expression to the developing pretectum, neural retina, and olfactory region [58].

In addition to the eye and CNS, other tissues to which GFP expression is directed by our PAX6-associated elements include the blood islands (PAX6_9–10, 36% of expressing embryos; PAX6_1, 16% of expressing embryos [Figure 5B]) and the median fin fold (PAX6_1, 55% of expressing embryos; Figures 5B and 6G); these tissues have not been associated with endogenous expression of PAX6.

HLXB9-Associated Elements

We assayed six elements associated with a genomic region containing the HLXB9 and KIAA0010 genes (Table 1). Each of these elements induces GFP expression in a variety of tissues (data from four elements are shown in Figure 5C). Most notably, KIAA0010_1 directs GFP expression to the notochord in more than 87% of expressing embryos (Figures 5C and6H), KIAA0010_2 directs expression to the blood (38% of expressing embryos; Figures 5C and6I) and the pericardial region (36% of expressing embryos; Figure 5C), HLXB9_1 directs expression to the skin/enveloping layer (EVL; 52% of expressing embryos) and skeletal muscle (40% of expressing embryos; Figure 5C), HLXB9_3 directs expression to skeletal muscle (48% of expressing embryos; Figures 5C and 6J) and to skin/EVL (33% of expressing embryos; Figures 5C and 6K), and HLXB9_2 directs expression to the spinal cord (87% of expressing embryos).

HLXB9 is a Mnx-class homeobox gene associated with autosomal dominant caudal defects [59]. The zebrafish orthologue, *hb9,* is expressed in the notochord, hypochord, tail mesoderm, and tailbud [60], paralleling some of the domains of GFP expression induced by HLXB9/KIAA0010-associated elements.

SHH-associated elements.

Two of the four SHH-associated elements tested in this study (Table 1) direct GFP expression most frequently to muscle cells (SHH_1, 46% of expressing embryos; SHH_6, 83% of expressing embryos [Figures 5D and 6L]). All four elements also prominently direct GFP expression to the CNS (SHH_1, 64% of expressing embryos; SHH_2, 42%; SHH_4, 57%; and SHH_6, 48% [Figures 5D, 6M, and 6N]).

The SHH signalling molecule is crucial for a number of developmental processes, and is extensively implicated in disease (reviewed in [61]). In zebrafish, *shh* and its co-orthologue*twhh* are both expressed predominantly in midline structures, i.e., floorplate and notochord. Later expression domains include the branchial arches, pectoral fin buds, and the retina [62,63]. GFP expression directed by SHH-associated elements and *shh/twhh* expression overlap in the floorplate; however, most of the other domains of GFP expression (e.g., muscle and blood islands; Figure 6O) are not reflected by endogenous expression of hedgehog genes.

DISCUSSION

Understanding the intricate and finely tuned process of gene regulation in vertebrate development remains a major challenge facing post-genomic research. In order to begin to understand how genomic information can coordinate regulatory processes, we have adopted an approach integrating comparative genomics and a medium-throughput functional assay. Nearly 1,400 non-coding DNA sequence elements were identified that exhibit extreme conservation throughout the vertebrate lineage. Despite a degree of overlap, less than half of the non-coding ultra-conserved regions (109 out of 256) identified from the mouse and human genomes [21] are present in this set. Most, if not all, of the CNE sequences appear to be associated with genes involved in the control of development, many of them transcription factors. A significant proportion of genes identified in this study are homologous to those identified in the sea urchin and other invertebrates as master regulators of early development, leading us to believe that they interact in GRNs. Consequently, it is extremely likely that the CNEs identified compose at least part of the genomic component of GRNs in vertebrates, acting as critical regions of regulatory control for their associated genes. Such regions would mediate up- or down-regulation of expression, effecting a cascade of downstream events.

In agreement with current GRN models, and given the function of many of the genes we have identified in our analysis, it is logical to speculate that CNEs consist of modules of binding sites for transcription factors. However, the model of CNEs as transcription factor binding sites, even for large numbers of transcription factors, does not fully explain their high sequence identity across vertebrates, given that transcription factor binding sites are generally rather short and exhibit a level of redundancy. Consequently, we have not ruled out the possibility that the CNEs may have a completely different mode of action or act in numerous different ways.

The relative positions and order of CNEs within a cluster is completely conserved in all vertebrate genomes we have analysed (generally mouse, rat, human, and *Fugu*) together with some degree of proportional compaction in the *Fugu* genome. This suggests that the CNEs might play a role in structuring the genomic architecture around *trans-dev* genes, which in turn may lead to an additional level of transcriptional control. Further evidence that genomic architecture may be important comes from the fact the *trans-dev* genes are generally located in regions of low gene density.

Alternatively, despite the lack of EST data, it is possible that CNEs are transcribed and work at the RNA level. A number of other ideas on the evolutionary mechanisms responsible for "ultra-conservation" have been suggested [21,64], involving decreased mutation rate, increased DNA repair, and multiply-overlapping transcription factor binding sites, but without more functional studies such hypotheses remain speculative. Whatever their mode of action, the striking degree of conservation displayed by this set of CNEs suggests they play critically important functional roles.

Having established a "map" of the major locations of CNEs in the genome, we were able to take a more sensitive alignment approach in a number of these regions in order to identify additional CNEs (rCNEs). The distinction between CNEs and rCNEs is purely a bioinformatics one, based on our search parameters, and we have no reason to believe that there is any functional distinction between the two sets of elements. We selected a number of elements (both CNEs and rCNEs) as candidates for functional analysis. Data from our functional assay of 25 elements from four different developmental genes demonstrate that a significant proportion can act as enhancers, inducing expression of a GFP reporter gene in a tissue-specific manner. The observed expression patterns differ among elements, but are reproducible for individual elements. Enhanced GFP expression domains frequently coincide with endogenous expression domains of the *trans-dev* gene most closely associated with a particular element, although in several instances, expression of GFP was induced in a tissue in which the most closely associated developmental gene is not normally expressed. This is not surprising because we are assaying elements out of context and individually. Thus, in our assay, we may have excluded another regulatory sequence in the region that under normal circumstances acts to silence the enhancer activity of an element in a specific tissue. Indeed GRN models would predict that a number of different regulatory regions must interact in order to precisely effect a particular spatiotemporal pattern of expression. One of our future directions will therefore be to assay the combinatorial effects of injecting a number of elements together. Alternatively, we may have associated a CNE with the wrong gene, particularly where there are two or more *trans-dev* genes in the same region (see below).

Whilst it is straightforward to assign CNEs unequivocally to the SOX21 and PAX6 genes based on their location in the genome, the situation is more complex for elements in the vicinity of the SHH and HLXB9 genes, which are situated in close proximity to each other in the human, rodent, and *Fugu* genomes. This is exacerbated by the fact that some CNEs may also be found within or around neighbouring genes. This phenomenon has been described for both the PAX6 [65] and PAX9 [32] genes, as well as for the SHH gene [30], where a long-range enhancer in the intron of a neighbouring, unrelated gene regulates SHH expression in developing limb buds and demonstrates the large genomic distances over which regulatory regions may act. This enhancer is identified as a CNE in our dataset and, despite its established mode of action, is located much closer to the HLXB9 gene (200 kb in human and 12 kb in *Fugu*) than to SHH (1,000 kb in human and 60 kb in *Fugu*). Furthermore, a number of elements are located directly 5′ of the HLXB9 gene, whilst others are found located further upstream, in introns of the next gene, KIAA0010. Although we strongly suspect that all these elements are associated functionally with the HLXB9 gene (e.g., KIAA0010_1 directs expression prominently to the notochord, an expression domain of the zebrafish HLXB9 orthologue), we cannot rule out the possibility that they may associate with the SHH gene (also expressed in the notochord), which lies a few genes downstream. There are a number of cases where a CNE cluster is located close to more than one *trans-dev* gene, illustrating the value of correlating endogenous expression pattern with CNE enhancer activity. However, it should be noted that in order to build GRN maps for the elements, it is desirable but not essential to know which element associates functionally with which gene.

Our confidence in the correctness of our gene assignment for the elements tested in this study is borne out by the results of our functional analysis. For the elements that we have associated with PAX6 and SOX21, there is a good correlation between tissues that express the gene endogenously and tissues induced by the associated co-injected elements to express GFP, i.e., the major sites of endogenous gene expression are highly represented in our mosaically expressing embryos (e.g., eye, hindbrain, and spinal cord for PAX6; forebrain, midbrain, hindbrain, and spinal cord for SOX21; see Figure 5). However, for elements in the vicinity of the HLXB9, KIAA0010, and SHH genes, GFP expression overlaps less often with expression domains of the associated gene to which the element has been assigned. As mentioned above, this reduced correlation with endogenous expression of their "associated" genes is probably due to the difficulty of assigning genes to elements in this region of relatively high *trans-dev* gene density.

It is likely that we have missed some developmental regulators in our whole-genome analysis owing to the stringency of our search parameters. Both the *RUNX2* [66] and *WNT1* [26] genes, for instance, share conserved non-coding sequences in humans and fish but were excluded because they failed to satisfy our stringent whole-genome search parameters. We may also have missed some elements because they were inadvertently hidden during the process used to mask coding sequence. Nevertheless, this is the first comprehensive attempt to identify the most highly conserved non-coding sequences common to all vertebrates. The use of the compact *Fugu* genome sequence, with its large evolutionary divergence from mammals, was critical in providing an exceptionally low degree of background noise in comparisons at the level of whole-genome and genomic regions.

As with any high-throughput approach, our functional screen has limitations. Since there is a negligible background level of GFP expression from our reporter construct alone, as well as from our other negative controls (see Table 1), the expression we see is most likely to be directly attributable to the enhancer properties of the CNEs. However, since GFP is a relatively stable protein [67], down-regulation of expression will not be detected during the time course of this screen; instead, expression of GFP by a particular cell indicates that expression was stimulated at some previous point in that cell's lineage. False negatives are a further limitation of the assay, e.g., tissues that develop from few cells will be under-represented and late-developing tissues or cell types (after 24 h) will be missed completely in this screen, since there is a delay between the time of onset of GFP transcription and the time when GFP fluorescence is detectable.

The proportion of screened embryos that showed GFP expression varied from around 4% (SOX21_21) to around 44% (SHH_6); this is probably due to many factors, e.g., variations in the embryonic stage at the time of injection and stochastic variations from embryo to embryo with regard to which cells the injected DNA is segregated into during cleavage. However, by combining expression data from a number of expressing embryos (an average of 30 embryos per positive element), we can gain insight into the overall pattern of reporter gene expression prescribed by each element.

In addition to seeing GFP expression in "expected" domains (with respect to the associated gene), GFP expression was also often detected in tissues in which the associated gene is not normally expressed (e.g., muscle cells for SHH_6 and notochord for SOX21_1; see Figure 5). This might be due to incorrect association of gene to element (see above); alternatively, it might reflect the importance of genomic context for function of CNEs and rCNEs. It is possible that certain regions of the genome function as silencers or suppressors, repressing the transcription-stimulating activity of enhancer elements. In our assay we are testing the autonomous enhancing function of our CNEs independent of their normal genomic context. Whilst this enables us to screen rapidly for function in an unconstrained context, it might also result in a loss of the endogenous negative constraints. It will be interesting to determine the combinatorial language of CNEs, and to uncover the importance of genomic context for their function.

Conserved non-coding sequences are likely to function as negative as well as positive regulatory elements. Indeed, it is possible for a conserved non-coding element to act as either an enhancer or repressor of transcription depending on what factors are bound to it [68]. Whether any of our CNEs can function as negative regulatory elements is an interesting question that is beyond the scope of the present study. Zebrafish are the ideal model vertebrate for this screen. These sequences are, by definition, highly similar between mammals and fish, and the data generated are therefore relevant to any vertebrate. Given that CNE DNA can easily be generated from any vertebrate species (given its high degree of sequence identity), subtle differences between CNE sequences may be tested functionally in this system. Zebrafish embryos are both readily produced and easily visualised, allowing convenient live screening throughout development. Their transparency makes the embryos ideally suited to GFP analysis and the problems associated with mosaicism in this screen are relatively easily overcome by injecting large numbers of embryos. Technical advances, such as the use of meganuclease injection, may facilitate this further.

The combination of a comparative genomics approach together with functional screening of conserved elements produces a large and complex dataset. Efficient access, integration, and interrogation of this bioinformatics and functional data is crucial, and of increasing interest to the scientific community, to begin to characterise GRNs in vertebrates. To this end, we have submitted all CNE DNA sequences from *Fugu* to the EMBL nucleotide database and are developing a publicly available relational database in order to store, curate, and analyse data from this study as well as data generated from ongoing identification and characterisation of rCNEs surrounding *trans-dev* genes.

We have identified an important set of highly conserved non-coding vertebrate sequences that associate with developmental regulators and have provided evidence that at least some of them demonstrate regulatory function. They are likely to be implicated in genetic disease, as has already been shown for the SHH gene [30]. Their distal location from coding sequence, often megabases away, makes them strong candidates as causative agents in position effect and breakpoint disorders [69,33]. They are amongst the most highly conserved of all sequences in vertebrate genomes yet they are completely unrecognisable in invertebrates. Given their strong association with genes involved in developmental regulation, they are most likely to contain the essential heritable information for the coordination of vertebrate development.

MATERIALS AND METHODS

Similarity Searching of Non-Coding Sequence between *Fugu* and Human Genomes

GENSCAN [70] (using a suboptimal exon probability cutoff of 0.1) and tRNA-scan-SE (release 1.1) [71] were used to predict coding exons and tRNA genes within the *Fugu* draft genome assembly (release 3.0; Rosalind Franklin Centre for Genomics Research Comparative Genomics Group; http://fugu.rfcgr.mrc.ac.uk/). These predicted sequences were then masked in the *Fugu* sequence by supplying them as a "repeat library" to Repeatmasker35. The masked sequence was similarity searched against human genomic sequence from the Ensembl [41] database v18.34.1 in 1-Mb sections using MegaBLAST [40] version 2.2.6 (word size 20 and mismatch penalty –2). Human and *Fugu* sequences with alignments of 100 bp or over were selected to form the initial CNE sequence dataset. All CNEs with a significant similarity to an expressed transcript in the EMBL database or protein sequence in Swiss-Prot/TrEMBL were removed from the dataset unless located within a UTR. CNEs with significant similarity to non-coding RNAs were also removed. These were located by comparing the CNEs to the microRNA Registry [72] and the Rfam database (version 5.0) [73] using BLASTn [74]. CNEs were also searched against Rfam using the INFERNAL software. This resulted in the detection of 1 microRNA, four U1 snoRNAs, six U2 snoRNAs, three U5 snoRNAs, one U6atac RNA, three 7S RNAs, one 7Sk RNA, and one 5S RNA. The CNEs were also searched against the UTRdb (http://www.ba.itb.cnr.it/BIG/UTRScan/, which is a collection of functional sequence patterns located in 5′ or 3′ UTR sequences, but no significant hits were found. We used the program QRNA [75] to see whether any of the BLAST matches had a pattern of mutation consistent with RNA secondary structure. However, the known RNAs detected above had the most significant hits from this analysis. QRNA uses the mutational pattern in a pairwise alignment to detect non-coding RNAs, but in general the sequence identity of the CNEs is too high for this to be of use.

Analysis of the Distribution of CNEs in the Human Genome

In order to test whether CNEs were randomly distributed, a new random location was allocated uniformly for each CNE within its chromosome. This process was repeated 1,000 times for each chromosome, and the average cluster sizes were calculated for the different distances given in Figure 1B. These cluster sizes were then compared to the cluster sizes of the CNEs. χ^2 tests were carried out comparing the number of clusters containing five or fewer CNEs with the number of clusters containing six or more CNEs. The p-values obtained from the χ^2 test statistics on one degree of freedom are also shown in Figure 1B. They give very strong evidence against the CNEs being randomly distributed.

Identification of Genes Associated with CNEs

The closest gene (using the transcription start site as defined in Ensembl) to the start of each CNE was determined from a list of all human genes supported by external evidence ("known" genes) downloaded using EnsMart, available from the Ensembl Web site (release 24.34e.1;http://www.ensembl.org/). The GOstat program was used to find statistically over-represented GOs in this group of genes [44], using the "goa_human" GO gene association database as a comparator. The minimum length of a considered GO path was five. The false discovery rate option was used to adjust for multiple comparisons.

MLAGAN Alignments

More sensitive global alignment of the CNE regions surrounding 25 orthologous genes in human, *Fugu,* and other vertebrate species was carried out using the MLAGAN alignment tool kit [50]. To locate the orthologous regions in mouse and rat, local similarity searches with BLASTn were carried out using the most outlying CNE associated with each gene. The relevant genomic regions were extracted from Ensembl for human, mouse, and rat. For *Fugu* the genomic regions were extracted from the Medical Research Council Rosalind Franklin Centre for Genomics Research *Fugu* Genomics Project Web site (http://fugu.rfcgr.mrc.ac.uk/) (where there is additional mapping information for scaffolds. All sequences were orientated prior to alignment so that the coding sequence of the gene was in positive orientation in all sequences. The MLAGAN alignment was visualised using the VISTA program [76], enabling the identification of conserved sequences. Because of the larger evolutionary distance between fish and mammals, conservation was measured using a 40-bp window and a cutoff score of 60% identity. *Fugu* was always used as the baseline sequence.

Similarity Searching of Human CNEs against Other Vertebrate and Invertebrate Genomes

To look for the presence of CNEs in other available vertebrate genomes, CNEs were similarity searched against Ensembl mouse (v19.32.2), rat (v21.3.2), chicken (v22.1.1), and zebrafish (v21.3.2) genome sequences using BLASTn with default parameters. All invertebrate sequences in the EMBL database were searched in the same way using BLASTn with non-stringent parameters (mismatch penalty –1, gap open penalty 1, word size 9, and soft masking). More sensitive alignment of flanking orthologous sequence around the SOX21 gene (up to the coding sequence of the genes on either side) from Ensembl C. elegans (v21.25), D. melanogaster (v21.3.1), and Anopheles gambiae (v21.2.2) was carried out using MLAGAN as above.

Fish Care

Zebrafish were raised and bred and embryos staged following standard protocols [77,78]; stages are described as the approximate number of hours post-fertilisation (hpf) when embryos are raised at 28.5 °C. To prevent pigment formation, some embryos were raised in 0.003% phenylthiocarbamide in embryo medium from tailbud stage.

Functional Assay

We assayed for enhancer activity in embryos co-injected with candidate enhancer elements or control DNA and a minimal promoter–reporter construct in a method adapted from Muller and colleagues [37] as described below:

For the preparation of DNA and micro-injection, CNEs, rCNEs, and negative controls were PCR-amplified from *Fugu* genomic DNA. The reporter construct consisting of EGFP (Clontech, Palo Alto, California, United States) under the control of a minimal promoter from the mouse β-globin gene, was PCR-amplified from a plasmid vector (available upon request). Amplified DNA was purified using the GFX PCR purification kit (#27–9602-01; Amersham Biosciences, Amersham, United Kingdom) or the QIAquick PCR purification kit (#28106; Qiagen, Valencia, California, United States). Element DNA or control DNA (at 150–300 ng/µl), reporter construct DNA (at 25 ng/µl), and phenol red (at 0.1%, used as a tracer) were combined and co-injected into embryos produced from natural matings between the one-cell stage and early cleavage stages, using an Eppendorf (Hamburg, Germany) FemtoJet pressure injection system. Any embryos developing abnormally were discarded before screening. For screening of embryos and data collection, on the second day of development (approximately 26–33 hpf), injected embryos were anaesthetised in Tricaine [77] and analysed for GFP expression by observation under fluorescence illumination using an Olympus (Tokyo, Japan) IX81 motorised inverted microscope. Images were captured using an FVII CCD monochrome digital camera and analySIS image-processing software.

GFP-expressing cells were classified according to the following tissue categories: forebrain, midbrain, hindbrain, spinal cord, eye, ear, notochord, muscle, blood (circulating)/ blood islands, heart/pericardial region (Please note: Some cells classified in this category may be circulating blood cells), epidermis/EVL, or fins. Cells that did not fall into one of these major expression categories (or that were not possible to unequivocally identify from morphology or localisation) were categorised as "other". The location and tissue category of each GFP-expressing cell for each embryo was recorded schematically using Adobe Photoshop software (Adobe Systems, San Jose, California, United States), by manually drawing colour-coded schematised cells in appropriate positions onto an overlay of a camera lucida drawing of a 31-hpf embryo (from staging series by C. Kimmel, downloaded from "Zebrafish: The Living Laboratory", courtesy of the Zebrafish CD Exchange Project; contact Mark Cooper at E-mail:mscooper@uwashington.edu;data relating to tissue category was also recorded on a spreadsheet.

GFP expression data were collected from between 25 and 55 expressing embryos per element injected. Cumulative overlaid schematised expression data for each element were compressed into a single JPEG file (displayed in Figure 5). Thus, the JPEG image for each element is designed to give an overall impression of the spatial pattern to which the element directs expression. Coupled with the accompanying graphs, the data present an overview of the spatial localisation of GFP expression as well as an idea of the number of cells per tissue in which GFP expression was detected, indicating the strength of the element›s enhancing properties or the size of the cell population to which expression is directed.

Anti-GFP Immunostaining

Embryos were fixed in 4% paraformaldehyde and stained with rabbit polyclonal anti-GFP (#TP401 at 1/1,000 dilution; AMS Biotechnology, Abingdon Oxon, United Kingdom) using standard protocols [79] and the ABC amplification system (Vectastain; Vector Laboratories, Burlingame, California, United States). Stained embryos were cleared in glycerol, flatmounted, and observed/imaged as above.

ACKNOWLEDGMENTS

This work was supported by the Medical Research Council and by private funds from Sydney Brenner.

AUTHOR CONTRIBUTIONS

MG and GE designed the study. AW, MG, DKG, PS, SFS, PN, and JEC performed the experiments. AW, MG, DKG, PS, GKM, TV, KK, JEC, and GE analysed the data. GKM, TV, HC, KW, IA, WG, and YJKE contributed reagents/materials/ analysis tools. AW, GKM, JEC, and GE wrote the paper.

REFERENCES

1. Davidson EH, Rast JP, Oliveri P, Ransick A, Calestani C, et al. (2002) A genomic regulatory network for development. Science 295: 1669–1678.
2. Davidson EH, McClay DR, Hood L (2003) Regulatory gene networks and the properties of the developmental process. Proc Natl Acad Sci U S A 100: 1475–1480.
3. Albert R, Othmer H (2003) The topology of the regulatory interactions predicts the expression pattern of the segment polarity genes in Drosophila melanogaster. J Theor Biol 223: 1–18.
4. Oliveri P, Davidson EH (2004) Gene regulatory network controlling embryonic specification in the sea urchin. Curr Opin Genet Dev 14: 351–360.
5. Lander ES, Linton LM, Birren B, Nusbaum C, Zody MC, et al. (2001) Initial sequencing and analysis of the human genome. Nature 409: 860–921.
6. Venter CJ, Adams MD, Myers EW, Li PW, Mural RJ, et al. (2001) The sequence of the human genome. Science 291: 1304–1351.
7. Aparicio S, Chapman J, Stupka E, Putnam N, Chia J, et al. (2002) Whole-genome shotgun assembly and analysis of the genome of Fugu rubripes. Science 297: 1301–1310.
8. Waterston RH, Lindblad-Toh K, Birney E, Rogers J, Abril JF, et al. (2002) Initial sequencing and comparative analysis of the mouse genome. Nature 420: 520–562.
9. Gibbs RA, Weinstock GM, Metzker ML, Muzny DM, Sodergren EJ, et al. (2003) Genome sequence of the brown Norway rat yields insights into mammalian evolution. Nature 428: 493–521.
10. Wasserman WW, Sandelin A (2004) Applied bioinformatics for the identification of regulatory elements. Nat Rev Genet 5: 276–287.
11. Hardison RC, Oeltjen J, Miller W (1997) Long human-mouse sequence alignments reveal novel regulatory elements: Reasons to sequence the mouse genome. Genome Res 7: 959–966.
12. Oeltjen JC, Malley T, Muzny D, Miller W, Gibbs RA, et al. (1997) Large-scale sequence comparison analysis of the human and murine Bruton's tyrosine kinase loci reveals conserved regulatory domains. Genome Res 7: 315–319.
13. Wasserman WW, Palumbo M, Thompson W, Fickett JW, Lawrence CE (2000) Human-mouse genome comparisons to locate regulatory sites. Nat Genet 26: 225–228.
14. Gottgens B, Barton LM, Gilbert JG, Bench AJ, Sanchez MJ, et al. (2000) Analysis of vertebrate SCL loci identifies conserved enhancers. Nature Biotechnol 18: 181–186.
15. Frazer KA, Sheehan JB, Stokowski RP, Chen X, Hosseini R, et al. (2001) Evolutionarily conserved sequences on human Chromosome 21. Genome Res 11: 1651–1659.
16. Frazer KA, Tao H, Osoegawa K, de Jong PJ, Chen X, et al. (2004) Noncoding sequences conserved in a limited number of mammals in the SIM2 interval are frequently functional. Genome Res 14: 367–372.
17. Dermitzakis ET, Reymond A, Lyle R, Scamuffa N, Ucla

C, et al. (2002) Numerous potentially functional but non-genic conserved sequences on human Chromosome 21. Nature 420: 578–582.

18. Dermitzakis ET, Reymond A, Scamuffa N, Ucla C, Kirkness E, et al. (2003) Evolutionary discrimination of mammalian conserved non-genic sequences (CNGs). Science 302: 1033–1035.
19. Dermitzakis ET, Kirkness E, Schwarz S, Birney E, Reymond A, et al. (2004) Comparison of human Chromosome 21 conserved nongenic sequences (CNGs) with the mouse and dog genomes shows that their selective constraint is independent of their genic environment. Genome Res 14: 852–859.
20. Tautz D (2000) Evolution of transcriptional regulation. Curr Opin in Genet Dev 10: 575–579.
21. Bejerano G, Pheasant M, Makunin I, Stephen S, Kent WJ, et al. (2004) Ultraconserved elements in the human genome. Science 304: 1321–1325.
22. Thomas JW, Touchman JW, Blakesley RW, Bouffard GG, Beckstrom-Sternberg SM, et al. (2003) Comparative analyses of multi-species sequences from targeted genomic regions. Nature 424: 788–793.
23. Gillemans N, McMorrow T, Tewari R, Wai AW, Burgtorf C, et al. (2003) Functional and comparative analysis of globin loci in puffferfish and humans. Blood 101: 2842–2849.
24. Chapman MA, Donaldson IJ, Gilbert J, Grafham D, Rogers J, et al. (2004) Analysis of multiple genomic sequence alignments: A web resource, online tools, and lessons learned from analysis of mammalian SCL loci. Genome Res 14: 313–318.
25. Marshall H, Studer M, Popperl H, Aparicio S, Kuroiwa A, et al. (1994) A conserved retinoic acid response element required for early expression of the homeobox gene Hoxb-1. Nature 370: 567–571.
26. Rowitch DH, Echelard Y, Danielian PS, Gellner K, Brenner S, et al. (1998) Identification of an evolutionarily conserved 110 base-pair cis-acting regulatory sequence that governs Wnt-1 expression in the murine neural plate. Development 125: 1735–1746.
27. Kammandel B, Chowdhury K, Stoykova A, Aparicio S, Brenner S, et al. (1999) Distinct cis-essential modules direct the time-space pattern of the Pax6 gene activity. Dev Biol 205: 79–97.
28. Bagheri-Fam S, Ferraz C, Demaille J, Scherer G, Pfeifer D (2001) Comparative genomics of the SOX9 region in human and *Fugu rubripes* conservation of short regulatory sequence elements within large intergenic regions. Genomics 78: 73–82.
29. Ghanem N, Jarinova O, Amores A, Long Q, Hatch G, et al. (2003) Regulatory roles of conserved intergenic domains in vertebrate Dlx Bigene clusters. Genome Res 13: 533–543.
30. Lettice LA, Heaney SJ, Purdie LA, Li L, de Beer P, et al. (2003) A long range Shh enhancer regulates expression in the developing limb and fin and is associated with preaxial polydactyly. Hum Mol Genet 12: 1725–1735.
31. Nobrega MA, Ovcharenko I, Afzal V, Rubin EM (2003) Scanning human gene deserts for long-range enhancers. Science 302: 413.
32. Santagati F, Abe K, Schmidt V, Schmitt-John T, Suzuki M, et al. (2003) Identification of Cis-regulatory elements in the mouse Pax9/Nkx2–9 genomic region: Implication for evolutionary conserved synteny. Genetics 165: 235–242.
33. Spitz F, Gonzalez F, Duboule D (2003) A global control region defines a chromosomal regulatory landscape containing the HoxD cluster. Cell 113: 405–417.
34. Kimura-Yoshida C, Kitajima K, Oda-Ishii I, Tian E, Suzuki M, et al. (2004) Characterization of the pufferfish Otx2 cis-regulators reveals evolutionarily conserved genetic mechanisms for vertebrate head specification. Development 131: 57–71.
35. Kumar S, Hedges SB (1998) A molecular timescale for vertebrate evolution. Nature 392: 917–920.
36. Brenner S, Elgar G, Sandford R, Macrae A, Venkatesh B, et al. (1994) Characterization of the pufferfish *(Fugu)* genome as a compact model vertebrate genome. Nature 366: 265–268.
37. Muller F, Williams DW, Kobolak J, Gauvry L, Goldspink G, et al. (1997) Activator effect of coinjected enhancers on the muscle-specific expression of promoters in zebrafish embryos. Mol Reprod Dev 47: 404–412.
38. Muller F, Chang B, Albert S, Fischer N, Tora L, et al. (1999) Intronic enhancers control expression of zebrafish sonic hedgehog in floor plate and notochord. Development 126: 2103–2116.
39. Dickmeis T, Plessy C, Rastegar S, Aanstad P, Herwig R, et al. (2004) Expression profiling and comparative genomics identify a conserved regulatory region controlling midline expression in the zebrafish embryo. Genome Res 14: 228–238.
40. Zhang Z, Schwartz S, Wagner L, Miller W (2000) A greedy algorithm for aligning DNA sequences. J Comput Biol 7: 203–214.
41. Hubbard T, Barker D, Birney E, Cameron G, Chen Y, et al. (2002) The Ensembl genome database project. Nucleic Acids Res 32: D468–D470.
42. Vandepoele K, De Vos W, Taylor JS, Meyer A, Van de Peer Y (2000) Major events in the genome evolution of vertebrates: Paranome age and size differ considerably between ray-finned fishes and land vertebrates. Proc Natl Acad Sci U S A 101: 1638–1643.
43. Force A, Lynch M, Pickett FB, Amores A, Yan YL, et al. (1999) Preservation of duplicate genes by complementary, degenerative mutations. Genetics 151: 1531–1545.
44. Beissbarth T, Speed TP (2004) GOstat: Find statistically overrepresented Gene Ontologies within a group of genes. Bioinformatics 20: 1464–1465.
45. Ashburner M, Ball CA, Blake JA, Botstein D, Butler H, et al. (2000) Gene ontology: Tool for the unification of biology. The Gene Ontology Consortium. Nat Genet 25: 25–29.
46. Mulder NJ, Apweiler R, Attwood TK, Bairoch A, Barrell D, et al. (2003) The InterPro database, 2003 brings

increased coverage and new features. Nucleic Acids Res 31: 315–318.

47. Sultana R, Yu CE, Yu J, Munson J, Chen D, et al. (2002) Identification of a novel gene on Chromosome 7q11.2 interrupted by a translocation breakpoint in a pair of autistic twins. Genomics 80: 129–134.

48. Miles C, Elgar G, Coles E, Kleinjan DJ, van Heyningen V, et al. (1998) Complete sequencing of the *Fugu* WAGR region from WT1 to PAX6: Dramatic compaction and conservation of synteny with human Chromosome 11p13. Proc Natl Acad Sci U S A 95: 13068–13072.

49. Chang CW, Tsai CW, Wang HF, Tsai HC, Chen HY, et al. (2004) Identification of a developmentally regulated striatum-enriched zinc-finger gene, Nolz-1, in the mammalian brain. Proc Natl Acad Sci U S A 101: 2613–2618.

50. Brudno M, Do CB, Cooper GM, Kim MF, Davydov E, et al. (2003) LAGAN and Multi-LAGAN: Efficient tools for large scale multiple alignment of genomic DNA. Genome Res 13: 721–731.

51. Margulies EH, Blanchette M, Haussler D, Green ED, NISC Comparative Sequencing Program (2003) Identification and characterization of multi-species conserved sequences. Genome Res 13: 2507–2518.

52. Argenton F, Giudici S, Deflorian G, Cimbro S, Cotelli F, et al. (2004) Ectopic expression and knockdown of a zebrafish sox21 reveal its role as a transcriptional repressor in early development. Mech Dev 121: 131–142.

53. Rex M, Uwanogho DA, Orme A, Scotting PJ, Sharpe PT (1997) cSox21 exhibits a complex and dynamic pattern of transcription during embryonic development of the chick central nervous system. Mech Dev 66: 39–53.

54. Rimini R, Beltrame M, Argenton F, Szymczak D, Cotelli F, et al. (1999) Expression patterns of zebrafish sox11A, sox11B and sox21. Mech Dev 89: 167–171.

55. Uchikawa M, Kamachi Y, Kondoh H (1999) Two distinct subgroups of Group B Sox genes for transcriptional activators and repressors: Their expression during embryonic organogenesis of the chicken. Mech Dev 84: 103–120.

56. Kleinjan DA, Seawright A, Schedl A, Quinlan RA, Danes S, et al. (2001) Aniridia-associated translocations, DNase hypersensitivity, sequence comparison and transgenic analysis redefine the functional domain of PAX6. Hum Mol Genet 10: 2049–2059.

57. Kleinjan DA, Seawright A, Childs AJ, van Heyningen V (2004) Conserved elements in Pax6 intron 7 involved in (auto)regulation and alternative transcription. Dev Biol 265: 462–477.

58. Griffin C, Kleinjan DA, Doe B, van Heyningen V (2002) New 3′ elements control Pax6 expression in the developing pretectum, neural retina and olfactory region. Mech Dev 112: 89–100.

59. Ross AJ, Ruiz-Perez V, Wang Y, Hagan DM, Scherer S, et al. (1998) A homeobox gene, HLXB9, is the major locus for dominantly inherited sacral agenesis. Nat Genet 20: 358–361.

60. Wendik B, Maier E, Meyer D (2004) Zebrafish mnx genes in endocrine and exocrine pancreas formation. Dev Biol 268: 372–383.

61. McMahon AP, Ingham PW, Tabin CJ (2003) Developmental roles and clinical significance of hedgehog signaling. Curr Top Dev Biol 53: 1–114.

62. Krauss S, Concordet JP, Ingham PW (1993) A functionally conserved homolog of the*Drosophila* segment polarity gene hh is expressed in tissues with polarizing activity in zebrafish embryos. Cell 75: 1431–1444.

63. Ekker SC, Ungar AR, Greenstein P, von Kessler DP, Porter JA, et al. (1995) Patterning activities of vertebrate hedgehog proteins in the developing eye and brain. Curr Biol 5: 944–955.

64. Boffelli D, Nobrega MA, Rubin EM (2004) Comparative genomics at the vertebrate extremes. Nat Rev Genet 5: 456–465.

65. Kleinjan DJ, Seawright A, Elgar G, van Heyningen V (2002) Characterization of a novel gene adjacent to PAX6, revealing synteny conservation with functional significance. Mamm Genome 13: 102–107.

66. Eggers JH, Stock M, Fliegauf M, Vonderstrass B, Otto F (2002) Genomic characterization of the RUNX2 gene of Fugu rubripes. Gene 291: 159–167.

67. Li X, Zhao X, Fang Y, Jiang X, Duong T, et al. (1998) Generation of destabilized green fluorescent protein as a transcription reporter. J Biol Chem 273: 34970–34975.

68. McDonald JA, Fujioka M, Odden JP, Jaynes JB, Doe CQ (2003) Specification of motoneuron fate in *Drosophila* Integration of positive and negative transcription factor inputs by a minimal eve enhancer. J Neurobiol 57: 193–203.

69. Kleinjan DJ, van Heyningen V (1998) Position effect in human genetic disease. Hum Mol Genet 7: 1611–1618.

70. Burge C, Karlin S (1997) Prediction of complete gene structures in human genomic DNA. J Mol Biol 268: 78–94.

71. Lowe TM, Eddy SR (1997) tRNAscan-SE: A program for improved detection of transfer RNA genes in genomic sequence. Nucleic Acids Res 25: 955–964.

72. Griffiths-Jones S (2004) The microRNA Registry. Nucleic Acids Res 32: D109–D111.

73. Griffiths-Jones S, Bateman A, Marshall M, Khanna A, Eddy SR (2003) Rfam: An RNA family database. Nucleic Acids Res 31: 439–441.

74. Altschul SF, Madden TL, Schaffer AA, Zhang J, Zhang Z, et al. (1997) Gapped BLAST and PSI-BLAST: A new generation of protein database search programs. Nucleic Acids Res 25: 3389–3402.

75. Rivas E, Eddy SR (2001) Noncoding RNA gene detection using comparative sequence analysis. BMC Bioinformatics 2: 8.

76. Mayor C, Brudno M, Schwartz JR, Poliakov A, Rubin EM, et al. (2000) VISTA: Visualizing global DNA sequence alignments of arbitrary length. Bioinformatics 16: 1046–1047.

77. Westerfield M (2000) The zebrafish book: A guide for

the laboratory use of zebrafish (*Danio rerio*). 4th ed. Eugene (Oregon): University of Oregon Press.

78. Kimmel CB, Ballard WW, Kimmel SR, Ullmann B, Schilling TF (1995) Stages of embryonic development of the zebrafish. Dev Dyn 203: 253–310.

79. Moens CB, Fritz A (1999) Techniques in neural development. Methods Cell Biol 59: 253–272.

CHAPTER 2

GENE EVOLUTION IN VERTEBRATES

Zhan Yu, David Morais, Mahine Ivanga and Paul M Harrison

BACKGROUND

Genes are subject to many different processes that give rise to novel sequences, such as segmental and local duplication, gene conversion, and retrotransposition. The extent to which these different processes contribute to gene evolution is unclear. In the present paper, we focus on the phenomenon of gene retrotransposition. Retrotransposition entails the reverse transcription of an mRNA transcript and the subsequent re-integration of the resulting cDNA into genomic DNA, in germ-line cells [1]. There is substantial genomic evidence for large-scale retrotransposition of mRNAs in mammalian genomes, from detection of thousands of apparent retropseudogenes in human, mouse and rat [2, 3, 4]. Such retropseudogenes (RψGs) are decayed or disabled gene sequence copies (typically bearing frameshifts and stop codons) that demonstrate the hallmark characteristics of retrotransposition, namely lack of introns of the parental gene, and also 3' polyadenine tails, if formed more recently [5]. Other features include short direct repeats flanking the sequence (for young retrotranspositions) [6], frequent 5' truncations, and genomic location different from that of the parent gene [2, 3]. It has been demonstrated experimentally that RψG s can be formed through the action of LINE-1 reverse transcriptases [7]. The computational comparison of LINEs and RψG s also supports the generation of RψG s by LINEs [8]. The poly(A) tails and frequent truncations found at the 5' end in the RψG s are typical for LINEs [2]. Moreover, they share similar structures, including a common TT|AAAA insertion motif [8].

Since the substantial majority of these retrosequences bear disablements (frameshifts and stop codons), or have codon substitution patterns indicative of decay [9, 5, 3], gene retrotransposition appears generally to lead to non-functional sequences that decay in the genomic DNA as evolution progresses [10, 2, 9]. However, even though the promoters of these gene retrosequences are not transferred, a small minority of them appears to be transcribed [11]. For the human genome, there is a small population of at least ~200 transcribed processed pseudogenes, which have the symptoms of a lack of coding ability despite evidence of transcription, and are significantly likely to be found near others genes (as would be expected if they are co-opting promoters) [11].

Generation of a new functional gene is also a possible outcome of retrotransposition [10]. There is an increasing number of transcribed, functionally characterized genes in

Citation: Zhan Yu, David Morais, Mahine Ivanga and Paul M Harison, gene evolution in vertebratesss, DOI: 10.1186/1471-2105-8-308

mammalian and invertebrate animal genomes reported to bear the characteristics of retrosequences [12]. Over ninety such retrogenes have been annotated in the human and mouse genomes [13]. Most of the functional retrogenes identified are expressed in testis and may have provided important raw material for rapid testis evolution in primates [12].

Here, to derive an overview of the role of gene retrotransposition in the genome evolution of vertebrates, and particularly mammals, we derive and apply a robust procedure to annotate gene retrotranspositions, built on our previous analyses of retropseudogenes [11, 3, 2, 14]. Our strategy incorporates a new rapid procedure for annotating retrocopies in the genomic DNA, in tandem with a pipeline to identify them in existing gene annotations. This PR annotation pipeline incorporates aging of the sequences through evolutionary rate analysis relative to putative parents and their orthologs, as well as analysis of the chromosomal milieus of these sequences and their putative parents. We find evidence for, on average, several hundred PRs in each proteome. Focussing on human and mouse, we find evidence for spurts of gene retrotransposition in both human and mouse, since divergence from dog. A small number (>50) of PRs have formed in both mouse and human since divergence form dog, that show signs of being under selection to maintain their coding sequences.

METHODS

Genome Data

The genome sequences and annotations of seven organisms analyzed in this paper (human, dog, cow, mouse, rat, chicken and Tetraodon nigriviridis), were downloaded from the Ensembl Web site [15], in January 2005.

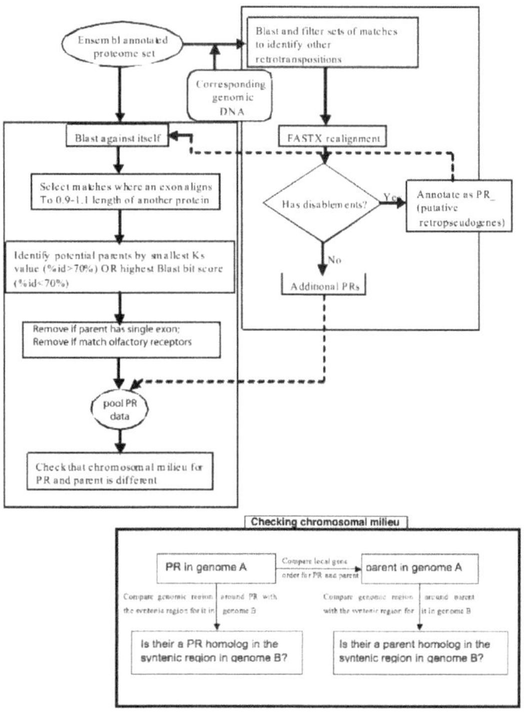

Figure 1 Pipeline summarizing the annotation of PRs and retropseudogenes. The pipeline for PR annotation is summarized. There is an inset at the bottom, that summarizes the tests for local gene order and chromosomal milieu.

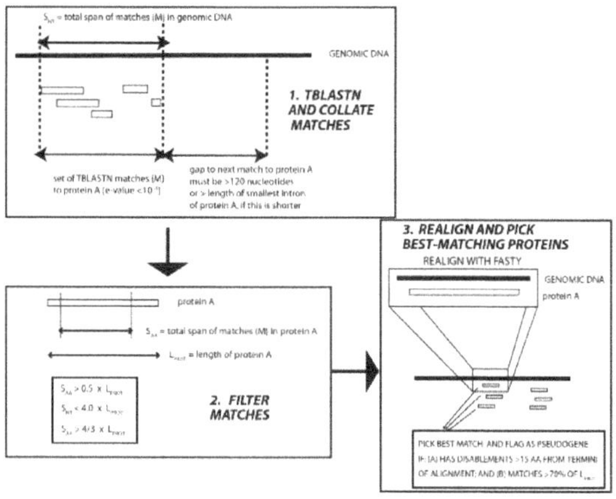

Figure 2 Rapid annotation of retropseudogenes. (1) TBLASTN matches (e-value $\leq 10^{-4}$) of the annotated proteome against the genomic DNA are sorted by coordinates and collated for each protein to form a set of matches {M}. (2) The sets {M} are filtered using length-based heuristics. (3) Each protein is realigned to the genomic DNA using FASTY, and the best-matching proteins at each point have disablements and that matches >70% of the length of the parent sequence are picked as retropseudogene annotations.

Version 2.1 of the chimpanzee assembly (downloaded in April 2006) was also used. Putative retrogenes (PRs) were identified in the annotated proteomes of eight vertebrates using the pipeline in Figure 1. This procedure is described in detail below. In tandem, putative retropseudogenes (RψGs), and additional PRs outside of current protein annotations, were assigned using a modification of previous procedures (Figures 1&2) [14, 2]. Genes from which PR s and RψG s are thought to have originated are called parent genes.

Rapid Identification of Retropseudogenes (*RψGs*)

Retropseudogenes were annotated on the Ensembl genome versions used in this present analysis, using the rapid improvement of previous procedures to identify retropseudogenes described above (summarised in Figure 2) [11, 5, 14, 2, 3].

Identification of putative retrotransposed genes (*PRs*)

- Homology detection: Each proteome was compared against itself using BLAST to find similarities with e-value $\leq 10^{-4}$ [16]. Any match to a potential pseudogene contaminant in the proteome annotations was removed (Figure 1).
- Exon seam analysis: Exon boundary information for each protein was extracted from the appropriate Ensembl genome annotation files. The positioning of exon boundaries in encoded protein sequences, i.e., 'exon seams', was then deduced [11]. Using the positioning of exon seams, the BLAST matches between proteins were filtered to pick out alignments between a protein encoded by a multiple-exon gene,

and a single exon of another gene. To define PRs, the length of the exon was required to be between 0.9 and 1.1 of the whole length of the multiple-exon protein. (This is stricter than the criterion of 0.7 of the length used for annotation of retropseudogenes [2]).

- Assignment of parent genes: To assign parent genes to PRs, we calculated substitution rates at synonymous codon sites (i.e., Ks values) for all matches to PRs using the package PAML, for all instances where the amino-acid sequence identity for the pair of sequences is ≥ 70%. The sequence with the smallest Ks value was chosen as the 'parent gene'. For sequence identities <70%, saturation of substitutions is likely [17], and so the sequence with the highest BLAST bitscore in alignment with the PR was chosen as the putative parent.
- Additional filtering: In addition, PRs were discarded if they matched olfactory receptors (ORs) with BLAST e-value ≤ 10^{-4} over ≥ 0.5 of the length of the OR, since recent olfactory receptors (ORs) have probably originated from different mechanism other than retrotransposition [2]. Olfactory receptor sequences were taken from ORDB [18].
- Local gene order test for similarity of the chromosomal milieus for PRs and parent genes: To check that the PRs did not arise from local or segmental duplication, we derived a 'local gene order' test. For this test, we compared the chromosomal milieus of PRs and parents for significant similarity, as follows. Proteins encoded by genes adjacent to PRs in the chromosomes were BLASTed against the corresponding proteins from genes adjacent to putative parents, for a given window (w_{genes}) of number of genes in either direction (5' and 3'). A w_{genes} size of 7 (the PR or parent, plus 3 genes in either direction), with an allowance for one gap between the positions of matches within w_{genes}, was found to be suitable. The number of significant homologous matches $N_{homologs}$ (BLAST e-value ≤ 10^{-4}, sequence identity >40%, and match ≥ 0.6 length of both PR and parent) between the milieus of PR and parent was tallied. An expected benchmark distribution for $N_{homologs}$ was derived for the chromosomal milieus of 1,000 randomly-sampled pairs of proteins that have any significant BLAST match to each other (e-value ≤ 10^{-4}). From examination of this distribution, we found that 80% of such random pairs have $N_{homologs}$ <1, and 87% have $N_{homologs}$ ≤ 1. We thus chose $N_{homologs}$ = 1, as a suitable threshold for local similarity arising from duplication of genomic DNA. However, the results differ little if a threshold of $N_{homologs}$ = 0 is used. This procedure was applied to the genomes of the human and mouse. Interestingly, application of this criterion resulted in the exclusion of many sequences with large individual exons (PRs with FLE_{parent} ≥ 0.8; 36/86 = 42% of those excluded), that may be false positives in our data set of PRs. A large fraction of these sequences (68%) tend to have long, tandem arrays of Zn-finger domains covering more than a third of their sequences.

Additional Filtering and Annotation

The following additional analysis was performed on the PRs:

(i) Fraction of largest exon in parent: We calculated the fraction of the length any parent gene that is taken up by its largest exon. This is denoted FLE_{parent}. We found that there is no peculiar tendency for the parents of PRs to have a single large exon (which would yield a tendency for high FLE values).

(ii) Overlap with retropseudogenes annotations: Retropseudogenes were annotated on the Ensembl genome versions used in this present analysis, using the rapid improvement of previous procedures to identify retro(pseudo) genes described above. Any PR that overlapped one of these annotations was flagged (Table 1).

Table 1 Overview of gene retrotransposition analysis for eight vertebrates

Species	**Number of genes** *	**Number of PRs**	**PR matches retrotransposed TE** **	**PR overlaps pseudogene annotation**	**FLE ≤ 0.8** ***	**Number of retropseudogenes (*RψGs*)**	**PRs passing local gene order test**	**Matching Refseq mRNA or Unigene consensus** ****
Human	22219	631 (3%)	78 (12%)	36 (6%)	504	2493	545	145/631 (23%)
Chimp	20980	476 (2%)	17 (4%)	5 (1%)	339	1889	----	
Dog	18199	409 (2%)	18 (4%)	25 (6%)	363	3505	----	
Cow	23147	790 (3%)	46 (6%)	104 (13%)	479	1996	----	
Mouse	25021	663 (3%)	31 (5%)	75 (11%)	518	2969	533	58/663 (9%)
Rat	22157	567 (3%)	21 (4%)	62 (11%)	492	4520	----	
Chicken	17707	321 (2%)	15 (5%)	26 (8%)	267	720	----	
Tetraodon	28005	227 (1%)	4 (2%)	10 (4%)	203	644	----	

* The number of gene annotations (both those labelled 'known' and 'novel') for the genome version downloaded from Ensembl [[15]; see Methods for details].

** TE = transposable element ; see Methods for details.

*** FLE = fraction of largest exon ; see Methods for details.

**** Number of PRs with complete Refseq mRNAs or complete Unigene consensus sequences (percentage of these in brackets); see Methods for details. PRs have significantly much less mapping of this transcription information than the whole annotated proteome for these organisms.

(iii) Filtering for potential transposable elements (TEs): Each proteome was compared using TBLASTN to libraries of transposable elements taken from the RepeatMasker distribution [19], using an e-value threshold of $\leq 10^{-4}$. Any proteins containing SINEs, or near-complete matches to LINEs (≥ 0.8 of their lengths), were labeled as potentially TE-containing.

(iv) Whether single-exon gene or multiple-exon gene: The PRs were labeled as either single-exon genes or part of multiple-exon genes.

Orthologs

Orthologs of parent genes were identified using the bi-directional best hits method, using BLAST (e-value $\leq 10^{-4}$, amino-acid sequence identity $\geq 40\%$ and requiring the alignment to cover ≥ 0.6 of the lengths of each sequence. The bi-directional best hits method is a common procedure for guarding against considering paralogs.

Analysis of K_s and K_a/K_s Values, and Derivation of Genome- and Lineage-Specific Gene Lists

The package PAML [20] was used to calculate maximum-likelihood Ka, Ks and Ka/Ks values for pairs of sequences (either PR versus computed ancestral sequences, or PR versus parent). In addition, branch-specific maximum-likelihood Ka/Ks values were calculated for three-way alignments of PR, parent and parent's orthologs from another close species.

We applied three different strategies based on analysis of Ks, to determine lists of genome-specific and lineage-specific PRs. For example, for the human genome, we calculated human-specific lists relative to the chimpanzee genome. Also, we calculated primate-specific lists for human plus chimpanzee, relative to a mammalian outgroup, such as dog or cow. To determine genome-specific lists of PRs, we investigated each of the following three methods ("parent's ortholog" refers to the ortholog of the parent in the most closely related genome):

- The distribution of Ks values for orthologous genes in the two organisms was calculated, and the median value m derived from this. If Ks [PR←→parent] <m and Ks [parent ←→parent's ortholog] >m, then a PR is labeled genome-specific;
- Secondly, a PR could be labeled genome-specific if Ks [PR ←→parent] < Ks [parent ←→parent's ortholog];
- Thirdly, a PR could be labeled genome-specific if, in a three-way tree of PR, parent and parent's ortholog, the branch-specific Ks [PR] is < (Ks [parent] + Ks [parent's ortholog])/2 ;
- Lineage-specific lists were derived in a similar fashion.

Analysis of Reading Frame Conservation

We assessed the reading frame conservation (RFC) in sequences using simulations of insertion and deletion governed by power-law insertion/deletion (indel) statistics [4]. Power-law statistics for indels were extracted from recently-formed RΨGs having $\geq 85\%$ amino-acid identity with their parent sequences. Power-law relationships were fitted, omitting points for any indels of size 3n, with n any positive integer. Expected ratios for insertions versus deletions were taken from this data; the expected number of indels per nucleotide substitution for several mammals was culled from the literature [21, 22]. The program DNADIST [23] was used to calculate the nucleotide-level divergence of the PRs from ancestral sequences (calculated using PAML [20]; see section on Ka/Ks analysis above). This divergence value is used as a target in simulations. For eachPR, repeated simulations of the evolution of the ancestral sequence towards present-day, for 1000 iterations, was performed using a Kimura two-parameter model. In each case, the resulting simulated protein coding sequence was marked for frame disablements (stop codons and frameshifts). PR sequences whose simulations yielded frame-disrupted sequences $\geq 99\%$ of the time were labeled as having significant RFC.

Assignment of Functional Categories

GO (Gene Ontology; [24]) functional categories were taken from the annotation files provided on the Ensembl [15] and Gene Ontology websites [25]. Further GO term annotations were derived by mapping functional GO annotations for the PDB (also downloaded from the GO website) onto Ensembl protein annotations, using 50% sequence identity and 0.8 fractional sequence coverage (for the protein domain) as thresholds, using alignment made by the program BLASTP (e-value ≤ 0.0001) [16]. These thresholds were benchmarked on the complete SCOP protein domain sequence database [26], to give a 2% false positive rate for GO term transfer.

Mapping of cDNAs/mRNAs

Refseq mRNAs and complete Unigene consensus sequences were downloaded from the NCBI website [27], for both human and mouse. These were mapped to the coding sequences of Ensembl gene annotations, using blastn (e-value $\leq 1 \times 10^{-10}$ for alignments ≥ 100 nucleotides) [16]. All mappings that match with $\geq 99\%$ sequence identity over ≥ 0.99 of the sequence length of the cDNA or mRNA, after removal of any polyadenylation, were retained. Further restriction of analysis of cDNA/mRNA mappings to those that do not match their putative parent sequences with >95% sequence identity, does not change the trends reported with regard to transcription evidence reported below.

RESULTS AND DISCUSSION

The pipelines for annotating the complement of gene retrotranspositions (both retropseudogenes (RψG s) and putative retrogenes (PRs)) were applied to eight vertebrates. In particular, we focused on the mammals, to analyse the ages of putative retrogenes (PRs), to derive genome- and lineage-specific lists and to check for spurts of gene retrotransposition activity. We then examined for evidence of transcription (mRNA and cDNA mapping), involvement in alternative splicing, selection pressures (significant Ka/Ks values and reading-frame conservation), and for functional categorizations of parent genes.

Overview of Gene Retrotranspositions in Vertebrates

Our analysis suggests that up to ~3% of the genes encoded in a vertebrate genome contain a PR (Table 1), with the smallest percentages in the chicken and puffer fish T. nigroviridis. By comparison, mammalian genomes have ~2,000–5,000 retropseudogenes (RψG s), that have at least 70% of the coding sequence of their parent genes, again with smaller numbers in non-mammal vertebrates (just 644 RψG s in T. nigroviridis) (Table 1). These results together indicate that there has been less, recent gene-retrotransposition activity in the two non-mammal vertebrates. These observations tally well with other evidence for less retrotransposition activity in chicken and Tetraodon. In chicken, there appear to be little or no SINEs [28], and only ~8% of the genomic DNA is comprised of the CR1 ('chicken repeat 1') LINE-1 [28, 29], whose reverse transcriptase is thought not to copy polyadenylated mRNAs [29]. In Tetraodon, <1% of the genome is comprised of retrotransposons, so gene retrotransposition should consequently be less likely [30]. For the eight genomes studied here, there are no significant linear correlations between the number of genes or proteins from a genome, versus the number of PRs, or RψGs (data not shown). Small percentages of the PRs could be classified as homologs of retrotransposed transposable elements, such as LINEs (2–12%), or as overlapping pseudogene annotations (4–13%).

As described in detail in Methods, we applied a 'local gene order' test, to set aside any PRs that may have arisen through local or segmental duplication, specifically for the human and mouse genomes (Table 1). This filter allows for at most one homologous protein encoded within a window of +/-3 genes along the genomic DNA (i.e., $N_{homologs} \leq 1$) (Table 1). The substantial majority of human and mouse PRs pass this filter (80–87%).

Ages of Primate and Rodent Gene Retrotranspositions

How old are these PRs? Is there any evidence for spurts of gene retrotransposition activity in mammalian evolution? To answer these questions, we examined the distribution of Ks values for PRs compared to their assigned parent genes, in the human and mouse genomes. (Only PRs passing the local gene order test, with threshold $N_{homologs} \leq 1$ were analysed.) Ks is the rate of synonymous substitutions per synonymous site in codons, and has been generally used to age coding sequences. From comparing Ks values for PRs, their parents, and orthologs of their parents, we have also been able to derive lists of genome-specific and lineage-specific PRs (Figure 4; see Methods for details).

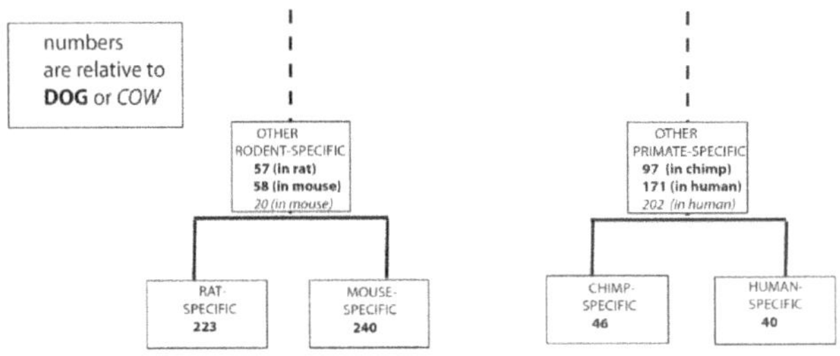

Figure 4 Lineage-specific lists of PRs: The number of species-specific PRs relative to other species. PRs specific relative to other species were obtained by comparison of Ks between the PR and its parent and the Ks between the parent (Ks_{PR_parent}) and the ortholog of the parent in the other species ($Ks_{parent\_ortholog}$). PRs with $Ks_{PR\_parent} < Ks_{parent\_ortholog}$ were defined as specific PRs relative to the other species. Only PRs which amino acid identity to parents is more than 70% and have an ortholog in other species were subjected to this calculation. Orthology criteria used are 40% identity over 60% length overlap. 'Human-specific' and 'Chimp-specific' PRs are those formed since the species diverged from each other; similarly, for 'Mouse-specific' and 'Rat-specific' PRs. 'Other primate-specific' are any other PRs formed in human or chimp since divergence from dog (in **bold** typeface), or from cow (in italic typeface); similarly, for 'Other rodent-specific'.

In human PRs, we see that there is a bimodal distribution of Ks (Figure 3A). The median Ks values for lists of PRs that are human-specific or that have otherwise been formed since divergence from dog, are labelled on the Ks histogram. The peak at Ks ~0.06–0.08 corresponds approximately to the median Ks value for PRs formed between human divergence from dog and from chimpanzee. This peak has been noted previously in analyses of retropseudogenes and total retrosequence populations [2, 3, 12], peaking at approximately the point of human lineage divergence from the New World Monkeys, some ~40 million years ago [3]. The peak at 0.0–0.02 (containing 21% of PRs) obviously corresponds to human-specific PRs. Some of these PRs may simply be too young to differentiate as a PR or retropseudogene (They have not been around long enough to acquire (apparent) reading-frame disablements.) Evidence for selection pressures on these sequences is discussed below.

By comparison, in the rodents, there is more, very recent gene retrotransposition activity. In mouse, we find proportionately more, genome-specific PRs (relative to rat), with 44% having Ks ≤ 0.02 (Figure 3B). In the two rodents, mouse and rat, there are >200 genome-specific PRs, compared to ~40 in each of the primates human and chimp. However, setting aside genome-specific examples, there are more gene retrotranspositions appearing in the primate lineage since its divergence from the dog or cow lineage (Figure4).

These observations are in keeping with the apparent maintenance of greater levels of LINE and SINE retrotransposition activity in the rodents [31, 32]; also, they tally well with previous observations for a general fall-off in such retrotransposition activity in the primate lineage [2, 33].

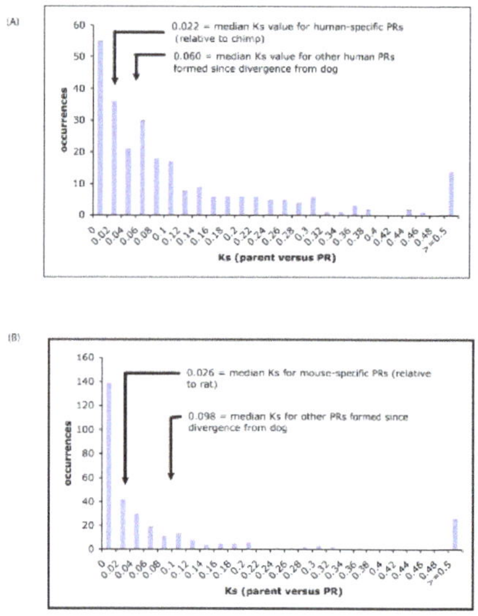

Figure 3 Ks distributions: **(A)** Ks distribution for human PRs meeting the local gene order test with threshold of $N_{homologs} = 0$, from comparison to their parent sequences. Labelled are the median values for the 'Human-specific' set, and those PRs formed between divergence from dog and from chimp [see panel (C)]. A similar distribution is observed with an $N_{homologs}$ threshold of ≤ 1 for the local gene order test. **(B)** Ks distribution for mouse PRs meeting the local gene order test with threshold of $N_{homologs} = 0$, from comparison to their parent sequences. Labelled are the median values for the 'Mouse-specific' set, and those PRs formed between divergence from dog and from chimp [see panel (C)]. A similar distribution is observed with an $N_{homologs}$ threshold of ≤ 1 for the local gene order test.

Transcription Evidence

Focussing on human and mouse, we examined the proportion of PRs that could be mapped to a complete Unigene consensus cDNA or a complete Refseq mRNA from the NCBI ([27]; see Methods for details). For both organisms, we found that the PRs have significantly less mapping of this transcription evidence ($P < 0.001$, using the z-score for distribution of the sample mean). For human, only 23% of human PRs mapped to a Refseq mRNA or Unigene cDNA consensus sequence (compared to 41% for the whole proteome). This may be due to lower transcription levels, because they are novel gene sequences using co-opted promoter elements at a site distal to the genomic location of their parent genes [11]. This general reduction in transcription is also to be expected, if some of the sequences are recent pseudogenes without disablements.

In addition, we examined how many PRs arise from alternatively-spliced genes. To do this, we cross-referenced the PR data with alternative splicings classified in the Alternative Splicing Database (ASD) at the EBI [34]. We found that 24% of genes for human PRs arose from an alternatively spliced gene, compared to 59% of genes overall (significantly less, $P < 0.001$ using the z-score for distribution of the sample mean). A significant reduction in representation from alternatively-spliced genes was also observed in mouse (4%, compared to 29% overall).

We examined the divergence of PRs from their putative parents, in the context of transcription evidence. This is illustrated in Figure 5 for those PRs that pass the local gene order test for both mouse and human, with $N_{homologs} \leq 1$. In human, there is a marked difference in the behaviour of transcribed PRs (purple bars in Figure5A), compared to those without transcription evidence (blue bars in Figure 5A). There are relatively very few transcribed PRs with high sequence identities (i.e., that formed relatively recently). The bimodal character of these plots may arise because some of the PRs: (i) are in a younger population of PRs that are not under selection pressures, but which have not accumulated deleterious mutations, simply by chance; i.e., they are pseudogenes without disablements; or (ii) are in a state of relaxed selection, and thus concomitantly have low transcription levels. Similarly, only a small fraction of the PRs calculated to have formed since divergence from the dog lineage (Figure 4), in either human (15/207, 7%) or mouse (20/292, 7%), are transcribed (significantly less, $P < 0.05$ using χ^2 tests for both cases).

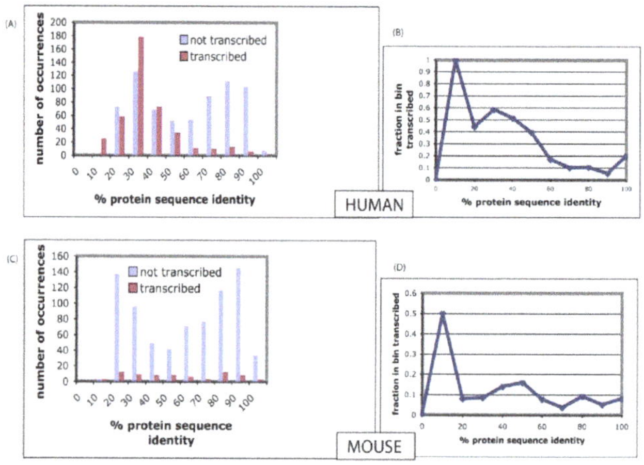

Figure 5 Distributions of percentage protein sequence identity between PRs and parents. **(A)** Distribution of % protein sequence identity for all human PRs that pass the local gene order test ($N_{homologs} \leq 1$). These are broken down into 'transcribed' and 'not transcribed'. **(B)** The fraction that are transcribed in each bin of the histogram in panel A. **(C)**Distribution of % protein sequence identity for all mouse PRs that pass the local gene order test ($N_{homologs} \leq 1$). These are broken down into 'transcribed' and 'not transcribed'. **(D)** The fraction that are transcribed in each bin of the histogram in panel C.

Ka/Ks and Reading Frame Conservation (RFC) Analysis

Is there any evidence for selection pressures on the PRs in human and mouse? We investigated this question for PRs that have been formed in human and mouse, since their divergences from dog. One standard indicator of selection pressures is the Ka/Ks ratio. This is the number of non-synonymous mutations per non-synonymous site, over the number of synonymous mutations per synonymous site, in codons. Negative (or 'purifying') selection in a specific lineage is indicated by a value significantly <1.0, where positive ('diversifying') selection is demonstrated by a value significantly >1.0. We calculated Ka/Ks values for PRs relative to ancestral sequences for the parents of the PRs (see Methods for details). We tested whether any of these Ka/Ks values were significantly < or >1.0 by generating 500 random pairs of sequences as diverged as the PR and ancestral sequence (to calculate expected means and standard deviations for the Ka/Ks values), and then deriving a P-value for the observed Ka/Ks.

Strikingly, when we correct for multiple hypothesis testing in the Ka/Ks calculations, we find only one PRsequence (formed since divergence from dog) that is under significant selection at the codon level in the human genome, and none in the mouse genome. (The one significant human example is a PR under purifying selection, from a family of proteins with the GTP-binding SAR1 domain.)

In addition, we calculated the distribution of Ka/Ks values from directly comparing the PRs versus their parents. From this specific sort of comparison, the neutral expectation for Ka/Ks is not ~1, because of non-synonymous mutations accumulating in the parent genes [2]. A significant excess of Ka/Ks values <0.5, however, may be indicative of purifying selection in the data set. For comparison, we also similarly calculated a Ka/Ks distribution for RψG s versus their respective parents, carefully parsing out disablements (frameshifts and premature stop codons) from the RψG sequences. This Ka/Ks distributional analysis is performed for both human and mouse (Figure 6). For human (Figure 6A), we find no significant excess of PRs with Ka/Ks values <0.5 relative to RψG s, either for the whole data set of PRs, or for the subset formed in the primate lineage, contrary to a previous report [12] (χ^2 test or Fisher's exact test). This distribution is thus consistent with a set of largely neutral retrotranspositions, behaving like RψG s. However, for mouse-specific PRs, there is a significant excess with Ka/Ks <0.5 ($P \leq 0.001$, χ^2 test and Fisher's exact test) (Figure 6B), indicating that some of these mouse PR sequences are under purifying selection.

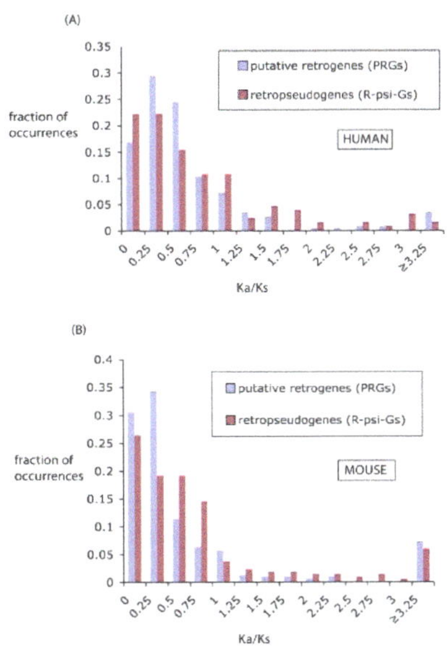

Figure 6 Ka/Ks distributions for PRs and for retropseudogenes (RψG s). **(A)** Distribution of Ka/Ks for human PRs (n = 262) meeting the local gene order test ($N_{homologs} \leq 1$), compared to Ka/Ks for the RψG s (n = 183). All sequences were required to have protein sequence identity ≥ 60.0% with their parent sequences. **(B)** As in (A), but for mouse PRs (n = 318) and RψG s (n = 220).

Conservation of open reading frames without disablements (frameshifts or stop codons), can also be an indicator of coding ability [12, 35, 36]. We derived a method for assessing significant conservation of open reading frames, using simulation with power-law insertion/deletion (indel) statistics [4]. Using simulations with calculated neutral rates of substitution, insertion and deletion, the likelihood of conservation of an open reading frame without interruption by frameshifts and stop codons, can be determined (see Methods for details). To give sufficient power, a P-value threshold of ≤ 0.01 was used as an indicator for significant reading-frame conservation (RFC). This calculation is complementary to the Ka/Ks analysis.

The results are listed Table 2. Even though there were no significant Ka/Ks values in mouse, we find over 30 mouse-specific PRs that have significant reading frame conservation, and a further 17 that were formed since divergence from dog (Table 2). In human, we find in total, 59 PRs with significant RFC, that have arisen since divergence from dog (Table 2). A phylogenetic tree for an example of one of the mouse PRs with significant RFC, which is homologous to citrate synthase.

Table 2 Results of analysis for reading-frame conservation (RFC)

Species	Set †	Number with significantly conserved reading frame ††
Human	Human-specific relative to chimp	10/40 (25%)
	Others in human, that were formed since divergence from dog	49/171 (29%)
	Other older PRs	162/378 (43%)
	TOTAL	221/589 (38%)
Mouse	Mouse-specific relative to rat	35/240 (15%)
	Others in mouse, that were formed since divergence from dog	17/58 (29%)
	Other older PRs	123/233 (53%)
	TOTAL	175/531 (33%)

† These are the sets illustrated in Figure 4. All of the PRs that pass the local gene order test, allowing for $N_{homologs} \leq 1$ were analysed for reading-frame conservation.

†† For PRs formed since divergence from dog, RFC simulations were performed using ancestral sequences calculated using PAML [20]. For PRs formed before divergence from dog, RFC simulations were conservatively performed to half of the nucleotide-level divergence between the PR and its assigned parent sequence.

These results are evidence for conservation of protein open reading frames, even though we found no evidence for purifying selection from examination of the sequences individually for Ka/Ks. This would arise if the PRs were generally under relaxed or positive selection pressures at the codon level. The existence of relaxed selection is consistent with the markedly low numbers of PRs found to be transcribed in both human and mouse, particularly those that were formed since divergence from dog.

Out of those with significant RFC, is there any evidence for non-neutral Ka/Ks trends? We checked for significant excess of PRs with Ka/Ks values <0.5, > 0.5, <1.0 or >1.0 for each of the subsets listed (Table 2). In the human lineage, we find a significant excess of PRs with Ka/Ks >0.5 (40/59, $P < 0.05$, χ^2 test and Fisher exact test, compared to an expectation from RψG sequences) formed between divergence from dog and from chimp. This non-neutral trend is suggestive of positive selection distributed throughout this specific subset population of RψG sequences. Out of the other subsets listed in Table 2, the only other significant non-neutral Ka/Ks tendency is for an excess of mouse-specific PRs to be under purifying selection (26/35 having Ka/Ks <0.5, compared to an expectation from RψG sequences, $P < 0.05$, χ^2 test and Fisher exact test).

Functional Categories

To assess whether the PRs and RψG s have any unusual functional associations, we assigned functional categories using the Gene Ontology (GO) functional classification (Table 3). As noted previously, 'Structural constituent of ribosome' is a prevalent functional category for RψG s [2, 3, 14]. Noticeably, for mouse, there are more retropseudogenes in metabolic categories, than in human (Table 3). A notable prevalence indicative of origin in retrotransposition, 'structural constituent of ribosome', occurs in the top ten of all PR (sub) sets, and is ranked number one for PRs formed since divergence from dog, for both mouse and human (Table 3).'Structural constituent of ribosome' is also the only Gene Ontology term that is statistically overrepresented in all of the retrotransposed gene sets listed (Table 3 legend; P' <0.05, using binomial statistics and a Bonferroni correction for multiple hypothesis testing [37]). The functional category preferences are not caused by over-representation of any one parent, since when representations of PR s on a parent-by-parent basis are tallied up, we find only a very small number of parents giving rise to five or more PR s; the substantial majority of parents have only one PR offspring (286/353 [81%] for mouse, and 279/347 [80%] for human).

Table 3 Most common Gene Ontology (GO) functional terms for different sets of sequences *

Human				
All genes (Total = 33930)	**retropseudogenes (Total = 2493)**	**All PRs (Total = 631)**	PRs ***formed since divergence from dog lineage (Total = 211)*** **	
GO:0005515, protein binding (2360)	***GO:0003735, structural constituent of ribosome (203)***	GO:0008270, zinc ion binding (49)	***GO:0003735, structural constituent of ribosome (11)***	
GO:0008270, zinc-ion binding (2069)	GO:0008270, zinc ion binding (189)	GO:0006355, regulation of transcription, DNA-dependent (35)	GO:0003677, DNA binding (10)	
GO:0006355, regulation of transcription, DNA-dependent (2029)	GO:0006355, regulation of transcription, DNA-dependent (166)	GO:0005509, calcium ion binding (25)	GO:0006355, regulation of transcription, DNA-dependent (9)	
GO:0005524, ATP-binding (1687)	**GO:0003676, nucleic acid binding (132)**	GO:0005525, GTP binding (21)	GO:0005525, GTP binding (5)	
GO:0003677, DNA binding (1339)	**GO:0003723, RNA binding (126)**	GO:0005515, protein binding (21)	GO:0003823, antigen binding (5)	
GO:0007165, signal transduction (1264)	GO:0005515, protein binding (114)	GO:0004842, ubiquitin-protein ligase activity (21)	GO:0003676, nucleic acid binding (5)	
GO:0016740, transferase activity (1263)	GO:0003677, DNA binding (110)	GO:0003677, DNA binding (20)	GO:0030145, manganese ion binding (4)	
GO:0004872, receptor activity (1242)	GO:0005524, ATP binding (93)	GO:0003676, nucleic acid binding (20)	GO:0020037, heme binding (4)	
GO:0016787, hydrolase activity (1171)	GO:0046872, metal ion binding (63)	***GO:0003735, structural component of the ribosome (16)***	GO:0016757, transferase activity, transferring glycosyl groups (4)	
GO:0003700, transcription factor activity (1052)	GO:0000166, nucleotide binding (57)	GO:0003723, RNA binding (13)	GO:0005509, calcium ion binding (4)	
Mouse				
All genes (Total = 32442)	**Retropseudogenes (Total = 2969)**	***PRs* (Total = 663)**	PRs **formed since divergence from dog lineage (Total = 298)** **	

GO:0005515, protein binding (2502)	**GO:0003676, nucleic acid binding (273)**	GO:0005515, protein-binding (17)	***GO:0003735, structural constituent of ribosome (16)***	
GO:0004872, receptor activity (1923)	**GO:0051287, NAD binding (243)**	GO:0003735, structural constituent of ribosome (16)	GO:0005524, ATP binding (8)	
GO:0006355, regulation of transcription, DNA-dependent (1571)	**GO:0008943, glyceraldehyde-3-phosphate dehydrogenase activity (243)**	GO:0008270, zinc ion binding (12)	GO:0005515, protein-binding (7)	
GO:0008270, zinc ion binding (1481)	**GO:0004365, glyceraldehyde-3-phosphate dehydrogenase (phosphorylating) activity (243)**	GO:0006355, regulation of transcription, DNA-dependent (12)	GO:0016740, transferase activity (6)	
GO:0005524, ATP binding (1252)	**GO:0008270, zinc ion binding (235)**	GO:0005524, ATP binding (12)	GO:0016491, oxidoreductase activity (6)	
GO:0016740, transferase activity (1036)	***GO:0003735, structural constituent of ribosome (201)***	GO:0005509, calcium ion binding (12)	GO:0006355, regulation of transcription, DNA-dependent (6)	
GO:0003677, DNA binding (1017)	GO:0005515, protein-binding (101)	GO:0016740, transferase activity (10)	GO:0016853, isomerase activity (5)	
GO:0016787, hydrolase activity (911)	**GO:0016491, oxidoreductase activity (94)**	GO:0016787, hydrolase activity (9)	GO:0016787, hydrolase activity (5)	
GO:0000166, nucleotide binding (873)	GO:0005524, ATP binding (78)	GO:0003677, DNA binding (9)	GO:0003677, DNA binding (5)	
GO:0003676, nucleic acid binding (872)	**GO:0004190, aspartic-type endopeptidase activity (77)**	GO:0003676, nucleic acid binding (9)	GO:0016874, ligase activity (3)	

* The most abundant Gene Ontology 'molecular function' terms are listed for each set of sequences, in decreasing order of abundance. The GO term number and a brief description are followed by the number of occurrences (in brackets). Significant overrepresentation of GO terms was calculated as described previously using binomial statistics, using a Bonferroni correction for multiple hypothesis testing ($P' < 0.05$) [37]. 'Structural constituent of the ribosome' (in italics) is the only term that is significantly overrepresented in all of the three putatively retrotransposed sequences.

** 'Structural constituent of the ribosome' remains the most abundant GO category in this column when PRs from parents with large exons (FLE>0.67) are removed, or a more stringent $N_{homologs}$ threshold of = 0 is used.

CONCLUSION

We have developed two parallel pipelines for the annotation of putative retrogenes (PRs) and retropseudogenes (RψG) in whole genomes. The new pipeline for retropseudogene annotation employs length-based heuristics to speed up the processing of sequence alignment data. We used these pipelines on the vertebrates here, but they are readily applicable to any genome and its set of gene/protein annotations. Genome analysis is constantly in flux, and so obviously, as vertebrate genome assemblies and their annotations are streamlined further, we will be further able to refine our retrotransposition analyses, to remove any errors from missing gene annotations, small genome assembly gaps, etc.

We focussed on the annotation of retro(pseudo)genes in mouse and human. We were particularly interested in the retro(pseudo)genes formed since divergence from an 'outgroup' genome, that of the dog. We found evidence for excess, recent gene-retrotransposition activity in both human and mouse, since their divergences from the dog lineage. We find some evidence for selection on PRs at different phases of mouse and human genome evolution. In human, there is statistical evidence for non-neutral evolution (suggestive of positive selection), for population of PRs that have significantly conserved reading frames and that formed since divergence from the dog lineage. Also, we found that, human PRs formed since divergence from the dog lineage have significantly less transcription evidence, which is consistent with the possibility that they are pseudogenes, or in some intermediate phase of relaxed selection. Such a state of low expression coupled with relaxed selection may also arise for alternatively-spliced exons [38, 39]. In summary, our genomic analysis suggests that some human PRs, formed since divergence from the dog lineage, are undergoing a form of non-neutral evolution, but the majority are either young pseudogenes (that are undisabled simply by chance), or lowly-expressed coding sequences in a state of 'relaxed' selection.

Further information on the PRs and RψGs is available at the website: [40].

ACKNOWLEDGEMENTS

This work was funded by a Discovery Grant awarded to P.M.H., from the National Science and Engineering Research Council of Canada.

AUTHORS' CONTRIBUTIONS

ZY developed the pipelines, performed most of the data analysis, and wrote initial drafts of the manuscript. DM performed some data analysis of alternative splicing events. MI performed phylogenetic analysis. PH conceived of the project, performed evolutionary analyses, and wrote later drafts of the manuscript.

REFERENCES

1. D'Errico I, Gadaleta G, Saccone C: Pseudogenes in metazoa: origin and features. Briefings in functional genomics & proteomics 2004, 3(2):157–167. 10.1093/bfgp/3.2.157
2. Zhang Z, Harrison P, Liu Y, Gerstein M: Millions of years of evolution preserved: a comprehensive catalog of the processed pseudogenes in the human genome. Genome Res 2003, 13(12):2541–2558. 10.1101/gr.1429003
3. Zhang Z, Harrison P, Gerstein M: Identification and analysis of over 2000 ribosomal protein pseudogenes in the human genome. Genome Res 2002, 12(10):1466–14482. 10.1101/gr.331902
4. Zhang Z, Gerstein M: Patterns of nucleotide substitution, insertion and deletion in the human genome inferred from pseudogenes. Nucleic acids research 2003, 31: 5338–5348. 10.1093/nar/gkg745
5. Harrison P, Gerstein M: Studying genomes through the aeons: protein families, pseudogenes and proteome evolution. J Mol Biol 2002, 318(5):1155–1174. 10.1016/S0022-2836(02)00109-2
6. Betran E, Thornton K, Long M: Retroposed new genes out of the X in Drosophila. Genome research 2002, 12(12):1854–1859. 10.1101/gr.6049
7. Esnault C, Maestre J, Heidmann T: Human LINE retrotransposons generate processed pseudogenes. Nature genetics 2000, 24(4):363–367. 10.1038/74184
8. Pavlicek A, Paces J, Elleder D, Hejnar J: Processed pseudogenes of human endogenous retroviruses generated by LINEs: their integration, stability, and distribution. Genome research 2002, 12(3):391–399. 10.1101/gr.216902. Article published online before print in February 2002
9. Torrents D, Suyama M, Zdobnov E, Bork P: A genome-wide survey of human pseudogenes. Genome research 2003, 13(12):2559–2567. 10.1101/gr.1455503
10. Brosius J: RNAs from all categories generate retrosequences that may be exapted as novel genes or regulatory elements. Gene 1999, 238(1):115–134. 10.1016/S0378-1119(99)00227-9
11. Harrison PM, Zheng D, Zhang Z, Carriero N, Gerstein M: Transcribed processed pseudogenes in the human genome: an intermediate form of expressed retrosequence lacking protein-coding ability. Nucleic acids research 2005, 33: 2374–2383. 10.1093/nar/gki531
12. Marques AC, Dupanloup I, Vinckenbosch N, Reymond A, Kaessmann H: Emergence of young human genes after a burst of retroposition in primates. PLoS biology 2005, 3: e357. 10.1371/journal.pbio.0030357
13. Emerson JJ, Kaessmann H, Betran E, Long M: Extensive gene traffic on the mammalian X chromosom. Science 2004, 303(5657):537–540. 10.1126/science.1090042
14. Harrison PM, Hegyi H, Balasubramanian S, Luscombe NM, Bertone P, Echols N, Johnson T, Gerstein M: Molecular fossils in the human genome: identification and analysis of the pseudogenes in chromosomes 21 and 22. Genome Res 2002, 12(2):272–280. 10.1101/gr.207102
15. Ensembl database [http://www.ensembl.org] .
16. Altschul SF, Madden TL, Schaffer AA, Zhang J, Zhang Z, Miller W, Lipman DJ: Gapped BLAST and PSI-BLAST: a new generation of protein database search programs. Nucleic acids research 1997, 25(17):3389–3402. 10.1093/nar/25.17.3389
17. Ophir R, Graur D: Patterns and rates of indel evolution in processed pseudogenes from humans and murids. Gene 1997, 205: 191–202. 10.1016/S0378-1119(97)00398-3
18. ORDB [http://senselab.med.yale.edu/senselab/ORDB].
19. Repeatmasker [http://www.repeatmasker.org] .
20. Yang Z: PAML: a program package for phylogenetic analysis by maximum likelihood. Comput Appl Biosci 1997, 13: 555–556.
21. Cooper GM, Brudno M, Stone EA, Dubchak I, Batzoglou S, Sidow A: Characterization of evolutionary rates and constraints in three mammalian genomes. Genome Res 2004, 14: 539–548. 10.1101/gr.2034704
22. Britten RJ, Rowen L, Williams J, Cameron RA: Majority of divergence between closely-related DNA samples is due to indels. PNAS 2003, 100: 4665–4670. 10.1073/pnas.0330964100
23. Felsenstein J: PHYLIP - Phylogeny Inference Package (Version 3.2). Cladistics 1989, 5: 164–166.
24. Consortium GO: The Gene Ontology (GO) database and informatics resource. Nucleic acids research 2004, 32: D258-D261. 10.1093/nar/gkh036
25. Gene Ontology [http://www.geneontology.org] .
26. Chandonia JM, Hon G, Walker NS, Lo Conte L, Koehl P, Levitt M, Brenner SE: The ASTRAL Compendium in 2004. Nucleic acids research 2004, 32: D189-D192. 10.1093/nar/gkh034
27. NCBI [http://www.ncbi.nih.gov] .
28. Consortium ICGS: Sequence and comparative analysis of the chicken genome provide unique perspectives on vertebrate evolution. Nature 2004, 432: 695–716. 10.1038/nature03154
29. Haas NB, Grabowski JM, Sivitz AB, Burch JB: Chicken repeat 1 (CR1) elements, which define an ancient family of vertebrate non-LTR retrotransposons, contain two closely spaced open reading frames. Gene 1997, 197: 305–309. 10.1016/S0378-1119(97)00276-X
30. Roest Crollius H, Jaillon O, Dasilva C, Ozouf-Costaz C, Fizames C, Fischer C, Bouneau L, Billault A, Quetier F, Saurin W, Bernot A, J. W: Characterization and repeat analysis of the compact genome of the freshwater pufferfish Tetraodon nigroviridis. Genome Res 2000, 10: 939–949. 10.1101/gr.10.7.939
31. Consortium RGSP: Genome sequence of the Brown Norway rat yields insights into mammalian evolution. Nature 2004, 428(6982):493–521. 10.1038/nature02426
32. Consortium MGS: Initial sequencing and comparative

analysis of the mouse genome. Nature 2002, 420(6915):520–562. 10.1038/nature01262

33. Genomics C: The sequence of the human genome. Science 2001, 291(5507):1304–1351. 10.1126/science.1058040
34. Consortium ENCODEP: The ENCODE (ENCyclopedia Of DNA Elements) Project. Science 2004, 306(5696):636–640. 10.1126/science.1105136
35. Harrison PM, Carriero N, Liu Y, Gerstein M: A "polyORFomic" analysis of prokaryote genomes using disabled-homology filtering reveals conserved but undiscovered short ORFs. J Mol Biol 2003, 333: 885–892. 10.1016/j.jmb.2003.09.016
36. Kellis M, Patterson N, Endrizzi M, Birren B, Lander ES: Sequencing and comparison of yeast species to identify genes and regulatory elements. Nature 2003, 423: 241–254. 10.1038/nature01644
37. Harrison PM: Exhaustive assignment of compositional bias reveals universally prevalent biased regions: analysis of functional associations in human and Drosophila. BMC Bioinformatics 2006, 7: 441. 10.1186/1471-2105-7-441
38. Modrek B, Lee CJ: Alternative splicing in the human, mouse and rat genomes is associated with an increased frequency of exon creation and/or loss. Nat Genet 2003, 34(2):177–180. 10.1038/ng1159
39. Wang W, Zheng H, Yang S, Yu H, Li J, Jiang H, Su J, Yang L, Zhang J, McDermott J, Samudrala R, Wang J, Yang H, Yu J, Kristiansen K, Wong GK, J W: Origin and evolution of new exons in rodents. Genome Res 2005, 15: 1258–1261. 10.1101/gr.3929705
40. http://biology.mcgill.ca/faculty/harrison/retro [http://biology.mcgill.ca/faculty/harrison/retro] .

CHAPTER 3 COLORATION AND THE GENETICS OF ADAPTATION

Nicholas I. Mundy

How do organisms adapt to new environments? The popular conception of the adaptive process generally runs like this: A mutation arises that leads to an improved phenotype under novel environmental conditions (for example, a beak modification suited to exploiting an untapped food source). Because of the enhanced reproductive success of individuals carrying the mutant allele (or gene variant), the frequency of the mutant allele increases. The mutant allele eventually becomes fixed in the population—that is, every member of the population has two copies of the allele—and the population has increased its fitness. The whole process can then repeat itself when the environment changes again.

This is the simplest possible model of adaptation, and is enshrined in well-known examples such as the evolution of melanism in the peppered moth and of insecticide resistance in mosquitoes. However, it has long been known that there are many potential complications to this tidy scenario involving a single mutant allele at a single genetic locus, and that the situation is frequently more complex in the real world. For example, more than one locus may be involved, mutant alleles may have effects on more than one aspect of the phenotype (a phenomenon known as pleiotropy), and the effects of alleles on the phenotype may be dependent on the environment (genotype × environment interactions). The timing of the occurrence of a mutation in relation to an environmental change in which the mutation is favored—that is, whether the beneficial mutation arose de novo in the new environment or was part of the standing genetic variation—also turns out to be critically important. Recent theoretical [1] and empirical work [2–4] has greatly increased our understanding of the adaptive process, but there are many fundamental questions remaining. In particular, we have little knowledge of the relative importance of different factors in actual cases of adaptation, and there are many details of the molecular basis of adaptation that are poorly understood.

If two or more loci are involved in adaptation to a new environment, then this leads to the possibility that the phenotypic effects of the two loci are not additive, that is, they are not independent of one another. For example, the effect on the phenotype of the presence of allele A rather than allele B at one locus may depend on whether allele X or allele Y are present at a second locus. Such epistatic effects may have a profound influence on the evolutionary dynamics of adaptation. One possibility is that genetic divergence among different populations that becomes fixed for alternate alleles at one or more loci may lead to different propensities for future

Citation: Mundy NI (2007) Coloration and the Genetics of Adaptation. PLoS Biol 5(9): e250. doi:10.1371/journal.pbio.0050250

evolution [5]. Epistasis also plays an important role in causing genetic incompatibilities that underlie reproductive isolation and speciation. However, in contrast to the growing number of examples in which the molecular genetics of single loci underlying adaptation have been investigated, we have little information on how epistasis works at the single gene level in adaptation in the wild.

Coloration is rapidly becoming established as a phenotype that is highly amenable to studying the molecular basis of adaptation. There are several reasons for this. Color variation is abundant in natural populations and can be readily quantified. The selective advantages of coloration are widely studied and include physical protection (e.g., against ultraviolet light), thermoregulation, concealment (including background color matching and camouflage), and a large variety of signaling functions both within and between species. Finally, genetic studies on model organisms such as fruit flies and mice have led to an excellent understanding of the gene networks involved in generating certain types of coloration, especially for those genes involved in regulating the production of melanin pigments.

Recent studies on organisms ranging from flowering plants [6] to fruit flies [7,8], butterflies [9], fish [10], lizards [11], birds [12,13], and mammals [14–17] have identified single loci contributing to color variation in natural populations. Perhaps the most notable conclusion from these studies is that, with some exceptions, it is surprisingly common for the same single loci or genomic regions to be responsible for similar color changes in different species. In vertebrates, many studies have found convergent association between variation in the coding sequence at the *Mc1r* locus and melanin-based coloration. *Mc1r* encodes the melanocortin-1 receptor, which has a critical function in determining the type of melanin synthesized in hair or feather melanocytes. However, these studies may present a biased picture of the overall importance of *Mc1r* in evolution since the locus is relatively easy to assay (it has a single coding exon of ~1 kb). In laboratory mice, the *Agouti* locus regulates pale ventral coloration and pale bands on dorsal hairs [18]. *Agouti* encodes *agouti* signaling protein, an inhibitor of*Mc1r*, and ventral-specific or hair-cycle specific expression of *Agouti* diverts the melanin synthesis pathway from brown or black eumelanin to yellow phaeomelanin. Many alleles of *Agouti* have been described, but unlike *Mc1r* the great majority of them are in the large 5′ regulatory regions of the *Agouti* gene, and thus far more difficult to assay. A priori, the *Agouti* gene is as good a candidate gene as *Mc1r* for involvement in evolutionary change of coloration, but until now association between *Agouti* variation and color in wild populations has not been described.

New research published in *PLoS Biology* from the lab of Dr. Hopi Hoekstra uncovers a persuasive example where color variation in natural populations is partly controlled by epistasis among two loci, and also provides the first case where *Agouti* has been implicated in natural color variation. Steiner et al. studied two subspecies of the oldfield mouse in the southeastern United States, Peromyscus polionotus subgriseus, a mainland subspecies with dark dorsal and pale ventral coloration, and P. p. leucocephalus, which inhabits white sand dunes and lacks visible coloration over much of the body, with a narrow band of darker coloration dorsally. The color of the mice matches the color of the substrates they inhabit, and previous work has shown that this provides crypsis against avian predators.

The authors use a powerful combination of methods to identify genes underlying the phenotypic variation. Using crosses between the two subspecies in captivity, they performed a quantitative trait locus (QTL) analysis with a total of 124 genetic markers. QTL analysis is an extension of standard genetic linkage mapping (originally designed for single locus Mendelian traits) to quantitative traits and can identify genomic regions contributing to heritable variation in a trait as well as the proportion of the trait variance attributable to a particular genomic region. In Steiner et al.'s analysis, the genomic regions identified were consistent across the seven color traits they measured: two regions were significantly associated with all seven traits while a third region was associated with four traits but not the remaining three.

All three QTL regions identified contained pigmentation genes (*Mc1r, Agouti*, and *Kit*), which are therefore candidates for functional involvement in the color differences. For *Mc1r* this was no surprise since the group had already shown that an amino acid substitution at *Mc1r* is associated with coloration and that there is a functional effect of the mutation [19]. Steiner et al. investigated *Agouti* further, and, interestingly, while there was no variation in the amino acid sequence, they did find significantly higher expression of *Agouti* mRNA transcript in several body regions of the paler P. p. leucocephalus than in P. p. subgriseus. This finding is consistent with the action of *Agouti* and, although more complicated scenarios cannot be ruled out, the simplest explanation is that there are one or more mutations in the large upstream regulatory regions of the *Agouti* locus that underline the QTL discovered.

The most novel part of the study is the demonstration of epistasis occurring between the phenotypic effects of alleles at *Mc1r* and *Agouti*. At both loci, parental alleles can be classed as dark (D) in P. p. subgriseus and light (L) in P. p. leucocephalus. For all seven traits, the effects of D and L alleles at *Mc1r* are dependent on the genotype at *Agouti*. The authors present the most dramatic case, which involves cheek pigmentation. For this trait, variation at*Mc1r* has no effect on a DD genetic background at *Agouti*, whereas on a LD or LL *Agouti* background, the LL *Mc1r* genotype is paler than DL or DD (see their Figure 5). This could have strong evolutionary significance, because it means that for cheek color the L *Mc1r* allele is neutral on a DD *Agouti* background, and only becomes visible to natural selection on a pale environmental substrate in an LD or DD *Agouti* genotype. There is interesting variation in the pattern of epistasis among traits—for example, for tail color, the LL *Mc1r* genotype is visible against a DD *Agouti* background, complicating attempts to reconstruct the evolutionary sequence of events of selection on *Agouti* and *Mc1r* genotypes. This study should stimulate further research into a number of questions. Is it generally the case that variation in color traits can be attributed to a relatively small number of loci, and is coloration different from other traits in this regard? How common is epistasis among color traits, and is it possible to predict the evolutionary consequences of particular epistatic interactions? What is the relative importance of different loci in evolutionary color change, and why? For *Agouti*, we finally have a case where the locus appears to be controlling evolutionary change in coloration—an exciting finding, given its important function in color patterning.

ACKNOWLEDGMENTS

I thank Chris Jiggins, Brenda Bradley, and Mohamed Noor for helpful comments. My work on pigmentation genetics has been supported by the Leverhulme Trust, the Biotechnology and Biological Sciences Research Council, and the Natural Environment Research Council.

REFERENCES

1. Orr HA (2005) The genetic theory of adaptation: A brief history. Nat Rev Genet 6: 119–127.
2. Colosimo PF, Hosemann KE, Balabhadra S, Villarreal G, Dickson M, et al. (2005) Widespread parallel evolution in sticklebacks by repeated fixation of ectodysplasin alleles. Science 307: 1928–1933.
3. Terai Y, Seehausen O, Sasaki T, Takahashi K, Mizoiri S, et al. (2006) Divergent selection on opsins drives incipient speciation in Lake Victoria cichlids. PLoS Biol 4(12): e433. doi:10.1371/journal.pbio.0040433.
4. Geffeney SL, Fujimoto E, Brodie ED III, Brodie ED Jr, Ruben PC (2005) Evolutionary diversification of TTX-resistant sodium channels in a predator–prey interaction. Nature 434: 759–763.
5. Weinreich DM, Watson RA, Chao L (2005) Perspective: Sign epistasis and genetic constraint on evolutionary trajectories. Evolution 59: 1165–1174.
6. Whibley AC, Langlade NB, Andalo C, Hanna AI, Bangham A, et al. (2007) Evolutionary paths underlying flower color variation in Antirrhinum. Science 313: 963–966.
7. Gompel N, Prud'homme B, Wittkopp PJ, Kassner VA, Carroll SB (2005) Chance caught on the wing: cis-regulatory evolution and the origin of pigment patterns in Drosophila. Nature 433: 481–487.
8. Wittkopp P, Williams B, Selegue J, Carroll S (2003) Drosophila pigmentation evolution: Divergent genotypes underlying convergent phenotypes. Proc Natl Acad Sci U S A 100: 1808–1813.
9. Joron M, Papa R, Beltrán M, Chamberlain N, Mavárez J, et al. (2006) A conserved supergene locus controls colour pattern diversity in Heliconius butterflies. PLoS Biol 4(10): e303. doi:10.1371/journal.pbio.0040303.
10. Protas ME, Hersey C, Kochanek D, Zhou Y, Wilkens H, et al. (2006) Genetic analysis of cavefish reveals molecular convergence in the evolution of albinism. Nat Genet 38: 107–111.
11. Rosenblum EB, Hoekstra HE, Nachman MW (2004) Adaptive reptile color variation and the evolution of the *Mc1r* gene. Evolution 58: 1794–1808.
12. Mundy NI, Badcock NS, Hart T, Scribner K, Janssen K, et al. (2004) Conserved genetic basis of a quantitative plumage trait involved in mate choice. Science 303: 1870–1873.
13. Theron E, Hawkins K, Bermingham E, Ricklefs RE, Mundy NI (2001) The molecular basis of an avian plumage polymorphism in the wild: A melanocortin-1-receptor point mutation is perfectly associated with the melanic plumage morph of the bananaquit,Coereba flaveola. Curr Biol 11: 550–557.
14. Nachman MW, Hoekstra HE, D'Agostino SL (2003) The genetic basis of adaptive melanism in pocket mice. Proc Natl Acad Sci U S A 100: 5268–5273.
15. Ritland K, Newton C, Marshall HD (2001) Inheritance and population structure of the white-phased "Kermode" black bear. Curr Biol 11: 1468–1472.
16. Eizirik E, Yuhki N, Johnson W, Menotti-Raymond M, Hannah S, et al. (2003) Molecular genetics and evolution of melanism in the cat family. Curr Biol 13: 448–453.
17. Gratten J, Beraldi D, Lowder BV, McRae AF, Visscher PM, et al. (2007) Compelling evidence that a single nucleotide substitution in *TYRP1* is responsible for coat-colour polymorphism in a free-living population of Soay sheep. Proc R Soc Lond B Biol Sci 274: 619–626.
18. Siracusa LD (1994) The *agouti* gene: Turned on to yellow. Trends Genet 10: 423–428.
19. Hoekstra HE, Hirschmann RJ, Bundey RA, Insel PA, Crossland JP (2006) A single amino acid mutation contributes to adaptive beach mouse color pattern. Science 313: 101–104.

CHAPTER 4 EVOLUTION OF AQUATIC INVERTEBRATES AND VERTEBRATES

Rahul Bale, Izaak D. Neveln, Amneet Pal Singh Bhalla, Malcolm A. MacIver, Neelesh A. Patankar

INTRODUCTION

How would life look if it evolved again on Earth, or for that matter, on any other habitable planet? The question of the role of chance versus necessity in evolution is a foundational issue in biology [1–3]. Gould gave us the metaphor of the "tape of life" for the evolution of life and argued that if it were somehow rewound and started again, life would have taken a very different course [4]. Conway Morris has argued that, on the contrary, the laws of physics limit the number of good solutions that are within reach of evolution, and that therefore we should expect life to take a similar course upon rewinding [5]. Examples of convergent evolution, such as wings on insects, birds, and mammals, are considered supporting evidence for this hypothesis. But our understanding of convergent evolution, as reflecting the dominance of natural selection plus variation over factors such as developmental constraints, pleiotropy [6], phylogenetic inertia, genetic drift, and other stochastic processes [3], is held back by a lack of quantitative arguments. Such arguments would expose the links from physical principles to the biological phenomena and help us understand where evolution is likely to converge to the same result or diverge to a wide variety of solutions [7].

Here, we present just such arguments for a phenomenon that unifies a vast diversity of swimming organisms, from invertebrates, like cuttlefish, to vertebrates, like cartilaginous and bony fish. Unlike the case of the convergently evolved wing, a morphological feature, here the evolved feature is a pattern of movement [8–12] that occurs across a morphologically diverse set of moving appendages on aquatic animals. These animals swim by undulating elongated fins while keeping their body relatively rigid (Fig 1). Their fins run lengthwise with the body rather than crosswise to the body like the tailfin of a trout. The fins occur along the midline of the body, either along the belly or back (Fig 1N–1S, knifefish), as a dorsoventral pair (Fig 1L, triggerfish), or as a left–right pair (Fig 1C–1K, cuttlefish, skates, and rays). Movement by means of these fins is therefore called median/paired fin swimming, in contrast to the more common swimming style of fish like trout, in which a portion of the body and the caudal tailfin is moved, termed body/caudal fin swimming [13].

Citation: Bale R, Neveln ID, Bhalla APS, MacIver MA, Patankar NA (2015) Evolution of Aquatic Invertebrates and Vertebrates . PLoS Biol 13(4): e1002123. doi:10.1371/journal.pbio.1002123

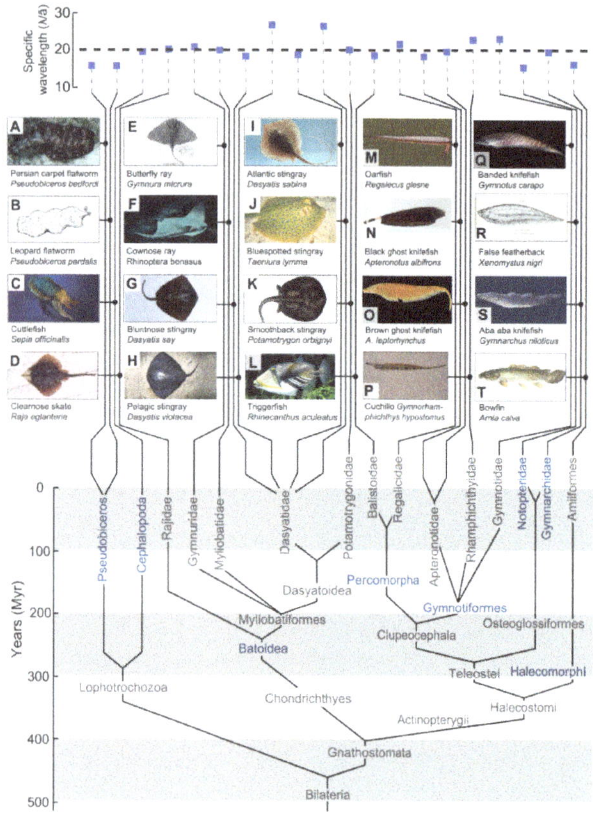

Figure 1. Undulatory median/paired fin swimmer phylogenetic relationships and SW = λ / *ã*, where λ and *ã* are wavelength and mean amplitude of undulations present along the fin, respectively. The eight instances of independent emergence of elongated median/paired fin swimming are highlighted in blue. Not shown here because of space constraints is the SW for the ray Dasyatis americana, which has an SW of 25.1, and the weakly electric knifefish Eigenmannia virescens, which use two counter-propagating waves on their fin during slow speed swimming and have an average SW of 17.7. The following images are licensed under CC-BY: C) Sepia officinalis image courtesy of Hans Dappen. D) Raja eglanteria image courtesy of George Burgess. F) Rhinoptera bonasusimage courtesy of Juan Aguere. J) Taeniura lymma image courtesy of Nicolai Johannesen. M) Regalecus glesne image courtesy of Sandstein. N) Apteronotus albifrons image courtesy of Clinton and Charles Robertson. O) Apteronotus leptorhynchus image courtesy of the Harvard Museum of Comparative Zoology. P) Gymnorhamphichthys hypostomus image courtesy of Mark Sabaj with support from IXingu Project (NSF DEB-1257813). S) Gymnarchus niloticus image courtesy of Masashi Kawasaki. All remaining images are public domain.

We find that in all cases where the wavelength of the traveling wave of median/paired fins has been documented during steady swimming, or can be extracted from readily available videos, the wavelength is about twenty times the mean amplitude (ã) of movement (mean = 19.5, standard deviation (STD) = 3.0, n = 22, Fig 1). This includes cases, such as swimming of the cownose ray, in which the wavelength is greater than the fin length. These fins appear to only oscillate up and down to a casual observer but in fact undulate in the same manner as described earlier.

The ratio of the wavelength of an undulation to its mean amplitude is called the specific wavelength (SW). This parameter has been used in the past to study undulatory swimming [14–17] under water and under sand [15,16]. However, the finding in this work that there is a particular value of the SW that is observed across a diverse set of animals is new. We call this the optimal specific wavelength, or OSW. The OSW identifies the optimality in kinematics associated with undulatory patterns, in contrast to prior work [18], in which the focus was on quantifying bends solely due to propulsor flexibility.

Many of the organisms that swim with the OSW, including invertebrates such as cuttlefish and flatworms, jawed vertebrates such as rays and skates, and ray-finned bony fishes such as gymnotiformes, gymnarchids, and notopterids have no known ancestor that swam by means of median/paired fins. Swimming through undulation of median/paired fins with the OSW has therefore independently arisen at least eight times in three phyla (Chordata, Mollusca, and Platyhelminthes) among the Bilateria (Fig 1).

RESULTS

The values of SW, shown in Fig 1 for a variety of median/paired fin undulatory swimmers, converge to a narrow range (mean = 19.5, STD = 3.0, n = 22). Is there a mechanical optimum due to the physics of undulatory swimming that has driven evolution, in multiple instances, toward this narrow range of values of SW? An initial clue was obtained by our group from simulations of undulating neotropical knifefish fins [19] and measurements of the swimming speed of an undulatory fin robot [20]. Prior experimental work and current simulations show that swimming speed is highest when the number of undulations present along the fin (the fin length divided by the wavelength of the traveling wave progressing down the fin) is around two, similar to what is found in measurements of live knifefish [21,22]. The simulations were carried out on two sizes of fins. The first fin was similar in size and shape as the ventral elongated ribbon fin of a specific type of knifefish, the black ghost knifefish (A.albifrons, Fig 1N), where the fin measures 10 cm long by 1 cm tall. The second fin was a smaller fin measuring 2 cm long by 0.4 cm tall. A smaller fin was used because simulating many cases of a knifefish-sized fin for a more detailed parametric study, which we report later in the paper, was not practical due to limited computational resources. The experiments were with a robotic fin that was approximately three times larger in length and height (32.6 cm long by 3.6 cm tall) compared to the knifefish-sized fin. A larger robotic fin was built because of limitations of motor packing density for the smallest motors that could provide sufficient torque while providing an adequate number of artificial fin rays (32 rays [20], compared to 148 found in A. albifrons [23]). In Fig 2, we show the free swimming speed of the robot [20] and a simulated knifefish-sized fin [24,25] as a function of number of undulations along the fin. We also plot the propulsion force generated by undulating robotic and simulated fins that are prevented from translating along their long axis. We find that swimming speed and propulsive force are well correlated. This correlation is expected, since maximizing propulsive force or thrust is likely to enable the fastest swimming speed. Hereafter, we will present our analysis only in terms of propulsive force.

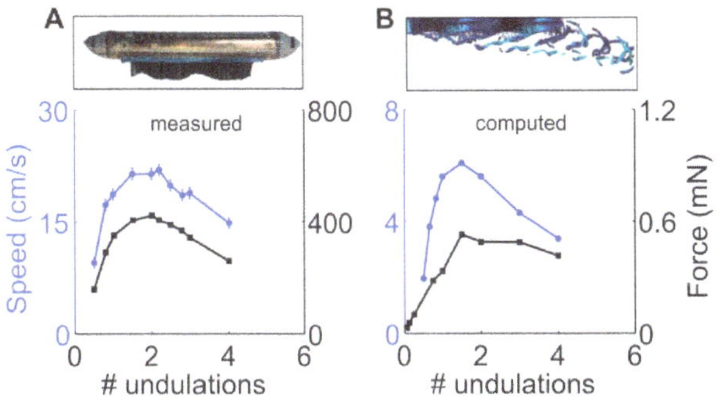

Figure 2. A) Experimentally measured axial swimming speed and propulsive force of a robotic knifefish as a function of the number of undulations along the fin (fin length divided by undulation wavelength). Modified from Fig 4B and Fig 5B of [20]. The error bars show STD from five trials of speed and force measurement experiments. The STD of error in measuring force is less than 1.5%, hence the error bars are barely visible at the scale of the plot. The fin was 32.60 cm long and 3.37 cm tall. For these experiments, f = 4 Hz and θ_{max} = 30°. Further details of measurements and error estimation can be found in our prior work [20]. Image of the robot taken by the authors. B) Axial swimming speed and propulsive force computed from simulations of a knifefish-sized fin (ten cm long, one cm tall, f = 2 Hz, θ_{max} = 30°) immersed in water, plotted against the number of undulations along the fin. In each plot, the left axis (in blue) and the blue curve corresponds to swimming speed, and the right axis (in black) and the black curve corresponds to propulsive force. Simulation code can be found at [26].

As we varied the length of the simulated fin and the robotic fin for this study, while keeping other kinematic parameters the same, we were surprised to find that the number of undulations that resulted in maximal propulsive force (correlating to maximal free swimming speed) changed, as shown in Fig 3A. In this work, we show that if we examine force versus SW, instead of force versus number of undulations, the peaks of propulsive force all align at around a SW of twenty (Fig 3B). We find that the same is true when other kinematic and morphological parameters (amplitude of undulations, frequency, fin height, and fin shape) of the fin are varied. Because the propulsive force was maximal at an SW of around twenty across these examined parameters, we refer to SW ≈ 20 as the optimal SW or OSW.

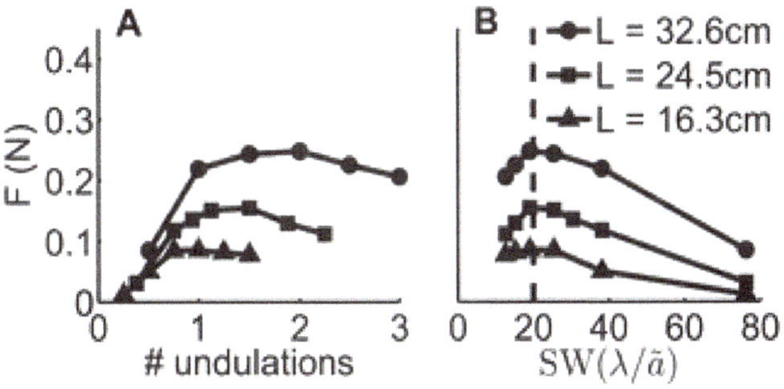

Figure 3. Effect of changes to fin length of a robotic knifefish on the optimal number of undulations and OSW. A) As the fin length is reduced from 32.6 cm to 16.3 cm, the optimal number of undulations reduces from 2 to 1. B) With the same changes in fin length, the OSW remains unaffected. Experimental parameters: θ_{max} = 20° (peak angle from midsagittal plane), fin height h = 5 cm, frequency of fin ray oscillation f = 3 Hz.

Robustness of OSW to Variations in Kinematics or Morphology

The axial propulsive force F depends on several physical variables of the problem:

$$F = fn(\rho, \mu, f, \lambda, L, h, a), \qquad (1)$$

where fn denotes "function of," ρ is the density of water (0.9956×10^3 kg/m³ at 25°C, varying less than 5% across the temperatures and salinities of aquatic organisms), μ is the viscosity of water (0.89×10^{-3} N · s/m² at 25°C), f is the frequency of oscillation of the fin, λ is the wavelength of undulation of the fin, L and h are the length and the height of the fin, respectively, a = h sinθ_{max} is the amplitude of undulation at the distal edge of the fin, and θ_{max} is the peak angle of excursion of the fin from the midsagittal plane. We assume a condition of steady temporally periodic flow in Equation 1. In our experiments and simulations, the force was generated by a undulating but non-translating fin. Hence, the fin velocity is not a physical variable of the problem. If the fin is attached to a swimming body then the swimming velocity is another dimensional variable of the problem. The steady swimming velocity also depends on the kinematic parameters listed in Equation 1.

Other physical variables of interest that depend on those listed in Equation 1 are noted next. A is the surface area of the fin, which is approximately given by $A = Lh(1 + (a / \lambda)^2)$. For the parameter space of interest in this work, $a / \lambda \ll 1$, implying A ≈ Lh. A measure of the lateral velocity of oscillation of the fin is $V_l = f\tilde{a}$, where ã is the average amplitude of oscillation of the fin. In the definition of SW, we use the average amplitude ã rather than the amplitude a. The reason is as follows: the propulsive force generated by an elongated fin depends on the amplitude of oscillation of the fin. For a given angular excursion, the amplitude of oscillation depends on the height of the fin, which typically varies along the length of the fin (see Fig 1 for a schematic). The contribution to propulsive force from the part of the fin with greater height would be larger than that part of the fin which has relatively smaller height. Therefore, we use the average amplitude ã to define SW = λ / ã (and other parameters), where $\tilde{a} = h_{\text{mean}} \sin(\theta^{\text{avg}}_{\text{max}})/2$. h_{mean} is the mean height of the fin and $\theta^{\text{avg}}_{\text{max}}$ is the mean angle of excursion of fin rays.

Equation 1 shows that there are eight physical variables including the propulsive force. The viscosity and density of water are known. Thus, the propulsive force F depends on the five remaining physical variables of which the wavelength, or equivalent SW, is one variable. We find that although F depends on multiple variables listed in Equation 1, the maximum is always at SW ≈ 20, irrespective of the other parameters. This is demonstrated in Fig 4. It is seen that the OSW did not change over the range of frequencies, angles of excursion, fin heights, and fin lengths shown. In addition to the effect of rectangular fin height, the effect of different nonrectangular fin shapes on the OSW was also investigated. Fin shapes ranging from rounded convex to triangular (such as those used in a robotic cownose ray [27]) had no effect on the value of the OSW.

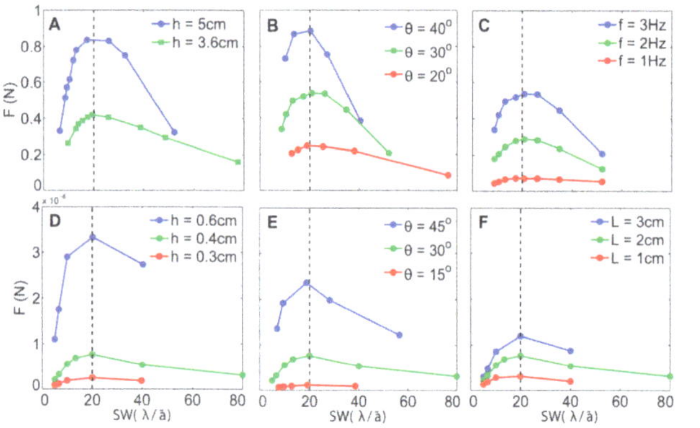

Fig 4. Results of a parametric study of the robotic fin plotted against SW in (A) to (C). (A) Fin height is varied, L = 32.6 cm, f = 4 Hz, and θ_{max} = 30°. (B) Angle of excursion is varied, L = 32.6 cm, h = 5 cm, and f = 3 Hz. (C) Frequency of undulations is varied, L = 32.6 cm, h = 5 cm, and θ_{max} = 30°. Results of a parametric study of the simulated fin plotted against SW in (D) to (F). (D) Fin height is varied, L = 2 cm, f = 1 Hz, and θ_{max} = 30° (E) Angle of excursion is varied, L = 2 cm, h = 0.4 cm, and f = 1 Hz. (F) Fin length is varied, h = 2 cm, f = 1 Hz, and θ_{max} = 30°. The fluid was water in all cases. The simulation code can be found at [26].

For the parameter space we examined, the qualitative trend of F with respect to SW is independent of the size of the fin. Consequently, the OSW is independent of size over the range examined here. Although our experiments and simulations are limited to fins between 2 cm and 32 cm in length, data from 22 species of swimming animals in Fig 1 show that the OSW holds over a wide range of lengths that span two orders of magnitude, from a 2 cm longPseudobiceros pardalis to a 255 cm long R. glesne. Beyond our sample of 22 species, there are over 1,000 median/paired fin species (either within the same genera or in higher taxonomic categories) that we predict will exhibit the OSW.

Exclusion of Seahorse Data

We have chosen not to include the seahorse Hippocampus hudsonius within our results. This species is the sole exception we could find to the OSW, according to fin kinematics reported by Blake [28]. We decided not to include this case for the following reasons: the fin of the seahorse discussed in [28] was 2.1 cm long and around 1 cm deep. Blake reported that the fin moved with an amplitude of 0.63 cm at a frequency of 40 Hz while having 3.3 undulations. We cannot be confident that the reported kinematics are physically achievable by animals. To understand why, consider the kinematics of the black ghost knifefish [22]. The fin height is the same for the seahorse and the knifefish. The length of the seahorse fin is five times less than that of the knifefish. If the knifefish were to undulate its fin in an equivalent scenario, it would need to have around 15 undulations. The highest number of undulations along the fin of a knifefish is around four [22]. Thus, for conditions equivalent to the seahorse, the fin membrane would be subjected to extremely high strains and stresses, which seem physically unrealizable.

DISCUSSION

Convergent Evolution and the OSW

Convergent evolution, the repeated emergence of a trait from an ancestor that did not have the trait, is the most compelling evidence for what has been called the "robust repeatability thesis" [3], the view that the evolution of animals can be explained by the comparative selective advantage of evolutionary survivors over their inferior counterparts. This is in contrast to Gould's view—that quite different animal forms would emerge if evolution were to rerun its course. This view of macroevolution, the "radical contingency thesis," [3] is that early patterns of animal evolution were generated by a selection process that was unrelated to the comparative fitnesses of the selected lineages.

What is the mechanism for macroevolutionary repeatability? In the language of the calculus of variations, these examples of convergent evolution—if correctly identified as such—imply that there is a gradient in the fitness landscape toward some optimum with respect to trait in question, and this gradient is large enough to overcome competing factors such as developmental constraints, pleiotropy [6], phylogenetic inertia, genetic drift, cases where optimality in one trait results in suboptimality in another trait (e.g., [29]), and approximations of the trait which provide local but not global optimality. With a sufficiently steep gradient in fitness in place and evolutionary dynamics capable of achieving near-optimal solutions, it is only a matter of time before the mechanism of selection with variation can arrive at the optimum. As the derivative of the trait with respect to fitness is stabilizing, departures from the optimum would be self-correcting over evolutionary time.

What is the evidence of a gradient in the case of the OSW? To address this question, we examine how variation in the SW affects force production (and therefore swimming speed) across the observed range of SW (Fig 1). We then compare the results to the variation in force production for SW outside of the observed range.

The effect of any decline in propulsive force—even less than one percent—from what it is at the OSW is amplified over the vast number of undulations an animal may make in its life. Thus, we can only speculate that the 7.5% decline in force occurring over the observed variation in SW is not large enough to overcome the many causes of suboptimality listed above, whereas the 25% decline we find beyond this range is large enough to cause selection pressure toward the OSW. Additional research is needed to establish whether this hypothesis is true.

Finally, we note another aspect of the stability of the OSW, which is its insensitivity to changes in key kinematic variables such as frequency and amplitude, as well as to changes in fin geometry such as height and shape.

Is the OSW an Optimal Regime for a Suboptimal Type of Swimming?

While the previous discussion shows that once an elongated fin species has arisen it may be costly to depart from the OSW, elongated fin swimmers constitute a minority amongst all swimmers. For example, there are approximately 1,000 median/paired fin swimming species within the jawed fishes (based on ascending to the nearest non-median/paired fin swimmer at each node of the tree presented in Fig 1) compared to 33,000 total fish species currently listed within FishBase. The fastest fish, and therefore the fish that are able to produce the highest amount of force, are body/caudal fin swimmers

[30,31]. The question is therefore why the slower forms of swimming exhibited by median/paired fin animals would emerge and thrive despite the prevalence of body/caudal fin swimming in ancestral species.

A similar question has arisen in simulation studies that show that a light-sensitive patch of skin can evolve through several intermediate forms into an advanced camera-type lens eye in only a few hundred thousand years [32]—why, then, are there so many existing animals with intermediate forms of eyes? Nilsson and Pelger's answer is that camera-type lens eyes are only the best solution for certain animal–ecosystem combinations [32]. Our answer is similar: body/caudal fin swimming makes little sense in isolation. It is only within particular ecological contexts that some types of animals are able to survive better with this type of swimming than with alternative approaches.

In particular, median/paired fin swimming appears to be a low speed, low cost of transport specialization [21,30,33]. The lower amplitudes of fin movement that are possible in median/paired fin swimmers, compared to the very high amplitudes possible when the high power axial musculature is used in body/caudal fin swimmers, is therefore an advantage instead of a liability due to the lower energetic cost of transport of median/paired fin swimming [30]. The fact that median/paired fin swimming is used at lower speeds should not be confused, however, with the concept of maximizing speed by swimming at the OSW. Even when swimming at lower speeds (or whatever speed for that matter, which is determined by frequency and amplitude), for a given set of parameters (amplitude of undulations, frequency, fin height, and fin shape), if an animal swims with elongated median/paired fins, then its speed can be maximized for that set of parameters by swimming at the OSW.

The ecological circumstances in which median/paired fin swimming will be favored over the typically faster, higher cost of transport body/caudal fin mode are not clear. If we consider one group specifically, such as the Gymnotiformes, a number of ecological features stand out as possible factors. First, they live in habitats of murky water and are active at night. Therefore vision—and high speed predation based on visual guidance—is not a factor. Second, periodic reductions in oxygen levels of the rivers they inhabit [33] may favor the lower cost of transport of median/paired fin swimming. Third, species within this group sense using electric fields, a mode of sensing where there is fourth-power attenuation of signal with distance [34]. This quartic attenuation implies a penalty on gross movement in sensory acquisition tasks, since typically sensing range is extremely short [34–36]. Therefore, precise, small, and slow movements are most effective. Finally, because the field generator is in the trunk, movement of the body causes large distortions of the emitted signal, possibly favoring holding the trunk rigid [37,38] rather than using it for propulsion. Given these constraints, the elongated fins that are universally present within the more than 150 species comprising Gymnotiformes may be favored, but clearly a tremendous amount of work would need to be done to assess the relative importance of all of these factors in giving rise to this one group of median/paired fin swimmers.

While the existence of body/caudal fin swimmers and the existence of median/paired fin swimmers may or may not be subject to robust repeatability, what is clear is that if median/paired fin swimming with elongated fins and semirigid trunks emerges—as it has independently on multiple occasions according to Fig 1—it is very probable that the specific trait of swimming at the OSW will also emerge. This is because of the thrust-maximizing property of the OSW, the existence of the gradient in thrust efficacy with a stabilizing derivative mentioned above, and the insensitivity of the OSW to variations in key fin movement parameters such as frequency and amplitude and to variations in fin height and shape.

Maximizing thrust ensures that speed can be maximized for a given set of parameters, irrespective of whether the animal swims energetically efficiently during cruising, or possibly less efficiently during escape or attack maneuvers. Thus, it seems likely that aquatic animals could benefit from adhering to the OSW at all—or at least at most—times. An example of using the OSW for thrust maximization even when swimming speed is not a concern is provided by the phenomenon of counterpropagating waves along a single fin. Gymnotid fish such as the black ghost knifefish A. albifrons [22,39] and the glass knifefish E. virescens will generate undulations from head to tail (called the head wave) and undulations from tail to head (called the tail wave) along its median anal fin, producing antagonistic forces [40]. We and collaborators have shown that this enhances stability and maneuverability [40]. Published videos show indications of this pattern being more widespread than gymnotid fish, such as the oarfish R. glesne [41] and cuttlefish [42]. For E. virescens, where counterpropagating waves have been analyzed more fully, while the angular amplitude is relatively constant between the two waves, there is a distinct difference in the wavelength. As the fin is deeper rostrally, for a similar angular excursion of each fin ray, the resulting mean lateral amplitude $\tilde{a}$ of the fin will be greater in the head wave than in the tail wave. The fish compensates for these differences in mean amplitude with a longer wavelength for the head wave than the tail wave, keeping the SW closer to 20 than if a constant wavelength were present throughout the fin. A. albifrons follows a similar trend, though more data is needed for a full analysis. Maintaining OSW for the head and tail counterpropagating waves maximizes the counteracting thrust forces which lead to increased stability and maneuverability [40].

Despite these uses of the OSW for force maximization, it should be noted that median/paired fin fish are able to switch to the higher force production mode enabled by body/caudal fin swimming when absolutely needed, such as C-starts in knifefish [43], gait changes in median/paired fin swimmers to body/caudal fin swimming as speed increases [30], and oarfish transitioning to anguilliform swimming for higher speed (for reference see the videos from [41]). A natural question is whether body/caudal fin swimmers, or those that undulate their entire body, such as eels and lamprey—or even terrestrial animals like snakes—abide by an OSW of ≈ 20. It is likely that some related criteria exist but the optimal value of the SW may be different, or it may not be a constant as in case of the animals considered here that swim by means of elongated fins while keeping their trunk semirigid. In this context, we note that in the characterization of sandfish swimming (effectively, a body/caudal fin swimmer in sand), Maladen et al. [14,15] find that the swimming speed of the sandfish is optimal at SW = 5. The issue of the applicability of an OSW-like optimum to other modes of swimming, although not within the scope of this work, merits future investigation.

How the OSW Is Implemented

One obvious way in which swimming animals could adhere to the OSW is through neural control. Whether at the level of spinal circuitry, or through descending control, this control would enforce the OSW through changing amplitude and wavelength so as to maintain a fixed ratio of around 20, as discussed below in remarks discussing the implications of the OSW for controlling robots. While outside the scope of this study, this hypothesis warrants further investigation, perhaps using the weakly electric fish model system, which is heavily used within certain domains of neuroscience [44].

Another approach is to build the OSW into the material properties of the structures that comprise the fin. The fin is made of a collagenous membrane that connects adjacent fin rays in teleosts [45] and a muscle and collagen matrix in cuttlefish [46,47]. This connective tissue limits the maximal strain, and thus the curvature of the fin cannot increase beyond a certain value. Suppose the amplitude of movement is increased for a given wavelength. This would increase the curvature and strain in some parts of the membrane. If the maximal allowable strain is being approached, then the material would relax to a larger wavelength for the new larger amplitude. A demonstration of this principle is seen in experiments in which the anterior end of a long, flexible, free-swimming foil is oscillated [48]. When the amplitude of a traveling wave along the foil increased by slightly over a factor of two, the wavelength increased by a factor of 1.6 (see Fig 4, [48]). Interestingly, the primary structural protein of aquatic animal fins, collagen, is an ancient protein predating the divergence of the three phyla where we have observed the OSW [49].

Whether by neural control or by material properties, there is an additional aspect of swimming by median/paired fins that could facilitate genetic aspects of the implementation of the OSW. For the more prevalent body/caudal fin swimming style, there are a host of other functions served by the body besides propulsion, such as holding internal organs, or providing a mirror-like surface to facilitate camouflage [50,51]. Having to move the mass of the trunk for propulsion, including such things as internal organs, is a disadvantage for the body/caudal fin style of swimming.

In contrast to most body/caudal fin swimmers, median/paired fin swimmers use their fins purely as thrust elements. In moving fluid to generate thrust, they are mostly putting energy into moving fluid instead of the mass of the membranous fin. Therefore, the collection of genes that give rise to axial elongation [52,53] and the associated elongated fins used for thrust production may have few other functions, perhaps facilitating their approach to an optimum in comparison to the genes controlling body shape in swimmers that move their trunk for swimming [6]. Further evidence in support of this idea is the finding of convergently evolved anatomical specializations for one type of body/caudal fin swimmer that minimizes trunk movement during tail fin movement, resulting in a similarly pure thrust element [12]. These are the thunniform swimmers such as tuna and lamnid sharks which largely restrict movement to their caudal fin propulsor, and in doing so are able to swim faster than most other fish. We hypothesize that regulation of the OSW, axial elongation, and fin elongation in the phyla in which we observe these traits may reflect the action of common genetic networks, as has been shown to underlie convergent evolution in the case of the electric organ [54].

Predicting the Number of Fin Undulations

Based on the OSW and the measured mean undulation amplitude ã, the number of undulations that should be present along the fin of a swimming animal, cruising at any velocity, can be predicted. This is given by

$$NU_{predict} = \frac{L}{OSW\,\tilde{a}}. \qquad (2)$$

We plot the measured number of undulations along with the predicted number of undulations in Fig 5 for all the same animals shown in Fig 1, using measurements of mean amplitude and assuming an OSW = 20. The number of undulations predicted is in excellent agreement with the observed number.

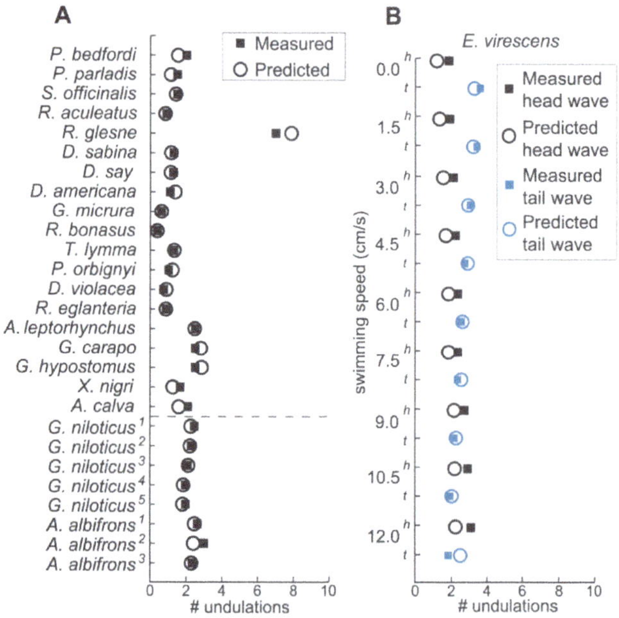

Figure 5. Predicted number of undulations along a fin (open circles) compared to measured number of undulations (filled squares) for the 20 species shown in Fig 1, plus D. americana in (A) and E. virescens in (B) (same data sources as listed in Fig 1). The order on the left plot is the same as Fig 1, except we have moved G. niloticus andA. albifrons to below the dashed line. Below the dashed line, the superscripts indicate data at different swimming speeds: For G. niloticus, 1 = 19.7 cm/s, 2 = 21.2 cm/s, 3 = 22.3 cm/s, 4 = 23.8 cm/s, 5 = 26.6 cm/s. For A. albifrons, 1 = 14.9 cm/s, 2 = 19.7 cm/s, and 3 = 32.6 cm/s. When the swimming velocity increases, amplitude is increased and the number of undulations is decreased to maintain an SW that is close to the OSW. (B) compares the predicted and measured number of undulations of both the head and the tail waves of the counterpropagating undulations in E. virescens averaged over five fish. The trends seen are largely due to the change in the lengths of the head and tail waves as the point where the two waves meet shift caudally with increased forward swimming speed. The STD across these data ranges from 0.15 to 0.67 and is not shown for clarity.

These predictions are for a single swimming velocity for each species. However, in two cases we have fin undulation data across several swimming velocities. Here too, the fin follows the pattern dictated by the OSW. The data for the African aba aba knifefish (G. niloticus) and South American black ghost knifefish (A. albifrons) below the dashed line in Fig 5A are at different swimming speeds. The superscripts on G. niloticus and A. albifrons relate to swimming speed, with the superscript number [1] corresponding to the slowest speed and higher numbers to higher speeds. The data show that as these fish swim faster, they reduce the number of undulations along their fins. This reduction in undulations along the fin is most likely because this is necessary to continue swimming at the OSW. To swim faster, the knifefish have to either increase the amplitude of fin undulations, or their frequency, or both. An important question for future investigation is whether the motor systems of undulatory animals provides for independent control of key kinematic parameters such as amplitude and frequency. For example, in a biologically inspired model of how coupled oscillators generate traveling waves along the anal fin of knifefish, the drive to the initial oscillator controlled frequency, while the gradient in drive across oscillators controlled the number of undulations [22], a situation in which there is no independent control of these key kinematic parameters. For the data plotted in Fig 5, to swim faster, the knifefish increase both frequency and amplitude [22,55]. Consequently, to compensate for the increase in amplitude, the wavelength is decreased to maintain the OSW.

Because the SW is independent of frequency, if an animal is swimming at the OSW and only frequency is varied to change speed, then the organism does not have to change the number of undulations along the fin to maintain the OSW. It appears that frequency control dominates at lower speeds for knifefish [21,22], which likely underlies the observation of a relatively constant number of undulations along the fin as swimming speed changes in a variety of electric and nonelectric knifefish [21,22]. Rosenberger [56,57] also mentions how some species of rays such as Gymnura micrura and T. lymma exhibit changes in fin motion with swim speed. As G. micrura and T. lymma increase their swimming speed, the number of undulations goes down. Rosenberger [57] also reported that T. lymma used smaller amplitudes at lower speeds. Consequently, to maintain the OSW, it would have to use more undulations when swimming slowly compared to when it is swimming rapidly using larger amplitudes. The same change in amplitude with speed has been mentioned qualitatively for G. micrura [56], and we would predict that quantitative measures would show that the OSW-prescribed pattern holds for this animal as well during changes in speed.

A final example of the correlation between amplitude and number of undulations is provided by the phenomenon of counterpropagating waves along a single fin discussed above. Despite traveling waves passing along the fin in opposite directions along fin membrane of differing depth (resulting in different amplitudes for similar angular excursions across fin rays), fish adjust the wavelength of undulations so as to maintain the OSW, as shown by Fig 5B.

The OSW and St

A nondimensional number commonly used to characterize the swimming of body/caudal fin swimmers is the St (St = 2fa / U, where U is the swimming speed, 2a is the maximum peak to peak amplitude of the lateral fin excursion, and f is the frequency of tail beating). It has been reported that body/caudal fin swimmers cruise at an St between 0.2–0.4 [8, 58]. It has also been reported, based on a meta-analysis of steady undulatory swimming, that the St varies over a much larger range of 0.2–1.8 [59]. Prior studies suggest that the basis of St between 0.2–0.4 in animals is high propulsive efficiency, which may relate to maintenance of attached flow [8] and hydrodynamic resonance [60], among other factors. Eloy [61], using Lighthill's elongated body theory [62], has shown that an St between 0.15 and 0.8 minimizes energy expenditure and maximizes Froude efficiency. In swimming live knifefish, the St ranges between 0.3 and 0.6 (from data in [22]), the same range found in a study reporting data from a robotic knifefish as well as computational fluid simulations of knifefish fins [63].

A quantitative relationship between the St and the SW follows from their definitions:

$$\mathrm{St}\ \eta_{wave}\ \mathrm{SW} = \gamma, \tag{3}$$

where $\eta_{wave} = \frac{U}{V_w}$ is the wave efficiency, which is the ratio of the velocity of a swimmer to the velocity of the traveling wave (frequency × wavelength) along the fin (see [64], where $1-\eta_{wave}$ is the "slip"). The wave efficiency can, in theory, take any value between zero and one. It has been reported that for high Froude efficiency, the wave efficiency should be in the range of 0.4–0.8 [61]. The wave efficiencies observed in swimming animals ranges from 0.1 to 0.9 [59]. The factor $\gamma = 2a / \tilde{a}$, where 2a is the maximum peak to peak amplitude, while $\tilde{a}$ is the mean amplitude as previously defined, is due to the difference in how the amplitude parameter is defined in the formula for St and how we define it for the SW. γ ranges from four, for a rectangular fin similar to those of knifefish, to around eight for a deep triangular fin similar to those of some batoid fishes.

Given St and SW, the wave efficiency, η_{wave}, is determined according to Equation 3 for a given γ. For the range of γ from 4–8, a SW of 20 combined with the reported range of St of 0.2–0.4 [8,58] results in high wave efficiencies. As an example, the relationship from Equation 3 for $\gamma = 6.7$, that is typical of batoid fish, is shown graphically in Fig 6. The wave efficiencies are all above 0.6. Therefore, while St and SW are complementary, the observed values are also consistent with each other.

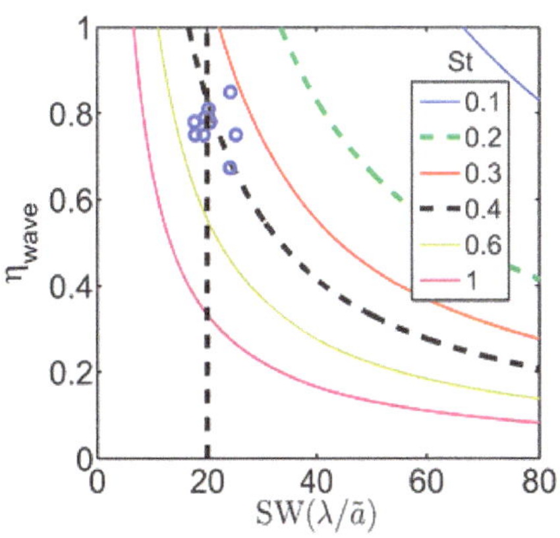

Figure 6. Isolines of St against SW on the x-axis and wave efficiency on the y-axis. St was computed using Equation 3, in which the mean value of $\gamma = 6.7$ for batoid fish, presented in Fig 1, was used. The OSW is highlighted by a dashed vertical line. The dashed contour lines correspond to the St range of 0.2–0.4. Also plotted are the data points (open circles) of the batoid fishes.

Although the St and the SW are related according to Equation 3, these two parameters are not equivalent. To appreciate this, three considerations should be noted. The first is that St and SW quantify different physical parameters of the problem. The St number quantifies information about the amplitude and frequency of undulations relative to the swimming speed, while the SW number quantifies the wavelength of undulations relative to amplitude. The difference between these two parameters is highlighted by situations in which median/paired fin swimmers undulate their fins while their swimming speed is zero to enhance stability and maneuverability [40], the phenomenon of counterpropagating waves mentioned above. In this situation, St approaches infinity, while the fish maintains the optimal SW.

The second consideration is that the optimal value of St (0.2–0.4) recommends the frequency that maximizes the propulsive efficiency (the ratio of thrust power to the power spent by the fin) of the fin [58] without any recommendation for the wavelength of undulations. A high propulsive efficiency case could be possible with large thrust or small thrust or other values. Of all the cases that have comparable propulsive efficiency, the case with maximum thrust is desirable since it will also lead to a fast swimming speed. For aquatic animals swimming by means of elongated median/paired fins, the OSW recommends an appropriate wavelength so that the propulsive thrust is also maximized. Thus, the St and the OSW are complementary and can help identify two physical parameters of the problem—frequency and wavelength—such that a median/paired fin animal swims with high propulsive efficiency and high thrust. Finally, the third difference between the St and the OSW arises from consideration of how these two parameters can enter into the control of advanced underwater robots that use undulating fins. The St incorporates swimming velocity, the time derivative of position, which—like orientation—is termed a group variable in mechanics. The SW only refers to variables internal to the body (amplitude and wavelength), frequently called shape variables. Often in mechanical systems there exists full control authority over shape variables, meaning any time evolution of the variables can be specified to arbitrary precision. The group variables then evolve through nonlinear dynamics involving the shape variables [65]. In the case of swimming velocity, for example, these dynamics are governed by the Navier-Stokes equation at intermediate Reynolds numbers. Control of robotic undulators for advanced underwater robots in the future may involve the relatively simple control of proportionately changing wavelength with amplitude for maximizing force production by maintaining the OSW. However, maintaining the St within a specific range would require estimating swimming velocity with a sensor or forward model, accompanied by control which accounts for the nonlinear effects that the shape variables such as frequency and amplitude have on the velocity. To the extent that the neural or mechanical control of a fin in an aquatic animal proceeds along similar lines, it may also be the case that adherence to the OSW is a relatively direct function of variables under the control of the animal.

A Hypothesis for Competing Mechanisms Underlying the OSW

What is the physical basis of the optimal value of the SW? The SW can be related to the peak steepness (slope) of the surface of an undulatory fin. The value of the peak slope corresponding to the OSW will depend on the fin geometry and the shape of the undulatory wave. For a 2D sinusoidal wave at the OSW, the angle made by the fin surface at the location of the peak slope is 17.44°. Our results indicate that for a given amplitude, the propulsive force is maximum at an intermediate peak steepness of the surface of the fin: very steep and very shallow are both suboptimal. We hypothesize that the optimal peak steepness may be rationalized in terms of two competing mechanisms (Fig 7). A steeper surface (smaller SW) of the fin should be able to trap and transport the fluid backward more efficiently. The increase in backward fluid momentum should lead to a stronger propulsive force at small SW. We call this the "friction mechanism" in Fig 7. A shallower surface (larger SW) of the fin leads to larger wave velocity resulting in the fluid being propelled backward at higher speed. Faster fluid velocity should increase backward fluid momentum and should lead to a stronger propulsive force at large SW. We call this the "velocity mechanism" in Fig 7. Thus, we speculate that SW influences two separate physical mechanisms with opposite trends. These competing mechanisms would lead to an optimal value of propulsive force at an intermediate value of SW.

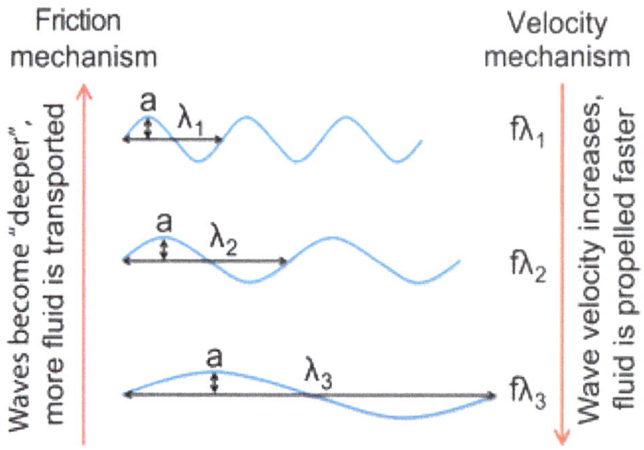

Figure 7. A schematic of a hypothesis for two competing mechanisms that could lead to the optimal propulsive force and therefore swimming speed. Keeping the amplitude, length, and frequency fixed, as the wavelength is increased (top to bottom), the wave velocity increases proportionally (indicated by the downward arrow on the right). As the wave velocity increases, the fluid is propelled backward at higher velocity. At the same time, the effectiveness with which the fin can trap and transport the fluid decreases (indicated by the upward arrow on the left).

METHODS

Experimental Setup

The experiments were carried out using an updated version of the "Ghostbot" used in a number of previous studies [20,39,66]. The Ghostbot generated propulsive force using an elongated fin inspired by the ventral fin of a knifefish. The fin was 32.6 cm long and 5 cm deep. The fin was composed of 32 discrete fin rays which were driven by individual motors housed in a cylindrical body. The rays were connected by a Lycra fabric which constitutes the fin membrane. A more detailed description of the Ghostbot can be found elsewhere [20].

Experimental Procedure

The axial force generated by the ribbon fin was measured similar to the methods presented in our earlier work [20]. The robot was submerged into a custom built flow tunnel and was suspended from above on a frictionless air bearing system allowing near-frictionless motion. The dimensions of the working area of the tunnel were 33 cm wide, 32 cm deep, and 100 cm long. The robot was oriented horizontally and fixed in place in the water. A calibrated single axis force transducer (LSB200, Futek, Irvine, CA, USA) was placed between the air-bearing platform and mechanical ground in the axial direction of the robot so that longitudinal forces could be measured. For each trial, voltages from the force transducer were recorded at 1,000 Hz over a period of approximately 20 seconds while the fin was undulating in stationary water. The last 8 s of data were averaged. A zero measurement was also taken while the robot's fin was at rest (not undulating) and was subtracted from the average voltage measurement from each trial. Voltage data were then converted into force units based on the calibration, which had a maximum nonlinear error of 0.034%. Details of the experiment in which free swimming velocity of the robot was measured can be found in prior work [20].

Error Measurements

The error bars in Fig 2 show STD of error from measurements of speed and force detailed in the following. The swimming speed of the robot was measured in five consecutive trials with the same kinematic parameters. The parameters chosen for these trials were f = 3 Hz, number of undulations = 2, and θ_{max} = 30°. The time gap between each trial was such that the flow in the tank was equal to zero before the start of the next trial. Similarly, the force measurements were also measured in five consecutive trials. In each trial the fin was actuated with f = 3 Hz, number of undulations = 2, and θ_{max} = 30°.

Numerical Method

Numerical simulations of the ribbon fin were carried out using an implementation of the constraint-based immersed body (IB) method (cIB) [24,25,67–70]. The cIB involves the solution of the mass and momentum conservation equations, the Navier-Stokes equations, for the combined fluid and solid domain with a constraint force in the solid domain to impose the solid motion on the fluid. The Navier-Stokes equations for the fluid flow are solved in the entire computational domain in an Eulerian framework. The motion and kinematics of the immersed body (ribbon fin in the present work) are represented in a Lagrangian frame. Information between the Lagrangian frame of the IB and Eulerian frame of the fluid is exchanged through the discrete delta function operator [71]. Further details of the numerical technique can be found in previously published work [24,25,63]. The cIB method has been incorporated within the IBAMR software [24,25]. IBAMR is an IB method implementation with support for adaptive mesh refinement and distributed memory parallelism.

Two kinds of fin simulations were carried out: free swimming and stationary. In free swimming fin simulations, the translational degree of freedom of the swimmer being simulated is free. The swimming speed of the swimmer is determined by the solution of the numerical simulation. In the stationary fin simulations, all the translational degrees of freedom of the swimmer are locked and the swimmer is held stationary while it undulates its body as it would during free swimming. These simulations are used to compute the propulsive force generated by the swimmer's undulations. Simulations were carried out on two sets of fins: a knifefish-sized fin (10 cm × 1 cm) and a smaller rectangular fin (2 cm × 0.4 cm). The computational cost of numerical simulations for the smaller fin is lower than that of knifefish sized fin. Because of this relatively lower computational cost, the smaller fin was used in the parametric study (Fig 4). The physical properties of the fluid in the simulations were for water at 25°C: Density $\rho = 0.9956\times10^3$ kg/m³, viscosity $\mu = 0.89\times10^{-3}$ N·s/m².

Fin Kinematics

In prior work, we have approximated the shape of the traveling waves along the fins of knifefish with a traveling sinusoid [19,20,39]. The sinusoidal approximation departs from measured wave shapes in several regards [22,55], but these differences appear to have negligible impact on our results. Consequently, the kinematics of the fin undulations in the experiments and numerical simulations are described by

$$\theta = \theta_{max} \sin 2\pi(x/\lambda - ft), \qquad (4)$$

where θ is the angular displacement of a fin ray located at an axial distance x from the rostral end of the fin, θ_{max} is the maximum possible angle of excursion, λ is the wavelength, and f is the frequency of undulation.

Computing Environment

The computational resources for the numerical simulations presented in this work were provided by the Quest high performance computing facility at Northwestern University. The Quest high performance cluster is composed of 252 nodes of Intel Westmere X5650 processors with 48 GB memory/node, 68 nodes of Intel Sandybridge E2670 processors with 68 GB memory/node, and 110 Intel IvyBridge E5-2680 processors with 128 GB memory/node.

ACKNOWLEDGMENTS

We thank Todd Murphey and Heather Eisthen for helpful discussions on an earlier draft, and Noah Cowan and Eric Fortune for ribbon fin kinematic data for E. virescens partially detailed in [40]. HDT Robotics (Evanston, IL USA) built the robotic knifefish with the support of Northwestern University.

AUTHOR CONTRIBUTIONS

Conceived and designed the experiments: RB NAP MAM. Performed the experiments: RB IDN. Analyzed the data: RB NAP. Wrote the paper: MAM NAP RB IDN. Led the numerical simulation and analysis effort: NAP. Developed the numerical tool: APSB. Carried out numerical simulations: RB. Led the experimental effort: MAM.

REFERENCES

1. Jacob F (1977) Evolution and tinkering. Science 196: 1161–1166. pmid:860134 doi: 10.1126/science.860134
2. Carroll SB (2001) Chance and necessity: the evolution

of morphological complexity and diversity. Nature 409: 1102–1109. pmid:11234024 doi: 10.1038/35059227

3. Powell R (2012) Convergent evolution and the limits of natural selection. Euro Jnl Phil Sci 2: 355–373. doi: 10.1007/s13194-012-0047-9
4. Gould SJ (1989) Wonderful life: the Burgess Shale and the nature of history. New York: W.W. Norton, 1st edition, 347 pp.
5. Conway Morris S (2003) Life's solution: inevitable humans in a lonely universe. Cambridge UK; New York: Cambridge University Press, xxi, 464 pp.
6. Guillaume F, Otto SP (2012) Gene functional trade-offs and the evolution of pleiotropy. Genetics 192: 1389–1409. doi: 10.1534/genetics.112.143214. pmid:22982578
7. Khatri BS, McLeish TCB, Sear RP (2009) Statistical mechanics of convergent evolution in spatial patterning. Proceedings of the National Academy of Sciences of the United States of America, 106(24), 9564–9569. doi: 10.1073/pnas.0812260106.
8. Taylor GK, Nudds RL, Thomas AL (2003) Flying and swimming animals cruise at a strouhal number tuned for high power efficiency. Nature 425: 707–711. pmid:14562101 doi: 10.1038/nature02000
9. Ritzmann RE, Quinn RD, Fischer MS (2004) Convergent evolution and locomotion through complex terrain by insects, vertebrates and robots. Arthropod Struct Dev 33: 361–79. pmid:18089044 doi: 10.1016/j.asd.2004.05.001
10. Gleiss AC, Jorgensen SJ, Liebsch N, Sala JE, Norman B, et al. (2011) Convergent evolution in locomotory patterns of flying and swimming animals. Nature Communications 2: 352. doi: 10.1038/ncomms1350. pmid:21673673
11. Srinivasan M, Ruina A (2006) Computer optimization of a minimal biped model discovers walking and running. Nature 439: 72–75. pmid:16155564 doi: 10.1038/nature04113
12. Donley JM, Sepulveda CA, Konstantinidis P, Gemballa S, Shadwick RE (2004) Convergent evolution in mechanical design of lamnid sharks and tunas. Nature 429: 61–65. pmid:15129279 doi: 10.1038/nature02435
13. Webb P (1984) Body form, locomotion and foraging in aquatic vertebrates. Am Zool 24: 107–120. doi: 10.1093/icb/24.1.107
14. Taylor G (1952) Analysis of the swimming of long and narrow animals. P Roy Soc Lond B Mat 214: 158–183. doi: 10.1098/rspa.1952.0159
15. Maladen R, Ding Y, Li C, Goldman D (2009) Undulatory Swimming in Sand: Subsurface Locomotion of the Sandfish Lizard. Science 325: 314. doi: 10.1126/science.1172490. pmid:19608917
16. Maladen RD, Ding Y, Umbanhowar PB, Kamor A, Goldman DI (2011) Mechanical models of sandfish locomotion reveal principles of high performance subsurface sand-swimming. J Roy Soc Interface 8: 1332–1345. doi: 10.1098/rsif.2010.0678
17. McMillen T, Holmes P (2006) The dynamics of choice among multiple alternatives. J Math Psycol 50: 30–57. doi: 10.1016/j.jmp.2005.10.003
18. Lucas KN, Johnson N, Beaulieu WT, Cathcart E, Tirrell G, et al. (2014) Bending rules for animal propulsion. Nat Commun 5: 3923. doi: 10.1038/ncomms4923. pmid:24879355
19. Shirgaonkar AA, Curet OM, Patankar NA, MacIver MA (2008) The hydrodynamics of ribbon-fin propulsion during impulsive motion. J Exp Biol 211: 3490–3503. doi: 10.1242/jeb.019224. pmid:18931321
20. Curet OM, Patankar NA, Lauder GV, MacIver MA (2011) Mechanical properties of a bio-inspired robotic knifefish with an undulatory propulsor. Bioinspir Biomim 6: 026004. doi: 10.1088/1748-3182/6/2/026004. pmid:21474864
21. Blake RW (1983) Swimming in the electric eels and knifefishes. Can J Zool 61: 1432–1441. doi: 10.1139/z83-192
22. Ruiz-Torres R, Curet OM, Lauder GV, MacIver MA (2013) Kinematics of the ribbon fin in hovering and swimming of the electric ghost knifefish. J Exp Biol 216: 823–834, Correction (vol. 217, p. 3765). doi: 10.1242/jeb.076471. pmid:23197089
23. Albert JS (2001) Species diversity and phylogenetic systematics of American knifefishes (Gymnotiformes, Teleostei). University of Michigan Misc Publ Mus Zool 190: 1–127.
24. Bhalla APS, Bale R, Griffith BE, Patankar NA (2013) A unified mathematical framework and an adaptive numerical method for fluid-structure interaction with rigid, deforming, and elastic bodies. Journal of Computational Physics 250: 446–476. doi: 10.1016/j.jcp.2013.04.033
25. Bhalla APS, Bale R, Griffith BE, Patankar NA (2014) Fully resolved immersed electrohydrodynamics for particle motion, electrolocation, and self-propulsion. Journal of Computational Physics 256: 88–108. doi: 10.1016/j.jcp.2013.08.043
26. IBAMR: An adaptive and distributed-memory parallel implementation of the immersed boundary method. https://github.com/IBAMR/IBAMR. IBAMR-web-page
27. Yang SB, Qiu J, Han XY (2009) Kinematics modeling and experiments of pectoral oscillation propulsion robotic fish. J Bion Eng 6: 174–179. doi: 10.1016/s1672-6529(08)60114-6
28. Blake R (1980) Undulatory median fin propulsion of two teleosts with different modes of life. Can J Zool 58: 2116–2119. doi: 10.1139/z80-291
29. MacIver MA, Patankar NA, Shirgaonkar AA (2010) Energy-Information Trade-Offs between Movement and Sensing. PLoS Comput Biol 6: e1000769. doi: 10.1371/journal.pcbi.1000769. pmid:20463870
30. Korsmeyer KE, Steffensen JF, Herskin J (2002) Energetics of median and paired fin swimming, body and caudal fin swimming, and gait transition in parrotfish (Scarus schlegeli) and triggerfish (Rhinecanthus aculeatus). J Exp Biol 205: 1253–1263. pmid:11948202
31. Webb PW (1984) Form and function in fish swimming. Sci Am 251: 58–68. pmid:6084304 doi: 10.1038/scientificamerican0784-72

32. Nilsson DE, Pelger S (1994) A pessimistic estimate of the time required for an eye to evolve. Proc Biol Sci 256: 53–58. pmid:8008757 doi: 10.1098/rspb.1994.0048

33. Julian D, Crampton WGR, Wohlgemuth SE, Albert JS (2003) Oxygen consumption in weakly electric neotropical fishes. Oecologia 137: 502–511. pmid:14505027 doi: 10.1007/s00442-003-1368-3

34. Nelson ME, MacIver MA (2006) Sensory acquisition in active sensing systems. J Comp Physiol A 192: 573–586. pmid:16645885 doi: 10.1007/s00359-006-0099-4

35. MacIver MA, Sharabash NM, Nelson ME (2001) Prey-capture behavior in gymnotid electric fish: Motion analysis and effects of water conductivity. J Exp Biol 204: 543–557. pmid:11171305

36. Snyder JB, Nelson ME, Burdick JW, MacIver MA (2007) Omnidirectional sensory and motor volumes in an electric fish. PLoS Biol 5: 2671–2683. doi: 10.1371/journal.pbio.0050301

37. Lissmann H (1963) Electric location by fishes. Sci Am 208: 50–59. doi: 10.1038/scientificamerican0363-50

38. Chen L, House JH, Krahe R, Nelson ME (2005) Modeling signal and background components of electrosensory scenes. J Comp Physiol A 191: 331–345. pmid:15800793 doi: 10.1007/s00359-004-0587-3

39. Curet OM, Patankar NA, Lauder GV, MacIver MA (2011) Aquatic manoeuvering with counter-propagating waves: a novel locomotive strategy. J R Soc Interface 8: 1041–1050. doi: 10.1098/rsif.2010.0493. pmid:21177695

40. Sefati S, Neveln ID, Roth E, Mitchell TR, Snyder JB, et al. (2013) Mutually opposing forces during locomotion can eliminate the tradeoff between maneuverability and stability. Proc Natl Acad Sci USA 110: 18798–18803. doi: 10.1073/pnas.1309300110. pmid:24191034

41. Benfield M, Cook S, Sharuga S, Valentine M (2013) Five in situ observations of live oarfish Regalecus glesne (Regalecidae) by remotely operated vehicles in the oceanic waters of the northern Gulf of Mexico. J Fish Biol: 83(1). doi: 10.1111/jfb.12156. pmid:23808703

42. Ma S (2012). Cuttlefish swimming inside Ocean Park's Grand Aquarium, Hong Kong.http://youtu.be/zjwNxbdKoeI.

43. Domenici P, Blake R (1997) The kinematics and performance of fish fast-start swimming. J Exp Biol 200: 1165–1178. pmid:9319004

44. Krahe R, Fortune ES (2013) Electric fishes: neural systems, behaviour and evolution. J Exp Biol 216: 2363–2364. doi: 10.1242/jeb.091322. pmid:23761461

45. Lauder GV, Madden PGA (2006) Learning from fish: Kinematics and experimental hydrodynamics for roboticists. International Journal of Automation and Computing 4: 325–335. doi: 10.1007/s11633-006-0325-0

46. Kier WM, Smith KK, Miyan JA (1989) Electromyography of the fin musculature of the cuttlefish Sepia officinalis. J Exp Biol 143: 17–31. pmid:2732660

47. Johnsen S, Kier WM (1993) Intramuscular crossed connective tissue fibres: skeletal support in the lateral fins of squid and cuttlefish (mollusca: Cephalopoda). J Zool 231: 311–338. doi: 10.1111/j.1469-7998.1993.tb01921.x

48. Lauder G, Lim J, Shelton R, Witt C, Anderson E, et al. (2011) Robotic models for studying undulatory locomotion in fishes. Mar Technol Soc J 45: 41–55. doi: 10.4031/mtsj.45.4.8

49. Boot-Handford RP, Tuckwell DS (2003) Fibrillar collagen: the key to vertebrate evolution? A tale of molecular incest. Bioessays 25: 142–151. pmid:12539240 doi: 10.1002/bies.10230

50. Brady PC, Travis KA, Maginnis T, Cummings ME (2013) Polaro-cryptic mirror of the lookdown as a biological model for open ocean camouflage. Proc Natl Acad Sci USA 110: 9764–9769. doi: 10.1073/pnas.1222125110. pmid:23716701

51. Denton EJ (1970) Review lecture: on the organization of reflecting surfaces in some marine animals. Philos Trans R Soc Lond, B, Biol Sci 258: 285–313. pmid:22408830 doi: 10.1098/rstb.1970.0037

52. Ward AB, Mehta RS (2010) Axial elongation in fishes: using morphological approaches to elucidate developmental mechanisms in studying body shape. Integr Comp Biol 50: 1106–1119. doi: 10.1093/icb/icq029. pmid:21558262

53. Copf T, Schroder R, Averof M (2004) Ancestral role of caudal genes in axis elongation and segmentation. Proc Natl Acad Sci USA 101: 17711–17715. pmid:15598743 doi: 10.1073/pnas.0407327102

54. Gallant JR, Traeger LL, Volkening JD, Moffett H, Chen PH, et al. (2014) Genomic basis for the convergent evolution of electric organs. Science 344: 1522–1525. doi: 10.1126/science.1254432. pmid:24970089

55. Fei L, Tian-Jiang H, Guang-Ming W, Lin-Cheng S (2005) Locomotion of Gymnarchus niloticus: experiment and kinematics. J Bion Eng 2: 115–121.

56. Rosenberger LJ (2001) Pectoral fin locomotion in batoid fishes: undulation versus oscillation. J Exp Biol 204: 379–394. pmid:11136623

57. Rosenberger LJ, Westneat MW (1999) Functional morphology of undulatory pectoral fin locomotion in the stingray Taeniura lymma (Chondrichthyes: Dasyatidae). J Exp Biol 202: 3523–3539. pmid:10574730

58. Triantafyllou M, Triantafyllou G, Yue D (2000) Hydrodynamics of fishlike swimming. Annual Review of Fluid Mechanics 32: 33–53. doi: 10.1146/annurev.fluid.32.1.33

59. van Weerden JF, Reid DA, Hemelrijk CK (2013) A meta-analysis of steady undulatory swimming. Fish and Fisheries 15: 397–409. doi: 10.1111/faf.12022

60. Moored K, Dewey P, Smits A, Haj-Hariri H (2012) Hydrodynamic wake resonance as an underlying principle of efficient unsteady propulsion. J Fluid Mech 708: 329. doi: 10.1017/jfm.2012.313

61. Eloy C (2012) Optimal strouhal number for swimming animals. J Fluids Struct 30: 205–218. doi: 10.1016/j.jfluidstructs.2012.02.008

62. Lighthill J (1971) Large-amplitude elongated-body theory of fish locomotion. Proc Roy Soc B 179: 125–

138. doi: 10.1098/rspb.1971.0085

63. Neveln ID, Bale R, Bhalla AP, Curet OM, Patankar NA, et al. (2014) Undulating fins produce off-axis thrust and flow structures. J Exp Biol 217: 201–213. doi: 10.1242/jeb.091520. pmid:24072799

64. Gray J, Lissmann HW (1964) The locomotion of nematodes. J Exp Biol 41: 135–154. pmid:14161604

65. Bloch AM, Krishnaprasad PS, Marsden JE, Murray RM (1996) Nonholonomic mechanical systems with symmetry. Arch Ration Mech An 136: 21–99. doi: 10.1007/bf02199365

66. Sefati S, Neveln I, MacIver MA, Fortune ES, Cowan NJ (2012) Counter-propagating waves enhance maneuverability and stability: a bio-inspired strategy for robotic ribbon-fin propulsion. In: Fourth IEEE RAS/EMBS International Conference on Biomedical Robotics and Biomechatronics. Rome, Italy.

67. Patankar NA (2001) A formulation for fast computations of rigid particulate flows. In: Annual Research Briefs, Stanford University: Center for Turbulence Research. pp. 185–196.

68. Sharma N, Patankar NA (2005) A fast computation technique for the direct numerical simulation of rigid particulate flows. J Comput Phys 205: 439–457. doi: 10.1016/j.jcp.2004.11.012

69. Shirgaonkar AA, MacIver MA, Patankar NA (2009) A new mathematical formulation and fast algorithm for fully resolved simulation of self-propulsion. J Comput Phys 228: 2366–2390. doi: 10.1016/j.jcp.2008.12.006

70. Curet OM, AlAli IK, MacIver MA, Patankar NA (2010) A versatile implicit iterative approach for fully resolved simulation of self-propulsion. Computer Methods in Applied Mechanics and Engineering 199: 2417–2424. doi: 10.1016/j.cma.2010.03.026

CHAPTER 5 VERTEBRATE DIVERSITY

Walter Jetz, Paul V. A. Fine

INTRODUCTION

The uneven distribution of species diversity is a key feature of life on Earth and has myriad implications. While the scale-dependence of the determinants of the global variation in diversity is well acknowledged [1]–[6], to date a quantitative accounting of the roles of history and environment in generating and maintaining gradients in species richness is still lacking. Over the past three decades, increased data availability has facilitated analyses of contiguous geographic patterns in species richness at relatively fine spatial grains (100–200 km) at both continental [7]–[9] and global scales [10],[11]. At these spatial resolutions, environmental variables such as productivity or temperature have been shown to offer extremely strong statistical predictions of species richness [8],[11]–[18]. However, it has been difficult to connect these results directly with underlying evolutionary and ecological processes. One problem is that the ultimate drivers underpinning diversity, namely speciation and extinction[19], operate at scales much larger than the spatial resolution (e.g., 100 km grids) of most analyses. A number of studies have confirmed the strong effect of regional richness on local richness [1]–[3],[6],[20] and have speculated on the role of energy driving diversification at regional scales [21]–[24] as well as sorting at local scales [25]–[27]. But attempts to integrate them at the appropriate scale have been limited, and we know of no study that has quantified the effect of productivity on richness gradients jointly at regional and local scales and both in terms of evolutionary and ecological processes.

Another impediment to interpretations of gridded richness analyses has been that species' geographic ranges are generally much larger than, for example, 100 km×100 km grid cells, resulting in geographically non-random patterns of pseudo-replication, inflated spatial autocorrelation, and an overrepresentation of wide-ranging species and their respective climatic associations [8],[28]. These issues have to date precluded straightforward evolutionary and ecological interpretations of macroecological environment correlations of gridded richness patterns [5],[29]. While partly motivated by limits in the knowledge of fine-scale species distributions [30], macroecological analyses have also been conducted using, for example, ca. 800 ecoregions as spatial units [14],[31], but these regions still incur significant and geographically variable redundancy in species. We are not aware of a study on richness gradients that has successfully overcome this problem and thus truly have given each species equal weight.

Citation: Jetz W, Fine PVA (2012) Vertebrate Diversity . PLoS Biol 10(3): e1001292. doi:10.1371/journal.pbio.1001292.

Finally, while there is little doubt about the importance of time for diversification [32]–[34], attempts to date to invoke paleoclimate for understanding richness have been hampered by the lack of data, especially at deeper time scales. Several studies have linked relatively recent climatic oscillations, for example, those causing quarternary high-latitude glaciation, to geographic richness patterns [22],[35]–[37]. The geography of deeper time climate conditions and exactly how it relates to the tempo of past clade diversification is inherently difficult to estimate. But given deep conservatism in the environmental (e.g., biome) associations within clades [38],[39] compared to relatively dynamic geographic ranges [40], clades are expected to much more strongly track climatically defined regions, or biomes, rather than specific geographical locations over evolutionary history. The ages of biomes may thus offer a promising avenue for understanding the role of paleoclimate contributing to contemporary patterns of species richness and have recently been successfully correlated with both turtle and tree richness at the regional scale [41],[42]. To date, analyses connecting the age and area of regions to finer grain richness patterns have not been attempted.

Here, we aim to address these problems with a hierarchical framework that integrates the drivers of regional diversification of species with those of their sorting into finer grain assemblages at their respective scales of influence. We use this model to test the relative importance of past spatio-temporal variation of climatic conditions (specifically time-integrated area and productivity) versus contemporary environment for explaining both the regional and finer scale variation in the species richness of terrestrial vertebrates worldwide. Due to environmental niche conservatism, organisms are generally restricted to climatically defined bioregions, or "evolutionary arenas," characterized by in situ speciation and extinction. We expect differences in species richness between such regions to arise from different levels of net diversification (speciation – extinction over time). The number of speciation and extinction events should vary among regions due to differences in the sizes of populations over time and the opportunities for reproductive isolation for all resident taxa [32]. We expect these drivers to be associated with today's area [29],[32],[43] and energy availability (i.e., productivity) [8],[11]–[14] of bioregions but, critically, also with the past levels of these factors—that is, how bioregions have varied in areal extent and productivity over time [42]. Furthermore, regional rates of diversification have been hypothesized to vary with temperature and its effects on activity and biological rates such as rates of molecular evolution or species interactions [15]–[17], [44]. We expect all of these drivers in concert to shape broad-scale gradients of diversity and predict that in an integrative assessment of regional differences in diversity, (i) models accounting for the temporal availability of area or productivity will outperform those without (i.e., regions that are older and/or have in the past been larger in extent will support higher vertebrate species richness than younger and/or smaller regions), (ii) area times energy availability (net primary productivity) will be a stronger predictor of richness than area alone[13],[45], and (iii) average bioregion temperature will positively affect richness above and beyond the effects of productivity and have a stronger effect in ectotherms compared to endotherms [15]. We test these predictions for the 32 main subdivisions, or "bioregions," of the world based on vegetation type and major landmass (Figure 1) [46],[47]. We excluded montane regions (which exclusively harbor ca. 5% of vertebrate species and represent ca. 15% of global land area) due to their extremely steep environmental gradients and associated species turnover, which impedes reliable bioregional delineation and estimates of their extent over time. Over historical time-scales these climatically and geographically distinct bioregions have been characterized by similar environmental and climatic conditions, but have changed in size and shape over time within their respective realms [42]. All bioregions are within the range of scales over which allopatric speciation of terrestrial vertebrate speciation typically occurs (100–1,000 km scale [48]) and may thus be considered bio-climatically and geographically distinct "evolutionary arenas." After deriving time-integrated models of bioregion species richness, we then in a second step assess their ability to predict the variation in richness at the scale of 110 km grid cell assemblages. We make these finer scale predictions first under a model of simple random sorting of species from those predicted for the bioregion and, second, under a model of sorting mediated by the relative productivity of a grid cell. The goals of this second step include (i) an evaluation of the ability of this hierarchical model to make strong fine-scale richness predictions (while including paleoclimate and avoiding regional-level conflation of sample size) and (ii) a demonstration of the separate roles energy availability has at different temporal and spatial scales.

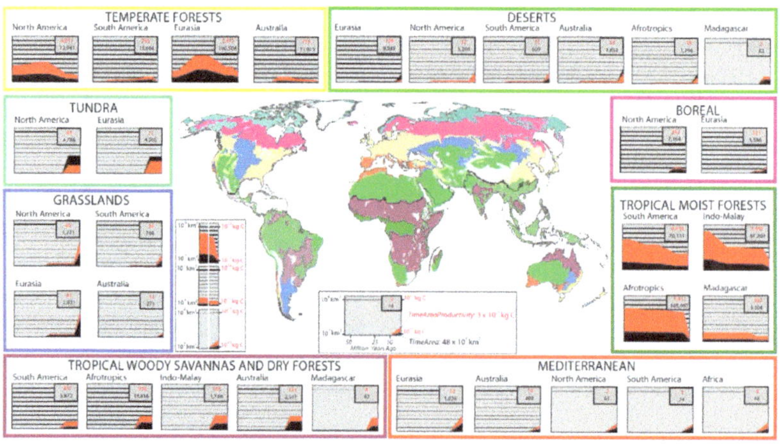

Figure 1. Map of study bioregions and their area and annual productivity dynamics. The variation in area (black) and annual productivity (red) over the last 55 million years forms the species richness predictors *TimeArea* (cumulative time-area, units $10^4 km^2 \times$million years) and *TimeArea-*

Productivity (cumulative total productivity, units 10^{17}kg Carbon), respectively (values in upper right box corner). Panel boxes have one of three different *y*-axis scales (note different line thicknesses and legend). For example, in tropical woody savannas and dry forests, the land area for the last few million years has been $\sim 1\times10^7$ km^2 in the Afrotropics, $\sim 2\times10^6$ km^2 in Australia, and $\sim 1\times10^5$ km^2 in Madagascar.

RESULTS AND DISCUSSION

Bioregion Historical Dynamics

Paleoclimatic data reveal dramatic variation in the age and spatial dynamics of different bioregions from the end of the Paleocene (55 MY bp) to the present day (Figure 1). For example, grasslands are not thought to have covered large areas on earth until 8 million years ago, resulting in a much smaller area over time than observed for, for example, temperate or tropical moist forests that have a longer history (Figure 1). Linking estimates of the extents of bioregions over time allows the calculation of "time-integrated area" (*TimeArea*)[42], a synthetic index of area available to the bioregion's biota over time, varying from just 48×10^4 km^2 integrated over 55 million years in the case of the Mediterranean bioregion at the southern tip of Africa to over $100{,}000\times10^4$ km^2 in Eurasian temperate and African moist tropical forests. Unlike bioregion extent and position, climatic conditions of bioregions are assumed to be relatively static over time [49], which allows the determination of average bioregion net primary productivity (*Productivity*) and *Temperature*. Summed *Productivity* over bioregion *Area* yields total bioregion productivity (*AreaProductivity*)—that is, total annual carbon flux measured in kg/year over a whole bioregion, a measure that exhibits joint dynamics with bioregion *Area*. But integrated over time in the form of *TimeAreaProductivity*, it exhibits very different geographic patterns than *TimeArea* (Figure 1) with, for example, African and IndoMalay tropical moist forests experiencing a flux of over $8{,}000\times10^{17}$ kg of carbon over the past 55 million years and the Mediterranean regions of the New World and Africa just under 3×10^{17} kg.

Bioregion Biotic Independence

We summarized terrestrial vertebrate richness per bioregion as *Total* (every species found in a bioregion), *Resident* (species for which a given bioregion contains the largest portion of the range), and *Endemic* (species that are restricted to a single bioregion). We find minimal overlap in *Total* species among bioregions (median Jaccard similarity among bioregions: 4% for birds, 0% for other taxa), which confirms their relative evolutionary isolation in addition to climatic and spatial independence and a consistently strong pattern of biome conservatism [38],[50],[51]. It also confirms that across all four vertebrate groups these selected bioregions represent useful spatial units that avoid the pseudo-replication of species: for *Resident* species richness every species enters a given analysis exactly once, and the number of distribution records is equal to the global richness of species (13,860 endothermic mammals and birds, 11,836 ectothermic amphibians and reptiles; montane endemics excluded). For *Endemic* species (total of 13,111 species and records) bioregions are even more likely to represent the true regions of origin compared to *Resident* species. We therefore expect a stronger correlation of area and productivity integrated over time (*TimeAreaProductivity*) with the diversity of *Endemic* species.

Bioregion Species Richness

All three predictions of our integrative model regarding the effect of time-area-productivity on richness are confirmed (Table 1). The models that account for time-integrated productivity and also include temperature as an additional predictor yield the strongest fits. For endotherms, the time-integrated measures of area outperform models that ignore time only for the *Endemic richness* dataset, which offered the more direct test of our hypotheses. Predictions of the two-predictor *TimeAreaProductivity+Temperature* model are consistently strong across all four vertebrate taxa, which represent independent replicates (Figure 2), explaining over 77% of the variation in richness (Figure 2, N=128. Models that fit*TimeArea* and *Productivity* as statistically separate terms do not on the whole yield stronger predictions. This lends support to Wright's [45] parallel findings for large islands, which represent similarly closed systems, and contrasts with previous results reported for 110×110 km grid cells [13]. The shape of the *Productivity*-richness relationship is linear (in Endotherm *Residents*) or positive accelerating in linear space (in Ectotherm *Residents* and both *Endemics* groups). In contrast, the slopes of the *AreaProductivity*- and*TimeAreaProductivity*-richness relationships, whether fitted with or without *Temperature*, are all positive saturating—that is, species richness tends to increase more steeply in the low than in the high productivity ranges (coefficients in ln-ln space vary between 0.4 and 1). We did not find evidence of a hump-shaped pattern for any measure of productivity and richness at the bioregional scale [52],[53].

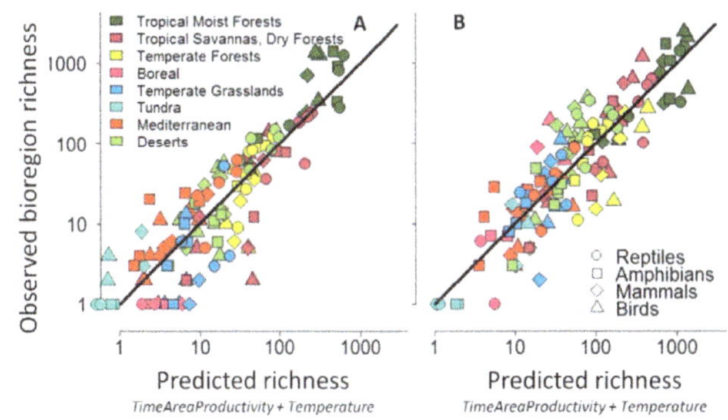

Figure 2. Observed versus predicted bioregion species richness of terrestrial vertebrates. Observed bioregion species richness (A, *Endemic* species, B, *Resident* species) is plotted against that predicted by the two-predictor *TimeAreaProductivity+Temperature*model fit separately for each of the four taxa (different symbols). Lines indicate least squares fit of regressions relating to observed predicted richness for each of the four taxa over the 32 bioregions (r^2 [*Endemic*]=0.78, r^2 [*Resident*]=0.78, N=128). For detailed results. Colors indicate biome membership (see the map in Figure 1 to match colors).

Table 1. Relative performance of integrated single- and two-predictor models of bioregion species richness.

Predictor Variables	*Endemic*				*Resident*			
	Endotherms		**Ectotherms**		**Endotherms**		**Ectotherms**	
	ΔAIC	r^2	*ΔAIC*	r^2	*ΔAIC*	r^2	*ΔAIC*	r^2
TimeAreaProductivity+Temperature	**0**	**0.67**	**0**	**0.82**	20	0.70	**1**	**0.86**
TimeArea+Temperature	**0**	**0.68**	**0**	**0.82**	14	0.75	**0**	**0.87**
AreaProductivity+Temperature	10	0.56	13	0.73	**0**	**0.84**	3	0.85
TimeAreaProductivity	20	0.36	46	0.22	31	0.56	55	0.20
TimeArea	23	0.29	49	0.14	32	0.54	58	0.13
AreaProductivity	26	0.22	50	0.10	22	0.66	57	0.16
Productivity	26	0.23	42	0.29	52	0.12	52	0.29
Area	31	0.09	53	0.02	35	0.49	61	0.04
Temperature	22	0.31	24	0.60	52	0.14	29	0.65

Endemic species are those restricted to a single bioregion and *Resident* counts species with the largest portion of their range in a given bioregion. Endotherms combine mammals and birds, ectotherms combine reptiles and amphibians. Best models (ΔAIC<2) are highlighted in bold, and r^2 refers to pseudo-r^2 values based on fitting model-predicted versus observed. Note that the results for both richness values are unaffected by the pseudo-replication that hampers the results of typical gridded analyses of species richness. Predictor variables: *Temperature*, average temperature of bioregion; *Area*, current-day extent of a bioregion; *Productivity*, average bioregion productivity; *AreaProductivity*, total bioregion productivity, that is, the product of *Productivity* and bioregion *Area*. *TimeArea*, time-integrated area, that is, the integrated areal extent of a bioregion over 55 million years; *TimeAreaProductivity*, time-integrated productivity, that is, the product of *Productivity* and *TimeArea*. For further details and results by taxon, see Methods and Tables S1, S2, S3, S4, S6, S7, S8, S9, S10.
doi:10.1371/journal.pbio.1001292.t001

As predicted, ectotherm richness increases much more steeply and strongly with temperature than endotherm richness, both when fitted singly and when controlled for *TimeAreaProductivity*. This supports at the global scale the significant and complementary effect temperature may contribute to levels of regional ectotherm diversity (see also [4],[14],[44]). For ectotherms, higher temperatures in tropical regions may be promoting higher rates of genetic incompatibilities among populations or faster rates of biotic interactions, further accelerating speciation rates [44],[54],[55]. Alternatively, the thermal dependence of activity represents a strong constraint on ectotherm distribution [56], likely imposing limits on clade origination and diversification in high-latitude regions. Third, in warm regions, ectotherms are released from physiological and behavioral adaptations to cold stress promoting a greater diversity of life histories and metabolic "niches" [57],[58]. These factors are not mutually exclusive, and more work is needed for understanding the potential role of temperature and thermal physiology in driving diversification. Preliminary results from phylogenetic analyses suggest increased diversification rates at lower latitudes in both amphibians [59] and mammals[60], but with a much weaker and more equivocal trend in the latter.

Overall, our bioregion results support the hypothesized interactions of environmental conditions and area over time in influencing the speciation and extinction and ultimately species richness of biota in bioregions. We suggest that the bioregional variation in time-integrated productivity successfully captures key factors affecting both cumulative population sizes over time as well as the different opportunities for reproductive isolation. Large, productive areas like the Neotropical moist/wet forest biome have been characterized by high productivity and a continuously large extent, and thus have supported large populations of each of the four vertebrate clades, since before the Eocene (700×10^{12} km^2 years and 663×10^{18} kg Carbon produced since 55 MY bp; Figure 1). Reproductive isolation has been facilitated by the large amount of time that vertebrate populations have had to encounter geographical barriers (such as rivers in non-volant mammals [61]) as well as heightened habitat heterogeneity related to the high productivity (i.e., multiple vertical forest strata) [12]. This contrasts with, for example, unproductive North American deserts, which have only come to cover a substantial area within the last few million years (12×10^{12} km years and 3×10^{18} kg Carbon; Figure 1) [62]. We suggest that the large *TimeAreaProductivity* seen in, for example, the Neotropical forest compared to the North American desert bioregion in Figure 1 reflects all factors affecting cumulative population sizes over time (which have affected both speciation and extinction probabilities) as well as opportunities for reproductive isolation. Together, these factors have led to the wide discrepancy in vertebrate diversity between these two bioregions.

Previous studies have employed phylogenies or sister-group comparisons to test whether the latitudinal diversity gradient derives from more evolutionary time [63], niche conservatism [38], or differences in speciation or extinction rates at different latitudes [22],[59],[60],[64]. Factors such as orbital forcing causing glaciation at high latitudes have been posited to elevate extinction rates and are expected to accentuate the observed disparities in species richness among bioregions, especially for endemics [35],[65]. The results reported here complement these studies and suggest that at the bioregion scale, and over an extremely large window of time (55 MY), diversification rates consistently vary with respect to the area, age, and productivity of a given bioregion (Figure 2). We thus view the time-integrated productivity of bioregions to be a general explanation for why so many clades originate at lower latitudes and correspondingly fewer have diversified into bioregions at higher latitudes. It is important to note that time alone is not sufficient to explain these patterns: temperate bioregions are just as ancient as tropical bioregions but strongly differ in their cumulative time-integrated area and productivity. In sum, the strong associations we find indicate a pathway toward first-order approximations of rates of net species production per bioregion, based on variation in area over time, productivity, and temperature. Future studies could integrate our approach with more detailed comparisons of clade-level diversification rates among bioregions or combine it with existing phylogenetic methods for quantifying correlates of diversification.

FINER SCALE SPECIES RICHNESS

Having addressed key evolutionary drivers affecting the broad-scale variation in vertebrate diversity, we next assess how each bioregion's species sort into grid cell assemblages and how both processes combine to explain the finer scale geographic variation in richness (Figure 3A). We perform this assessment for the 18,467 bird, mammal, and amphibian species in the bioregion analysis and their 2,966,137 occurrences across the 9,253 110×110 km terrestrial grid cells encompassed by the bioregions (Figure 3B). Strong effects of regional- on fine-scale richness have previously been demonstrated [1],[2], and here we provide a first test of their pervasiveness at a global scale by evaluating the performance of bioregion models for explaining grid cell richness. We find that the two-predictor *TimeArea Productivity+Temperature* model developed above (Table 1) alone explains 46%–60% and 32%–50% of the variation in *Resident* richness and *Total* richness, respectively (Figure 3B left column). This highlights how regional effects together with even simple null models of proportional sorting are able to explain much of the finer scale richness patterns. Fine-scale–regional richness relationships are known to be affected by spatial scale as well as by species' dispersal abilities [66]. In larger regions a grid cell of the same size represents a smaller portion of the regional area and, assuming similar levels of grid cell immigration/extinction, grid cell richness is expected to be smaller. This should apply whenever average species range sizes increase less than proportionally with bioregion size and should be particularly noticeable for taxa with relatively low dispersal rates or small within-bioregion range sizes (such as amphibians compared to birds or mammals), because with increasing bioregion size species will be progressively less likely to occupy a given grid cell. We find these expectations confirmed. Bioregion *Area* exhibits an additional negative effect and improves fine-scale predictions, especially for *Total* richness. It does so most strongly in Amphibians, whose greater dispersal limitation (and on average by a factor of four smaller geographic ranges) compared to mammals or birds has been previously suggested as contributing to their strong patterns of species turnover [67].

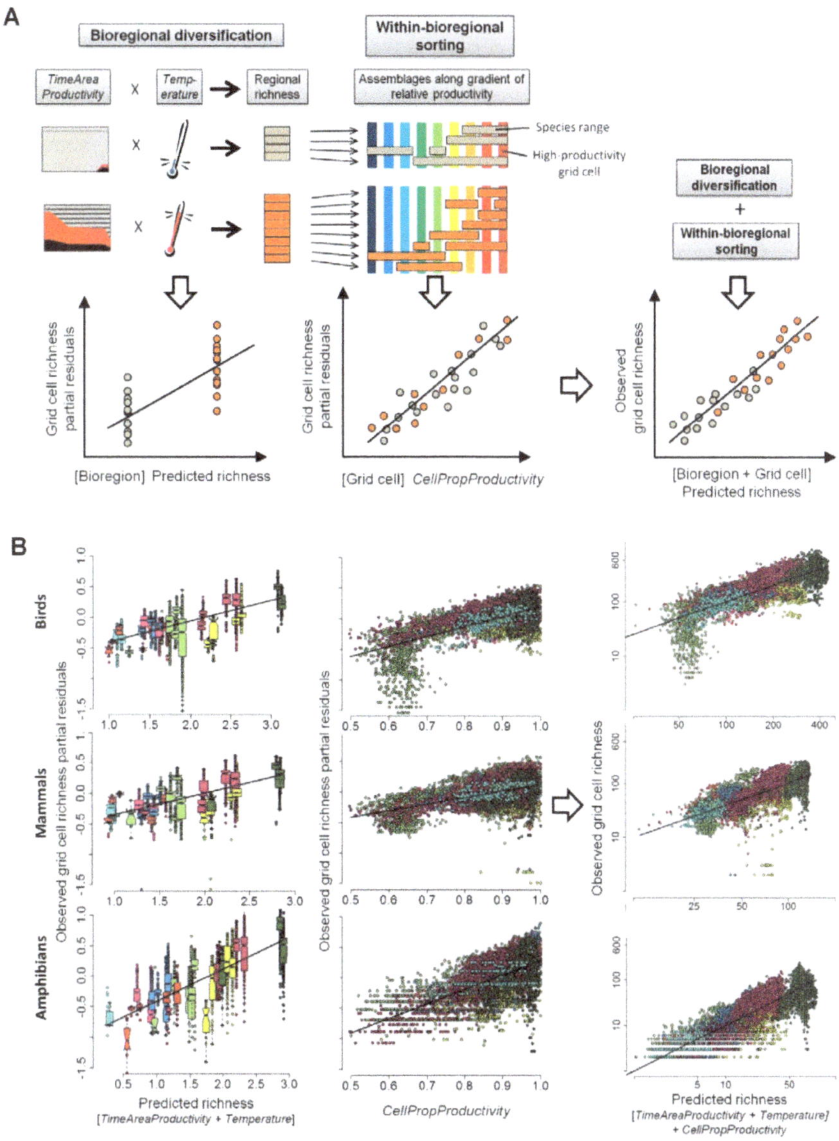

Figure 3. Hierarchical prediction of species richness at the scale of 110 km grid cells (N=9,253). (A) Conceptual outline of the model and (B) empirical evaluation for the 110 km grid cell *Total Richness* of Mammals, Birds, and Amphibians. The model first fits differences in grid cell

richness among bioregions based on the *Resident* richness model of bioregion-level diversification (*TimeAreaProductivity*, *Temperature*, see Table 1, Figure 2; additional effect of *Area* was also fitted and significant for Amphibians). Second, the effect of within-bioregion gradients in productivity (*CellPropProductivity*, i.e., proportion of bioregion grid cell maximum, a measure that standardizes productivity across bioregions) is fitted to predict subsequent sorting of each bioregions' species into grid cell assemblages. The resulting hierarchical prediction of grid cell richness accounts for the scale dependence of different effects and in the case of productivity addresses the different mechanisms of the same variable at different scales. In (B), lines indicate least squares model fits (r^2 values for observed–predicted; bioregion level, grid cell level, respectively: r^2 [Birds]=0.40, 0.61; r^2[Mammals]=0.45, 0.58; r^2 [Amphibians]=0.59, 0.77). Boxplots (left panels) summarize points for each of the 32 bioregions. Colors indicate biome membership (see Figure 2for legend). Partial residuals illustrate the relationship between a predictor and the response given other predictors in the model.

Species vary strongly in the number of assemblages they occupy and the species richness of grid cell assemblages is a function of the drivers that affect species' sorting and resulting overlap in geographic ranges. One variable strongly associated with the sorting into assemblages, particularly by wide-ranging species, is local energy availability [8],[25]. We find that relative productivity in a grid cell (*CellPropProductivity*, i.e., the proportion of the maximum grid cell productivity observed in a bioregion) predicts a substantial additional amount of observed variation in grid cell richness (Figure 3B middle column) and confirms the expected greater tendency of species within a bioregion to occupy high-productivity grid cells. Allowing the shape of the richness–productivity relationship to vary among regions improves predictions, but only slightly so, suggesting a within-regional role of productivity that is globally fairly consistent. Nevertheless, the total amount of variation explained by the *TimeAreaProductivity+Temperatur e* model (58%–77%) is remarkable and similar to that found in previously published broad-scale gridded richness regression analyses[8],[11]. Notably, however, the hierarchical approach avoids the dual problems of species pseudo-replication and conflation of among- and within-regional processes—issues that have seriously impeded interpretations of all previous gridded biogeographic or macroecological analyses at broad scales.

Our results largely corroborate past studies that have hypothesized that net primary productivity should be a dominant predictor of fine-grain assemblage richness [8],[11],[16]. However, our hierarchical model is able to separate how productivity influences species richness at different temporal and spatial scales. At the bioregional scale, productivity should increase the cumulative population size and opportunities for reproductive isolation over time, promoting higher species richness in high-productivity bioregions [12]. At the fine scale productivity affects the occupancy of assemblages in relation to the regional pool [27],[68]. In addition to the sampling effects inherent with larger assemblage-level population sizes, increased productivity may promote greater richness due to an increased number of niches facilitating species coexistence [12],[25].

Conclusions

We consider the contributions of this study to be conceptual in addition to empirical and hope that its framework will inspire further consideration of diversity gradients that aims to integrate ecological and evolutionary mechanisms across scales. Our global hierarchical approach represents an analytical paradigm shift away from the traditional analysis of fine-scale assemblages as independent spatial units. But there are obvious limits to our analysis. While the strong association of vertebrates with dominant vegetation types and the observed biotic independence of bioregions support their delineation as major evolutionary arenas, challenges remain surrounding the demarcation of the exact boundaries of such regions, the accuracy of past climate reconstructions, and their comparability across clades. Future availability of higher resolution phylogenies of the four vertebrate clades will allow more rigorous comparative approaches within and across lineages, but even comprehensive, strongly supported phylogenetic reconstructions are unlikely to provide vital information regarding the estimation of ancestral distributions (or ranges) and extinction rates [69]. Thus, our model can be viewed as a template on top of which other processes surely influence the origin and maintenance of diversity. For example, glaciation cycles influence speciation and extinction rates [36] and play an important role in driving recent speciation over broad scales [70]. Historical climate dynamics along elevation gradients in particular are known to create opportunities for rapid climate-associated parapatric or allopatric speciation and contribute strongly to the high richness of many tropical mountain areas [71]–[73]. Furthermore, a multitude of trophic interactions are likely to interact with these large-scale processes to cause positive, coevolutionary feedback loops, thus further increasing fine-scale and regional diversity [15].

Our findings show that energy availability has a large effect on both the regional pool and local sorting of richness. This highlights its importance for both evolutionary and ecological processes and the critical need to integrate these effects. This is especially crucial today, given the attention paid to recent models predicting the effects of climate change on the richness of whole gridded assemblages. The redundancy of information and conflation of ecological and evolutionary processes in smaller scale models impede interpretation in a way that is overcome in our analysis. Here we have shown how history can be integrated into a model predicting diversity with area, productivity, and temperature at the global scale. The separate consideration of drivers of diversification and finer scale occupancy and their joint effects on observed gradients of species richness should help pave the way for a more integrated macro-evolutionary and -ecological understanding of the origin and maintenance of global richness gradients.

MATERIALS AND METHODS

Bioregion Selection, Time-Integrated Area Calculations

We selected 32 well-established, geographically and climatically distinct bioregions (Figure 1). These bioregions correspond to the biomes (tundra, desert, grassland, boreal forest, temperate forest, tropical moist/wet forest, tropical dry forest/savanna, and Mediterranean forest/shrublands) within the world's main biogeographic realms (Neartic, Paleartic, Neotropical, Australian, IndoMalayan, and Afrotropics) as described by Olson et al. [46] and also used in the Wildfinder vertebrate distribution database (see below) [74]. Although we do not have detailed, fine-scale records throughout every interval of time for the past 55 million years, enough information exists regarding the age of all biomes and directionality of their expansion and contraction to make reasonable estimates of the measures of their area integrated over time. We excluded the "Mangroves" biome (Biome ID 14 in [46],[74]) and also the "Montane Grasslands & Shrublands" Biome (Biome ID 10 in [46],[74]). The latter was not included due to the difficulty in estimating areal and climate changes over their steep gradients over such a long time period. For example, in the Andes, different biomes occur at different elevations on the western and eastern slopes at different latitudes, and the available data are not sufficient to accurately estimate the elevations of the southern, central, and northern Andes at various time intervals since the Miocene, as each chain has uplifted at different rates and at different times [75]. This is critical information to be able to reconstruct the areal extent over time of each bioregion in the Andes and a general problem common to all of the world's mountain ranges, which is why they were excluded from our analysis.

The last 55 million years is an appropriate interval of time to measure the time-integrated area of the world's biomes within realms for two reasons. First, the beginning of the window of time is 10 million years after the massive extinction, which occurred 65 million years ago, causing major upheaval in the vertebrates. By 55 million years ago, the biosphere had recovered but its biota was very different from the plants and animals that had dominated the Cretaceous. Second, most of the "higher taxa"—that is, ancestors of modern lineages of vertebrates that now dominate the extant diversity of mammals, birds, amphibians, and reptiles (for example, fossils recognizable as extant genera)—are already represented in the fossil record by 55 million years ago [76]. Plant communities by the Eocene are, for the first time, composed of Angiosperms and Gymnosperms that are recognizable as the "genera" and "families" that are dominant in today's biomes [62],[77]. Thus, the biota in the Eocene has a "modern aspect"[76],[77].

The Earth's biomes have experienced large changes over the last 55 million years due to the consistent pattern of cooling and drying that has steadily taken place over this period of time[62],[78],[79]. Average global temperatures have plummeted from 27°C 55 million years ago to today's average of 15°C and precipitation has similarly dropped [62],[80]. For the moist/wet forest biomes (boreal forest, temperate forest, and tropical moist/wet forest) we used maps generated by Fine and Ree (2006) that were based on five sources: [49],[62],[81]–[83]. For the other biomes, our approach to estimate the time-integrated area of each biome was first to try to determine the paleobotanical consensus opinion for the age of each biome. Then, we took the extant area of that biome and backcasted in time over the years that it has been present, reasoning that as tropical forests have receded during the past 55 MY years, dry and cold biomes such as tundra, desert, Mediterranean, grassland, and dry forest/savanna must have increased in size from the date of their origin to today's area.

We made two interpretations—a "wet" and a "dry" interpretation. These two interpretations span the diverse opinions regarding the extent and age of the world's biomes over the last 55 million years and thus gauge the robustness of our results according to a range of expert opinions. For example, desert plants are absent in fossil records until about 2 Ma [77], even though it is hypothesized with molecular dating that plant lineages today found only in desert floras are at least 50 Ma old [62]. Thus, the consensus opinion is that deserts were probably present in the Eocene, but much restricted in size compared to today. For example, evaporite sediments point to extreme aridity in western Africa, Arabia, and central Asia in the late Miocene [82]. We thus made two estimates for the time-integrated area of desert biomes. The "wet" interpretation gives deserts an origin of 34 MYA but covering 10% of their current area from 34 MYA until 2 MYA, which is consistent with the lack of fossil evidence for any desert plant communities. The "dry" interpretation also gives the origin of deserts 34 MYA but has deserts covering the same areal extent as today since their origin, which is almost certainly an overestimate but is possible given the ancient age of some desert plant lineages and the difficulty of fossilization of desert environments.

Our wet and dry interpretations both yield qualitatively similar results, and for simplicity, we focus on the "wet" interpretation throughout the article. The current-day extent of a bioregion as given in [46] yielded our predictor variable *Area* (units km^2). Time-integrated area (*TimeArea*, in units year km^2) was given as the integrated areal extent of a bioregion over 55 million years, or simply the sum of the area estimated for each of the 55 one-million-year periods. We acknowledge that this offers only a first order approximation. While exact values will be subject to change as paleoecological knowledge advances, we expect these changes to refine the details rather than radically alter overall patterns, which would have relatively little effect on our analyses, and thus we do not expect systematic biases in our results.

While topographic heterogeneity is expected to also influence the potential for reproductive isolation [32], in this dataset (which excludes montane regions) it is largely captured by bioregion *Area* and does not yield improved predictions.

Bioregion Species Data

We aggregated existing eco-regional terrestrial vertebrate species lists for the selected 32 bioregions from the Wildfinder distribution database [74]. We excluded all eco-regions in biomes not selected for analysis (see above), including all montane eco-regions (which have a total of 1,015 terrestrial vertebrate species restricted to them). This resulted in 54,122 bioregion occurrence records for 25,696 species (9,229 birds, 4,607 amphibians, 4,631 mammals, and 7,229 reptiles). We calculated terrestrial vertebrate richness ("vertebrate richness") per bioregion in three different ways: *Total*, which includes every vertebrate species found within each bioregion; *Resident*, which only counts species in the bioregion with the largest proportion of its geographic range; and *Endemic*, which counts only species that are restricted to a single bioregion. Assigning each species only to its dominant bioregion to eliminate pseudo-replication yields a *Resident* richness pattern very similar to that of *Total* richness (r_S=0.85). For the analyses, vertebrates were divided into ectotherms (amphibians and reptiles) and endotherms (birds and mammals) and further separated into birds, mammals, reptiles, and amphibians. All richness values were natural log-transformed.

Finer Scale Species Data

Species occurrence data across grid cells were compiled from global expert opinion range maps extracted across a 110×110 km equal area grid in a Behrman projection. For mammals [84], and amphibians, sources were the IUCN assessment (http://www.iucnredlist.org). For birds, breeding distributions were compiled from the best available sources for a given broad geographical region or taxonomic group [85]. For reptiles, global-scale expert range maps have not yet been compiled, and they were therefore not included in the grid cell assemblage analyses. We excluded all cells that were not >50% inside the selected bioregion boundaries as described above (and shown in Figure 1). Only cells with >50% dry land and with at least one species from each of the three vertebrate groups were included in the analysis, resulting in 9,253 cells. For each grid cell we summarized richness of *Resident* species (i.e., species were counted if they occurred in several grid cells only within the same bioregion) and of *Total*species (i.e., species were counted whether they occurred in multiple grid cells within the same or in a different bioregion). Values were log10-transformed before analysis. For *Total*species, the full database consisted of a total of 2,966,137 grid cell records (birds 2,010,091; mammals 695,133; and amphibians 260,913).

Bioregion and Finer Scale Environmental Data

Bioregion-typical temperature estimates (*Temperature*) were based on average annual temperatures calculated from the University of East Anglia's Climatic Research Unit gridded climatology 1961–1990 dataset at native 10-min resolution [86]. For estimates of bioregion-typical annual net primary productivity, we used an average from 17 global models at a spatial resolution of 0.5 degrees latitude-longitude [87]. Average bioregion productivity (*Productivity*, units grams Carbon m^{-2} $year^{-1}$) was calculated from all 0.5×0.5 degree grid cells that predominantly fall inside a bioregion, and summed productivity (*AreaProductivity*, units grams Carbon $year^{-1}$) was then given by the product of this value and bioregion *Area*. With bioregions defined by their typical environmental conditions, we assumed average productivity characteristic of a bioregion to have been constant through time [49],[62]. Time-integrated productivity (*TimeAreaProductivity*, unit grams Carbon) was thus given as the product of *Productivity* and *TimeArea*. All response and predictor variables were natural log-transformed for analysis, except for temperature, which was 1/kT transformed (where k is the Boltzmann constant, see [44]). We used the same global net primary productivity dataset [87] to estimate productivity at the level of 110×110 km grid cells. First, we calculated average grid cell productivity (*NPP*) across all encompassing 0.5×0.5 degree grid cells. Second, we normalized each grid cell by dividing by the maximum productivity grid cell value observed in a bioregion, resulting in a measure of proportional productivity (*PropNPP*) varying from 0 to 1.

Bioregion Analyses

We performed a total of nine GLM models on the bioregion data and used the Akaike criterion to identify those offering the best fit [88]. Six models were given in the form of single predictors (*Temperature*, *Area*, *Productivity*, *AreaProductivity*, *TimeArea*, and*TimeAreaProductivity*). An additional three models were formed by the combination of the latter three variables with *Temperature*. We performed a separate set of analyses to assess the potential additional effect of elevation range within a bioregion, but adding this variable to any of the three two-predictor models did not improve model fit, and thus we excluded the variable from further consideration. Because of the strong independence of sampling units both in terms of response (no overlap in species) and predictor variables (by definition each bioregion is environmentally highly distinct from neighboring bioregions), the usual concerns about spatial autocorrelation affecting model results [89],[90] do not apply to this analysis, and additional spatial regression analysis was not performed.

Finer Scale Analyses

Having established models of bioregion richness, we assessed the success of predictions of resident bioregional richness to explain the species richness (*Total* and *Resident*, see above) of all 110×110 km grid cells within bioregions (for a conceptual overview of the analytical steps, see Figure 3). Note that unlike the bioregional tests described above, analyses at this scale do double-count species. In our study we make the simplifying assumption that diversification processes are sufficiently accounted for at the bioregional scale. The models at the within-bioregion scale then address the sorting of these species each into multiple grid cells, with multiple occurrences an integral part of the signal. We acknowledge that, depending on taxon and region, diversification processes may still exert influence on the within-bioregion patterns of distribution and richness, and we hope that our work will spur further research into additional approaches that can be integrated across all scales.

We first evaluated bioregion predicted resident richness alone (in essence testing for a random sorting of bioregion species into finer scale assemblages), then included bioregion *Area* as an additional predictor, and finally we added estimates of grid cell NPP as a finer scale predictor. We first performed simple GLM models with all 9,253 grid cells as sampling units, together with bioregion *Resident* richness as predicted by the *TimeAreaProductivity+Temperature* and*AreaProductivity+Temperature* models as a first predictor (*BioregPred*) and bioregion *Area* as a second predictor (Figure 3). In the same GLM we then added grid cell proportional net primary productivity (*CellPropNPP*, i.e., relative productivity within a bioregion, see above) as an additional predictor. In preliminary post hoc analyses with a number of environmental variables *CellPropNPP* remained by far the strongest, in line with recent work on within-regional richness filters that also find productivity-related variables to be dominant [26],[27]. Given the nested nature of these analyses we focus on pseudo-r^2 values (fit of observed versus predicted) and visual examination of results in the form of partial residual plots (Figure 3). For this first demonstration, focused on a single variable, we did not include further analyses additionally fitting the signal of spatial autocorrelation.

We performed a second set of analyses in an explicit mixed effects model setting, with bioregion as a random effect (R library lme4, Version 0.999375-32, function lmer). As in the GLM model, grid cell richness is first fitted by the predictions for regional resident species richness (*BioregPred*, see Table 1), and then by area of the region (*Area*), and grid-cell-level NPP (*NPP*). Region was fitted as a random effect, and the slope and strength of *BioregPred*and *BioregPred+Area* as fixed effects were assessed (model formula in R: lmer (y~*BioregPred+Area*+(1|*Bioregion*)). The additional effect of grid cell *NPP* was then evaluated by fitting it as an additional fixed effect with a globally constant slope (NPP_{const}) and by allowing the NPP–richness relationship to vary within regions as random slope (NPP_{var}) (model formula in R: lmer (y~*BioregPred+Area*+(1|*Bioregion*)+(NPP|*Bioregion*)).

ACKNOWLEDGMENTS

We would like to thank J. Belmaker, R. Bowie, G. Crutsinger, N. Cooper, P. Hull, F. La Sorte, T. Lee, D. Rabosky, R. Ricklefs, and D. Rosauer for discussions.

AUTHOR CONTRIBUTIONS

The author(s) have made the following declarations about their contributions: Conceived and designed the experiments: WJ PF. Performed the experiments: WJ PF. Analyzed the data: WJ PF. Contributed reagents/materials/analysis tools: WJ PF. Wrote the paper: WJ PF.

REFERENCES

1. Ricklefs R. E (1987) Community diversity: relative roles of local and regional processes. Science 235: 167–171.
2. Caley M. J, Schluter D (1997) The relationship between local and regional diversity. Ecology 78: 70–80.
3. McGill B. J (2010) Matters of scale. Science 328: 575–576.
4. Belmaker J, Jetz W (2011) Cross-scale variation in species richness–environment associations. Global Ecology and Biogeography 20: 464–474.
5. Gotelli N. J, Anderson M. J, Arita H. T, Chao A, Colwell R. K, et al. (2009) Patterns and causes of species richness: a general simulation model for macroecology. Ecol Lett 12: 873–886.
6. Willis K. J, Whittaker R. J (2002) Species diversity–scale matters. Science 295: 1245–1248.
7. Schall J. J, Pianka E. R (1978) Geographical trends in number of species. Science 201: 679–686.
8. Jetz W, Rahbek C (2002) Geographic range size and determinants of avian species richness. Science 297: 1548–1551.
9. Currie D. J (1991) Energy and large-scale patterns of animal species and plant species richness. Am Nat 137: 27–49.
10. Buckley L, Jetz W (2007) Environmental and historical constraints on global patterns of amphibian richness. Proc R Soc Lond B Biol Sci 274: 1167–1173.
11. Hawkins B. A, Porter E. E, Diniz J. A. F (2003) Productivity and history as predictors of the latitudinal diversity gradient of terrestrial birds. Ecology 84: 1608–1623.
12. Evans K. L, Warren P. H, Gaston K. J (2005) Species–energy relationships at the macroecological scale: a review of the mechanisms. Biol Rev Camb Philos Soc 80: 1–25.
13. Hurlbert A. H, Jetz W (2010) More than "more individuals": the non-equivalence of area and energy in the scaling of species richness. Am Nat 176: E50–E65.
14. Qian H (2010) Environment–richness relationships for mammals, birds, reptiles, and amphibians at global and regional scales. Ecological Research 25: 629–637.
15. Mittelbach G. G, Schemske D. W, Cornell H. V, Allen A. P, Brown J. M, et al. (2007) Evolution and the latitudinal diversity gradient: speciation, extinction and biogeography. Ecol Lett 10: 315–331.
16. Kaspari M, Ward P. S, Yuan M (2004) Energy gradients and the geographic distribution of local ant diversity. Oecologia 140: 407–413.
17. Wright S, Keeling J, Gillman L (2006) The road from Santa Rosalia: a faster tempo of evolution in tropical climates. Proc Natl Acad Sci U S A 103: 7718–7722.
18. Wright D. H, Currie D. J, Maurer B. A (1993) Energy supply and patterns of species richness on local and regional scales. In: Ricklefs R, Schluter D, editors. Species diversity in ecological communities. Chicago: University of Chicago Press. pp. 66–74.
19. Dobzhansky T (1950) Evolution in the tropics. American Scientist 38: 209–221.
20. Ricklefs R. E, Schluter D (1993) Species diversity: regional and historical influences. In: Ricklefs R. E, Schluter D, editors. Species diversity in ecological communities. Chicago: Chicago University Press.
21. Cardillo M, Orme C. D. L, Owens I. P. F (2005) Testing for latitudinal bias in diversification rates: an example using New World birds. Ecology 86: 2278–2287.
22. Jansson R, Davies T. J (2008) Global variation in diversification rates of flowering plants: energy vs. climate change. Ecol Lett 11: 173–183.
23. Buckley L. B, Davies T. J, Ackerly D. D, Kraft N. J. B, Harrison S. P, et al. (2010) Phylogeny, niche conservatism and the latitudinal diversity gradient in mammals. Proc R Soc Lond B Biol Sci 277: 2131–2138.
24. Rosenzweig M. L, Abramsky Z (1993) How are diversity and productivity related. In: Ricklefs R. E, Schluter D, editors. Species diversity in ecological communities. London: University of Chicago Press.
25. Storch D, Sizling A. L, Gaston A. J (2007) Scaling species richness and distribution: uniting the species–area and species–energy relationships. In: Storch D, Marquet P. A, Brown J. H, editors. Scaling biodiversity. Cambridge: Cambridge University Press.
26. Bonn A, Storch D, Gaston K. J (2004) Structure of the species-energy relationship. Proc R Soc Lond B Biol Sci 271: 1685–1691.
27. White E. P, Hurlbert A. H (2010) The combined influence of the local environment and regional enrichment on bird species richness. Am Nat 175: E35–E43.
28. Lennon J. J, Koleff P, Greenwood J. J. D, Gaston K. J (2004) Contribution of rarity and commonness to patterns of species richness. Ecol Lett 7: 81–87.
29. Ricklefs R. E (2007) History and diversity: explorations

at the intersection of ecology and evolution. Am Nat 170: S56–S70.

30. Hurlbert A. H, Jetz W (2007) Species richness, hotspots, and the scale dependence of range maps in ecology and conservation. Proc Natl Acad Sci U S A 104: 13384–13389.
31. Fritz S. A, Purvis A (2010) Phylogenetic diversity does not capture body size variation at risk in the world's mammals. Proc R Soc Lond B Biol Sci 277: 2435–2441.
32. Rosenzweig M. L (1995) Species diversity in space and time. Cambridge: Cambridge University Press.
33. Wallace A. R (1878) Tropical nature and other essays. London: Macmillan.
34. Willis J. C (1922) Age and area: a study in geographical distribution and origin of species. Cambridge: Cambridge University Press.
35. Araujo M. B, Nogues-Bravo D, Diniz-Filho J. A. F, Haywood A. M, Valdes P. J, et al. (2008) Quaternary climate changes explain diversity among reptiles and amphibians. Ecography 31: 8–15.
36. Jansson R, Dynesius M (2002) The fate of clades in a world of recurrent climatic change: Milankovitch oscillations and evolution. Annu Rev Ecol Syst 33: 741–777.
37. Hawkins B. A (2004) Summer vegetation, deglaciation and the anomalous bird diversity gradient in eastern North America. Global Ecol Biogeography 13: 321–325.
38. Wiens J. J, Donoghue M. J (2004) Historical biogeography, ecology and species richness. Trends Ecol Evol 19: 639–644.
39. Wiens J. J, Ackerly D. D, Allen A. P, Anacker B. L, Buckley L. B, et al. (2010) Niche conservatism as an emerging principle in ecology and conservation biology. Ecol Lett 13(10): 1310–1324.
40. Machac A, Zrzavý J, Storch D (2011) Range size heritability in carnivora is driven by geographic constraints. Am Nat 177: 767–779.
41. Stephens P. R, Wiens J. J (2003) Explaining species richness from continents to communities: the time for speciation effect in emydid turtles. Am Nat 161: 112–128.
42. Fine P. V. A, Ree R. H (2006) Evidence for a time-integrated species-area effect on the latitudinal gradient in tree diversity. Am Nat 168: 796–804.
43. Kisel Y, McInnes L, Toomey N. H, Orme C. D. L (2011) How diversification rates and diversity limits combine to create large-scale species–area relationships. Philos Trans R Soc Lond B Biol Sci 366: 2514–2525.
44. Allen A. P, Gillooly J, Brown J. H (2007) Recasting the species-energy hypothesis: the different roles of kinetic and potential energy in regulating biodiversity. In: Storch D, Marquet P. A, Brown J. H, editors. Scaling biodiversity. Cambridge: Cambridge University Press.
45. Wright D. H (1983) Species energy theory—an extension of species-area theory. Oikos 41: 496–596.
46. Olson D. M, Dinerstein E, Wikramanayake E. D, Burgess N. D, Powell G. V. N, et al. (2001) Terrestrial ecoregions of the worlds: a new map of life on Earth. Bioscience 51: 933–938.
47. Woodward F (1987) Climate and plant distribution: Cambridge Univ Pr.
48. Kisel Y, Barraclough T. G (2010) Speciation has a spatial scale that depends on levels of gene flow. Am Nat 175: 316–334.
49. Beerling D, Woodward F (2001) Vegetation and the terrestrial carbon cycle: modelling the first 400 million years: Cambridge Univ Pr.
50. Crisp M, Arroyo M, Cook L, Gandolfo M, Jordan G, et al. (2009) Phylogenetic biome conservatism on a global scale. Nature 458: 754–756.
51. Harrison S, Grace J. B (2007) Biogeographic affinity helps explain productivity-richness relationships at regional and local scales. Am Nat 170: S5–S15.
52. Mittelbach G. G, Steiner C. F, Scheiner S. M, Gross K. L, Reynolds H. L, et al. (2001) What is the observed relationship between species richness and productivity? Ecology 82: 2381–2396.
53. Waide R. B, Willig M. R, Steiner C. F, Mittelbach G, Gough L, et al. (1999) The relationship between productivity and species richness. Annu Rev Ecol Syst 30: 257–300.
54. Gillman L, Keeling D, Ross H, Wright S (2009) Latitude, elevation and the tempo of molecular evolution in mammals. Proc R Soc Lond B Biol Sci 276: 3353.
55. Rohde K (1992) Latitudinal gradients in species-diversity—the search for the primary cause. Oikos 65: 514–527.
56. Kearney M, Porter W (2009) Mechanistic niche modelling: combining physiological and spatial data to predict species' ranges. Ecol Lett 12: 334–350.
57. Clarke A, Gaston K. J (2006) Climate, energy and diversity. Proc R Soc Lond B Biol Sci 273: 2257–2266.
58. Anderson K. J, Jetz W (2005) The broad-scale ecology of energy expenditure of endotherms. Ecol Lett 8: 310–318.
59. Wiens J. J (2007) Global patterns of diversification and species richness in amphibians. Am Nat 170: S86–S106.
60. Davies T. J, Buckley L. B (2011) Phylogenetic diversity as a window into the evolutionary and biogeographic histories of present-day richness gradients for mammals. Philos Trans R Soc Lond B Biol Sci 366: 2414–2425.
61. Patton J. L, da Silva M. N. F, Lara M. C, Mustrangi M. A (1997) Diversity, differentiation, and the historical biogeography of non-volant small mammals of the neotropical forests. In: Lawrence W. F, Bieregaard J. T, editors. Tropical forest fragments: ecology, management, and conservation of fragmented communities. University of Chicago Press.
62. Willis K, McElwain J (2002) The evolution of plants. Oxford: Oxford University Press. 378 p.
63. Stephens P, Wiens J (2003) Ecological diversification and phylogeny of emydid turtles. Biological Journal of the Linnean Society 79: 577–610.
64. Weir J. T, Schluter D (2007) The latitudinal gradient

in recent speciation and extinction rates of birds and mammals. Science 315: 1574–1576.

65. Jansson R (2003) Global patterns in endemism explained by past climatic change. Proc R Soc Lond B Biol Sci 270: 583–590.
66. He F, Gaston K. J, Connor E. F, Srivastava D. S (2005) The local-regional relationship: immigration, extinction, and scale. Ecology 86: 360–365.
67. Buckley L. B, Jetz W (2008) Linking global turnover of species and environments. Proc Natl Acad Sci U S A 105: 17836–17841.
68. Huston M. A (1999) Local processes and regional patterns: appropriate scales for understanding variation in the diversity of plants and animals. Oikos 86: 393–401.
69. Rabosky D. L (2010) Extinction rates should not be estimated from molecular phylogenies. Evolution 64: 1816–1824.
70. Weir J. T, Schluter D (2004) Ice sheets promote speciation in boreal birds. Proc R Soc Lond B Biol Sci 271: 1881–1887.
71. Hughes C, Eastwood R (2006) Island radiation on a continental scale: exceptional rates of plant diversification after uplift of the Andes. Proc Natl Acad Sci U S A 103: 10334–10339.
72. Jetz W, Rahbek C, Colwell R. K (2004) The coincidence of rarity and richness and the potential signature of history in centres of endemism. Ecol Lett 7: 1180–1191.
73. Cadena C. D, Kozak K. H, Gómez J. P, Parra J. L, McCain C. M, et al. (2011) Latitude, elevational climatic zonation and speciation in New World vertebrates. Proc R Soc Lond B Biol Sci 279: 194–201.
74. WWF (2007) WildFinder: online database of species distributions, ver. July-07.
75. Mora A, Baby P, Roddaz M, Parra M, Brusset S, et al. (2010) Tectonic history of the Andes and sub-Andean zones: implications for the development of the Amazon drainage basin. In: Hoorn C, Wesselingh F. P, editors. Amazonia-landscape and species evolution: a look into the past. Chichester, UK: Wiley-Blackwell.
76. Carroll R (1988) Vertebrate paleontology and evolution. New York: W.H. Freeman and Company.
77. Graham A (2010) Late Cretaceous and Cenozoic history of Latin American vegetation and terrestrial environments. St Louis: Missouri Botanical Garden Press.
78. Miller K, Fairbanks R, Mountain G (1987) Tertiary oxygen isotope synthesis, sea level history, and continental margin erosion. Paleoceanography 2: 1–19.
79. Zachos J, Pagani M, Sloan L, Thomas E, Billups K (2001) Trends, rhythms, and aberrations in global climate 65 ma to present. Science 292: 686–693.
80. Frakes L, Francis J, Syktus J (1992) Climate modes of the Phanerozoic: the history of the Earth's climate over the past 600 million years. Cambridge; New York, N.Y., USA: Cambridge University Press.
81. Scotese C. R (2003) PALEOMAP project
82. Ziegler A, Eshel G, Rees P, Rothfus T, Rowley D, et al. (2003) Tracing the tropics across land and sea: Permian to present. Lethaia 36: 227–254.
83. Morley R (2000) Origin and evolution of tropical rain forests. Chichester, UK: Wiley.
84. Schipper J, Chanson J. S, Chiozza F, Cox N. A, Hoffmann M, et al. (2008) The status of the world's land and marine mammals: diversity, threat, and knowledge. Science 322: 225–230.
85. Jetz W, Wilcove D. S, Dobson A. P (2007) Projected impacts of climate and land-use change on the global diversity of birds. Plos Biol 5: 1211–1219. doi:10.1371/journal.pbio.0050157.
86. New M, Lister D, Hulme M, Makin I (2002) A high-resolution data set of surface climate over global land areas. Climate Research 21: 1–25.
87. Cramer W, Kicklighter D. W, Bondeau A, Moore B, Churkina C, et al. (1999) Comparing global models of terrestrial net primary productivity (NPP): overview and key results. Global Change Biology 5: 1–15.
88. Burnham K. P, Anderson D. R (2002) Model selection and multimodel inference. New York: Springer.
89. Lennon J. J (2000) Red-shifts and red herrings in geographical ecology. Ecography 23: 101–113.
90. Hawkins B. A, Diniz-Filho J. A. F, Mauricio Bini L, De Marco P, Blackburn T. M (2007) Red herrings revisited: spatial autocorrelation and parameter estimation in geographical ecology. Ecography 30: 375–384.

CHAPTER 6

EVOLUTION OF THE VERTEBRATE SKELETON: MORPHOLOGY, EMBRYOLOGY, AND DEVELOPMENT

Tatsuya Hirasawa and Shigeru Kuratani

Keywords: Vertebrate Skeleton Evolution Development Homology

INTRODUCTION

"Is histological development as complete a test of homology as morphological development?" (Huxley, 1864 [1]: 296)

The vertebrate skeletal system has paramount importance for analyses in evolutionary biology. Because vertebrate skeletons can be viewed as aggregates of apparently discrete units, namely bones, they have attracted the interest of comparative anatomists since even before the dawn of the concept of evolution [2]. In addition, because bones can be preserved as fossils, comparative research can include extinct vertebrates, thereby shedding light on evolutionary patterns and processes (e.g., [3]). In addition, the vertebrate skeletal system is well suited to biomechanical analyses, allowing both morphological and functional transitions throughout evolution to be reconstructed (e.g., [4]).

In any comparative study, homology is a conceptual basis for comparing equivalent units. There is, however, a difficulty in establishing homology—that is, "the apparent loose relationship between morphological characters and their genetic basis" [5]. Incongruities between morphologies and their genetic bases may lead to errors when homology is defined solely according to criteria of ontogeny.

Skeletal systems of vertebrates are intolerant of such incongruities (reviewed by [6]). Historical continuities of skeletal elements as step-wise morphological changes along a phylogenic lineage are inferable from detailed comparative analyses. However, within these continuities, discontinuities of genetic and developmental bases arise in which morphologically homologous bones are produced through different developmental processes [7,8].

Before the concept of evolution was established, two distinct types of bones were recognized in vertebrate skeletons and were thought to reflect their embryonic development; specifically, whether the bone arose from a cartilaginous precursor or not (e.g., [9,10]). Bone arising from precursor cartilage develops not only on the surface of the cartilage (perichondral ossification), but also within the cartilage mass as the cartilage template becomes degraded (endochondral ossification), thereby distinguishing this type of bone from that lacking a cartilaginous precursor in terms of developmental process, or histogenesis. This line of demarcation in histogenesis

Citation: Tatsuya Hirasawa and Shigeru Kuratani, Evolution of the vertebrate skeleton: morphology, embryology, and development, DOI 10.1186/s40851-014-0007-7

was later considered to reflect the evolutionary succession of bones. For example, Huxley (1864: 298) [1] wrote, "It is highly probable that, throughout the vertebrate series, certain bones are always, in origin, cartilage bone, while certain others are always, in origin, membrane bone." In addition, differences in the cell type of the osteoblast precursors—either mesodermal or neural crest cells—has historically been offered in support of the notion that these two histogenetically distinct types of bone generally evolved separately. However, here, we confirm, through a review of both classical and recent research, that both histogenesis and cell lineage are decoupled with the two independent lineages of skeletal systems, namely endo- and exoskeletons, the continuities of which are inferable from comparative morphology.

In this review, we first summarize various evolutionary continuities of vertebrate skeletal systems. We then describe their developmental bases at two hierarchal levels, namely histogenesis and cell lineage, according to recent studies in developmental biology. In light of this understanding, we discuss the loose relationship between morphology and developmental basis and suggest that a frame shift in character identity occurred across cell lineages during the evolution of vertebrate skeletal systems.

Morphological Divisions—Endoskeleton vs. Exoskeleton

From the perspective of comparative morphology, including paleontology, it has been suggested that two lineages of skeletal systems—the endoskeleton and exoskeleton—have succeeded in vertebrate evolution (Figure 1, Table 1) [7,11]. This mode of classification is defined exclusively by phylogenetic continuities, and thus differs from terminology based on ontogeny [7]. For example, the endoskeleton consists of bones preformed from cartilage and their evolutionary derivatives, or homologues (Table 1) [7]. Most endoskeletal bones, such as those in the axial and limb skeletons, are located together with muscles within a deep layer of the body. However, in the evolution toward turtles, the thoracic axial skeleton was exposed, owing to loss of the dorsal axial muscles, to form the carapace [12]. In this sense, the turtle carapace should be considered an "exposed endoskeleton." The distal tip of the distal (ungual) phalange is another example of an exposed endoskeleton that is recognized in vertebrates [12].

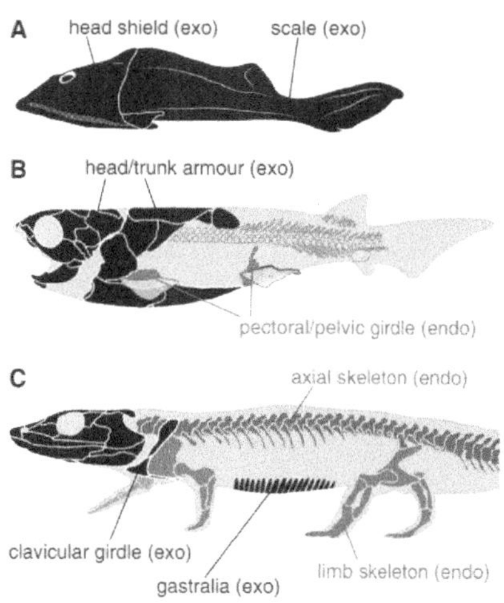

Figure 1 Distribution of endoskeletons (endo) and exoskeletons (exo) in the vertebrate body. **(A)** Osteostracan*Cephalaspis* (redrawn from [13]). **(B)** Basal jawed vertebrate *Compagopiscis* (redrawn from [14]). **(C)** Temnospondyl tetrapod *Dendrerpeton* (redrawn from [15]).

Table 1 Classification of skeletal systems

Skeletal system	Bone	Example
Endoskeleton	Cartilage bone	Vertebrae, ribs, limb bones
	Membrane bone	Centra of teleosts, sesamoid, orbitosphenoid of the Amphisbaenia
Exoskeleton	Dermal bone	Skull roof bones, dentary, clavicle, gastralia, scale of fishes, osteoderm

There is convincing evidence that cartilaginously preformed bones changed during evolution to become intramembranous bones. For example, the orbitosphenoid, a cranial skeletal element, of the Amphisbaenia (Reptilia: Squamata) develops intramembranously, although it clearly is homologous with the cartilaginously preformed orbitosphenoid of other tetrapods [16]. Patterson (1977) [7] proposed calling such intramembranous bones "membrane bones" and discriminated them from bones that developed within the dermis, or "dermal bone." According Patterson's terminology, the endoskeleton consists of cartilage and membrane bones (Table 1: Note that the above-mentioned Huxley's definition of "membrane bone" is different from that used in this paper, as he did not distinguish dermal bones from other intramembranously formed bones).

In contrast, the exoskeleton consists of dermal bones (*sensu* [7]), which are homologous with bony armor and are often coated with enameloid or dentine tissues in basal vertebrates (Figure 1, Table 1; [17]). Exoskeletal bones are located superficially in the body in ancestral conditions, but some exoskeletal bones, such as the dentary and clavicle of mammals, have shifted in their positions to a layer deeper than that of some muscles [18-20]. In this sense, the dentary and clavicle might be referred to as "sunken exoskeleton."

A possible intermediate condition between ancestral and sunken exoskeletons is represented by the gastralia (Figure 2). The gastralia are a series of segmental rod-like bones that cover the ventral aspect of the abdomen in crocodilians and the tuatara, among living forms. Based on fossil evidence, the gastralia are thought to have evolved from exoskeletal bony scales and thus are exoskeletal elements [21]. However, the gastralia embryonically develop in close association with the rectus abdominis muscle in a deep layer, whereas other trunk exoskeletal elements develop close to the epidermis [22,23] (see also Figure 2). Accordingly, Hay (1898) [22] distinguished the gastralia from other dermal bones and classified the gastralia as "fascia bone." Such a concept had been inherited in the distinction between "epithecal" and "thecal" ossifications, as used by Völker (1913) [24] and Zangerl (1939) [25], which indicate outer and inner dermal layers of ossification, respectively.

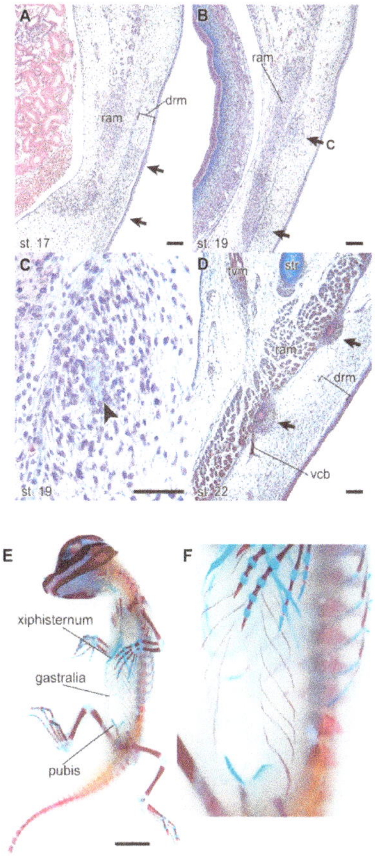

Figure 2 Gastralia of the American alligator (Alligator mississippiensis). The embryos were staged according to Ferguson (1985) [26]. **(A)** Transverse section of the ventral trunk of an embryo at stage 17. Formation of the gastralia begins with condensation of cells (arrows) in the dermis (drm). Alcian-blue, hematoxylin and eosin stains; scale bar, 100 µm. **(B)**Transverse section of the ventral trunk of an embryo at stage 19. The distance between the primordial gastralia and the rectus abdominis muscle (ram) decreases. Alcian-blue, hematoxylin and eosin stains; scale bar, 100 µm. **(C)** Enlarged image of the primordial gastralia, showing the matrix that is stained with Alcian blue (arrowhead), which appears transiently before the bony tissue is formed. Alcian-blue, hematoxylin and eosin stains; scale bar, 50 µm. **(D)**Transverse section of the ventral trunk of an embryo at stage 22. The gastralia contact the rectus abdominis muscle. The ventral cutaneous branch of the intercostal nerve (vcb) runs adjacent to the margin of the gastralium. Alcian-blue, hematoxylin, eosin and immunohistochemistry with anti-acetylated tubulin antibody (T6793, Sigma-Aldrich) stains; scale bar, 100 µm. tvm, transversus ventralis muscle. **(E)** Ventral view of a stage 25 embryo. Alizarin red and Alcian blue stains; scale bar, 1 cm. **(F)** Enlarged image of **E**.

The previously mentioned evolutionary shifts in the topographic positions of exoskeletal elements recalls the idea of Holmgren (1940) [27], who suggested that, in some cases, various exoskeletal elements evolved into endoskeleton as the result of a topographic shift (delamination theory). However, studies of comparative morphology provide no evidence of interchangeability between endo- and exoskeletons [7]; the two historical lines of endo- and exoskeletal systems are likely to have evolved quite independently from each other. It is true that, in some cases, exposed endo- and exoskeletal elements become fused into a single element during ontogeny, as seen in the ontogenetic fusion between endoskeletal costal plates and exoskeletal peripherals to form the carapace in turtles, and in the fusion between endoskeletal vertebrae and exoskeletal osteoderms to form a tail club in ankylosaurid dinosaurs [28]. However, the ossification centers maintain their separate entities, implying incompatibility between the endo- and exoskeletons. (Nevertheless, it is worth noting that a vestigial component of the cleithrum (exoskeletal element) on the scapula (endoskeletal element) in mammals has been suspected repeatedly [29,30]. This evolutionary change represents a "phylogenetic fusion" advocated by Patterson, 1977 [7]).

Some skeletal elements cannot always be traced back to the ancestral endo- or exoskeleton. There are some examples of newly acquired endo- or exoskeletons in various derived taxa. In special cases, bones are sometimes produced within musculotendinous tissues as neo-formations in specific taxa (e.g., the ossified tendon [31]; and sesamoid bones) or by pathologic ossification. Smith (1947) [32] called these bones "subdermal bones," whereas Patterson (1977) [7] classified them as membrane bones and components of the endoskeleton (Table 1).

As another example of newly evolved endoskeletal bones, the baculum is a cartilage bone that was newly acquired in the lineage of eutherian mammals [33]. Likewise, non-eutherian mammals have epipubic bones, which were newly acquired in the more basal mammalian lineage and lost in the crown eutherians [34]. It remains uncertain whether the baculum evolved from the epipubic bone of non-eutherian mammals [35], but examples of the baculum and epipubic bone are suggestive of a novel cartilage bone (a component of the endoskeletal system) that was acquired as an autapomorphy of a specific clade.

In addition, novel exoskeletal elements have been acquired in specific clades. The predentary and rostral bones are examples of such exoskeletal elements [36,37]. Osteoderms (the bony plates covering body contours) occur recurrently throughout vertebrate evolution [38-40]. Although morphological traits are distributed intermittently along the phylogeny, osteoderms are considered to share a developmental basis ("latent homology" *sensu* [40]), perhaps illustrative of the historical continuity of these bony elements [39,40].

Histogenesis: Endochondral and Intramembranous Ossifications

In contrast to the distinction of exo- and endoskeletons, adjectives such as 'endochondral', 'dermal' and 'intramembranous' are used exclusively for histogenetic aspects of skeletal tissues, and primarily unrelated to skeletal morphological identities [11]. In many cases, endoskeletal bones develop in association with preexisting cartilage, whereas exoskeletal bones develop solely intramembranously. However, some endoskeletal bones develop solely intramembranously, without any association with cartilage (membrane bones: Table 1), and some exoskeletal bones are likewise associated with cartilage. Comparative morphology studies have shown that cartilaginously preformed bone in the ancestral endoskeleton became intramembranously developed bone in derived taxa (e.g., the orbitosphenoid of the Amphisbaenia [16]). In contrast, cartilage (secondary or adventitious cartilage) develops on the periphery of exoskeletal bones that develops intramembranously, late in ontogeny of derived clades [7,41]. Cartilage has even been identified in the exoskeletal armor of the trunk (placodont sauropterygians [42]). Therefore, histogenetic modes with respect to the association of cartilage are interchangeable throughout evolution, as once suggested by De Beer (1937) [43].

Cartilaginously preformed bone is produced through both intramembranous (perichondral) and endochondral ossification. In perichondral ossification, the typical mode for periosteal bone formation, osteoblasts are differentiated from the perichondrium/periosteum surrounding the cartilage and subsequently produce the osteoid inside the periosteum. In the development of the costal plate of the turtle carapace, the periosteum expanded outward; therefore, osteoblasts produce outgrowths of the periosteal bone collar, or bony trabeculae [12].

Recent studies have shown that osteoblast cells derived from the perichondrium also support endochondral ossification [44]. In the early phase of this developmental process, osteoblastic precursors differentiate from perichondrial cells (Figure 3A) and subsequently migrate from surfaces in which the cartilage template is degraded into the primary ossification center of the endochondral bone (Figure 3B). Typically, blood vessels invade the cartilage from entrances of osteoblastic precursors and extend along their migration, suggesting intimate developmental relationship between vascularization and endochondral ossification [44]. The osteoblast precursors mature into osteoblasts to form bony trabeculae inside the cartilage (Figure 3C).

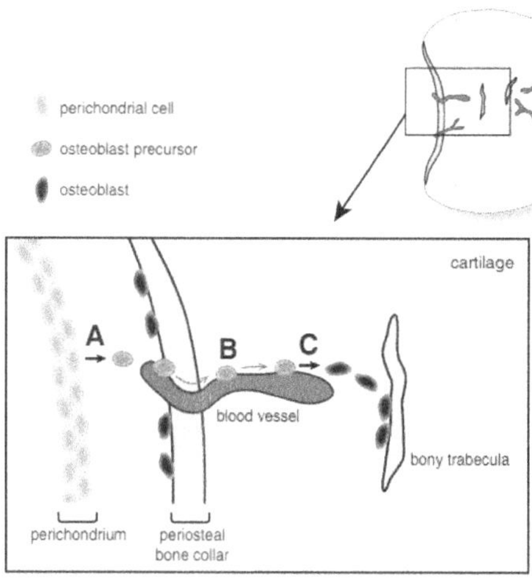

Figure 3 Process of endochondral ossification. **(A)** Differentiation of osteoblastic precursors from perichondrial cells. **(B)**Migration of osteoblastic precursors **(C)** Formation of bony trabeculae by mature osteoblasts.

According to histological analyses of fossils, perichondral ossification evolved in the clade containing osteostracans and jawed vertebrates, whereas the endoskeletons of galeaspids comprise calcified cartilages, not perichondral bones [45]. Endochondral ossification originated evolutionarily in osteichthyes—that is, later than the emergence of perichondral ossification [46].

In the development of the cranial exoskeletal bones of extant osteichthyans, osteogenic cells are differentiated from mesenchymal condensations in the dermis. During this process of intramembranous ossification, osteoblasts mature from a specific transitional cell type (chondrocyte-like osteoblast), which co-expresses both osteogenic and chondrogenic marker genes [47]. Postcranial osteoderms (exoskeletal bones) develop in the dermis, presumably regulated by an intimate interaction with the epidermis. For example, in armadillos, the osteoderm is produced by osteoblasts that are differentiated from the condensation of dermal cells, with the orientation of the primordial osteoderm parallel to that of the epidermis [48]. In contrast, the osteoderm of alligators develops beneath the keel of scutes, but no osteoblasts are morphologically recognizable during this process [49]. There remains much room for investigation regarding the development of reptilian osteoderms.

In some fishes, exoskeletal bones are coated with enameloid or dentine tissues, namely, odontogenic components (reviewed by [50]). These enameloid- and dentine-coated bones occur widely among stem-osteichthyans, and odontogenic components are present in chondrichthyans also. However, the odontogenic components seen in chondrichthyans are believed to represent the vestige of the enameloid- and dentine-coated bones of ancestral jawed vertebrates, in which the bony portion was lost secondarily [51]—the exoskeleton of stem-gnathostomes likely was composed primarily of bone. This view is supported by recent histological data from placoderms (a taxon of stem-gnathostomes), indicating that the condition seen in extant chondrichthyans is derived. In placoderms, bony components always contributed to the exoskeleton, whereas odontogenic components did not always contribute to the exoskeleton [52,53], suggesting that odontogenic components were not prerequisite for exoskeletal development in these taxa.

In addition to endochondral and intramembranous ossifications there is a disparate mode of bone formation, namely metaplastic bone formation [54], the process by which preexisting tissues change directly (i.e., through metaplasia) into bony tissues. Exposed endoskeleton [12,55,56] and exoskeleton [57] contain portions of metaplastic bone, in which the collagen fibers of the dermis are engulfed.

Collectively, comparisons of histogenesis in living and fossil vertebrates suggest the following scenario (Figure 4). In stem vertebrates basal to the clade of osteostracan-jawed vertebrates, the endoskeleton was composed purely of cartilage (Figure 4A). Osteostracans and non-osteichthyes jawed vertebrates evolved ossified endoskeletons (Figure 4B). In these animals, both endo- and exoskeletons developed purely through intramembranous ossification, although the endoskeleton developed on the surface of cartilage (perichondral ossification; as for perichondral ossification in chondrichthyes, see [58]). Osteichthyes acquired endochondral ossification, in which bony tissues are produced within (as well as on top of) cartilage (Figure 4C). During evolution, cartilage structures were occasionally lost and replaced in part by endoskeletal bones (membrane bones) and occasionally acquired in association with exoskeletal bones (secondary cartilages). Exoskeletal bones might be coated with enameloid and dentine tissues, but whether such a trait represents the ancestral or derived state is equivocal, on the sole basis of histological data. Alternatively, perhaps exoskeletal bones in the ancestral condition were not associated with enameloid and dentine tissues.

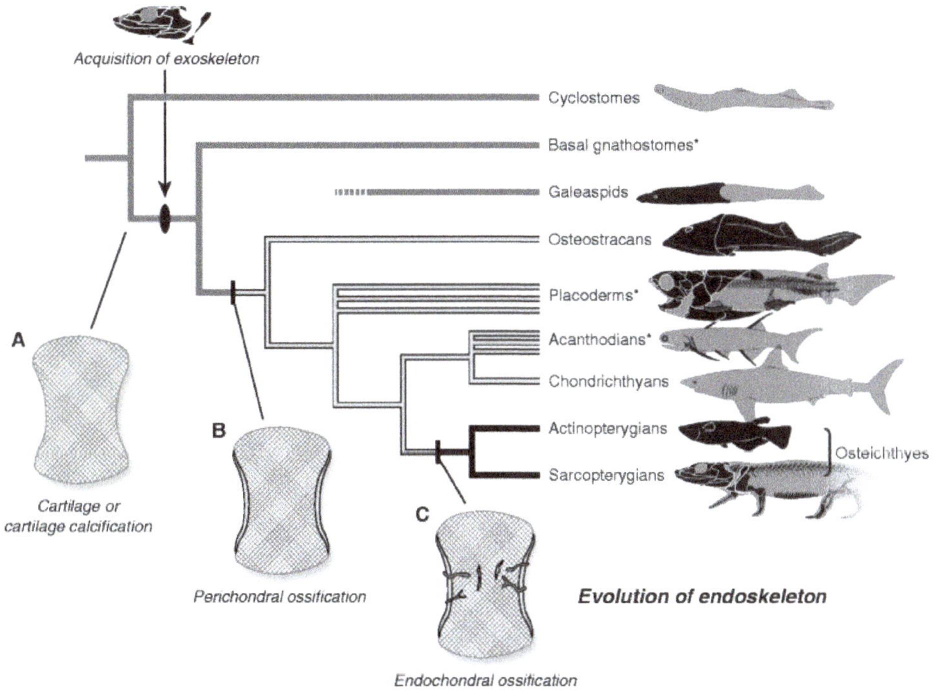

Figure 4 Evolution of the endoskeleton. Phylogenetic framework was adopted from [59]. Asterisks indicate paraphyletic groups. **(A)** Endoskeleton composed purely of cartilage. **(B)** Endoskeleton with perichondral ossification. **(C)**Endoskeleton with peri- and endochondral ossifications.

Developmental Origins and Cell Lineages—Mesoderm and Neural Crest in the Vertebrate Cranium

Developmentally, the skeletal tissues of vertebrates have dual origins—the mesoderm and neural crest. Platt (1893) [60] suggested that the ectodermally derived mesenchyme (that is, ectomesenchyme) contributes to the cranial skeleton in basal vertebrates. De Beer (1958, 1971) [61,62] later used Platt's notion to refute von Baer's germ layer theory [63], because mesoderm generally was believed to be the main source of skeletal tissue in animals.

The origination of part of the vertebrate cranium from the neural crest has been exemplified through several experimental embryologic analyses involving amphibian and avian models in which neural crest grafting experiments are possible (reviewed by [64,65]). Even in non-model vertebrate species, including lampreys, similar results have been obtained [66,67] (also see [68,69]). The use of transgenic techniques has revealed the contribution of the neural crest to the skull in teleosts and mammals (Figure 5) [70-73]. It was previously thought that the rostral neural crest (cephalic crest) yielded mesenchymal tissue throughout the bodies of vertebrates, whereas the posterior portion (that is, the trunk crest) typically gave rise to a more limited repertoire of tissues, including melanocytes and the peripheral nervous system [74-77]. In the head, it has generally been accepted that the visceral arch skeleton (see below) is derived from the neural crest [78] (reviewed by [79]), which however, is not yet completely exemplified for some of the visceral dermal bones at the genetic level in the mouse (reviewed by [8]; Figure 5C).

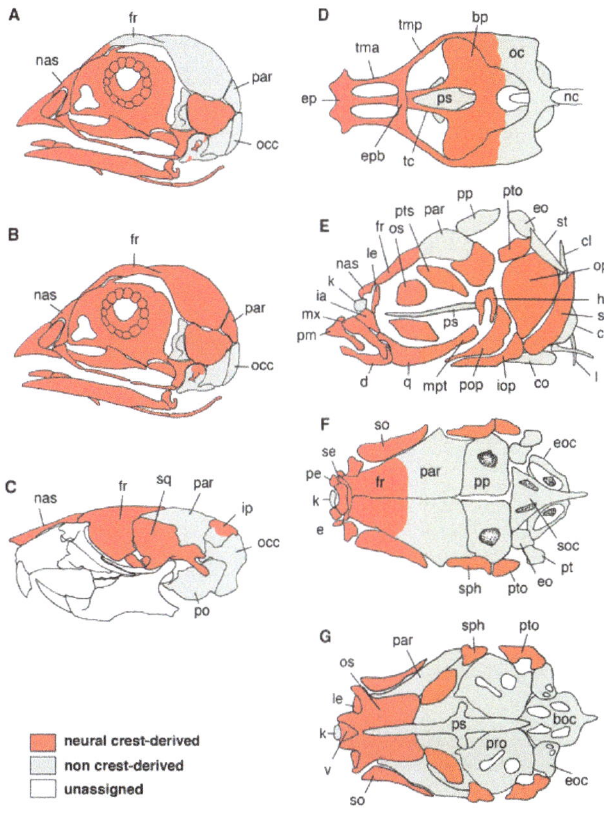

Figure 5 Developmental origins of the dermal skull roof. **(A** and **B)** Different views of the neural crest. Noden (1982, 1984) [80,81] placed the neural crest-mesodermal boudary in the dermal skull roof in the rostral part of the avian frontal **(A)**, whereas Couly at al. (1993) [82] reported that the entire dermal skull roof is derived from the neural crest. Note that the occipital represents an endoskeletal vertebral element secondarily assimilated to the cranium in gnathostomes. **(C)** Developmental origins of the dermal skull roof and the posterior cranium in the mouse, based on transgenic approaches by [70,72,83,84]. Neural crest-mesoderm boundary is located at the boundary between

the frontal and parietal. **(D-G)** Neural crest- and mesodermal origins of the cranial elements in zebrafish based on transgenic techniques by Kague et al. (2012) [73]. Names of the bones were revised based on comparative osteology by [85,86]. Dorsal view of the chondrocranium **(D)**, and left lateral **(E)**, dorsal **(F)**, and ventral **(G)** views of adult zebrafish. In these views, the elements colored grey are of mesodermal origin. Note tha the neural crest-mesodermal boundary of the dermal skull roof is found in the frontal of this animal. Abbreviations: boc, basioccipital; bp, basal plate; cl, cleithrum; co, coracoid; d, dentary; e, ethmoid; eoc, exoccipital; fr, frontal; hm, hyomandibula; ia, intercalar; iop, interopercle; ip, interparietal; k, kinethomoid; le, lateral ethmoid; mpt, metapterygoid; mx, maxilla; nas, nasal; nc, notochord; oc, otic capsule; occ, occipital; op, opercle; os, orbitosphenoid; par, parietal; pe, preethmoid; pm, premaxilla; po, periotic; pop, preopercle; pp, postparietal; pro, prootic; ps, parasphenoid; pto, pterotic; pts, pterosphenoid; q, quadrate; se, supraethmoid; soc, supraoccipital; so, supraorbital; soc, supraoccipital; sop, subopercle; sph, sphenotic; sq, squamosal; st, supratemporal; tc, trabecula; tma, taenia marginalis anterior; tmp, taenia marginalis posterior; Redrawn from [8] **(A-C)** and from [73] **(D-G)**.

In the context of comparative embryology and morphology, the cranium traditionally has been divided into several components, primarily the dorsal and ventral moieties (the neurocrania and viscerocrania, respectively) [43,79,87-94]. The neurocrania and viscerocrania are both recognized as endoskeletons over which a dermal covering, the dermatocranium, develops to encapsulate the entire endocranium. As noted earlier, the endocranium forms as a cartilage precursor and either ossifies through endochondral ossification to be replaced by bone, or degenerates (in cases in which dermal bones can perform the same functions). The cartilaginous skull roof in elasmobranchs is complete, but in animals in which the dermal skull roof is well developed that part of the cartilaginous neurocranium typically is absent.

Like the cranium, the dermatocranium can be divided into dorsal and ventral components corresponding to its neural and visceral elements. The cartilaginous neurocranium was initially recognized as a rostral continuation of the vertebral column, its elements being united and expanded to hold the enlarged brain. In contrast, the viscerocranium is composed of serial and metameric visceral arch skeletons surrounding the pharynx. In jawed vertebrates, one of the rostral elements is enlarged and divided dorsoventrally into the upper and lower jaws. The developmental origins of these cranial components have been, and remain, the focus of much debate.

According to Noden (1988)'s scheme [78], the neural crest-derived ectomesenchyme resides predominantly within the ventral part of the pharyngular head, in the region in which the craniofacial structures will form, whereas the majority of the cranial mesoderm is found more dorsally, lateral to the notochord and surrounding the brain primordium [78] (reviewed by [8]). This arrangement prompts the speculation that the distinction between neurocrania and viscerocrania will correspond to that of their embryonic cell lineages, i.e., mesoderm and neural crest. This seems reasonable, given that, like that of trunk somites, chondrification of the mesoderm is understood to require signals that emanate from the notochord. In contrast, the skeletogenesis of neural crest cells differs from that of the paraxial mesoderm, and is highly dependent on epithelial–mesenchymal interactions [82] (reviewed by [95]). Although this explanation holds true for part of the cranium, it is contradicted elsewhere.

First, the so-called cranial base is not entirely made of mesodermal cells—its rostral portion (rostral to the position of hypophysis) is preformed as paired rods of cartilages called trabeculae, which are derived from the neural crest [71,82,96] (reviewed by [97]; Figures 5D, 6A–C). Comparative embryologists have suggested that this structure represents visceral arch skeletons that had been ancestrally developing rostral to the mandibular arch (reviewed by [97-99]). Although trabeculae in the cyclostomes are not homologous with those in jawed vertebrates, it is now generally accepted that the rostral part of the neurocranium originates from the neural crest throughout the vertebrate species [79,100,101] (also see [68,102]). Therefore, in a developmental sense, the endoskeletal neurocranium is a composite structure, derived from both the mesoderm and cephalic neural crest. Its posterior part, which originates mainly from a pair of longitudinal plates called parachordals, is a mesodermal structure, except for the otic capsule, which is derived partly from the neural crest. The parachordals secondarily incorporate segmented somitic (vertebrae-like) materials to complete the posteriormost portion, the occipital region [43,90,103-106]. In the chicken, and in other sauropsids as well, this part of the neurocranium contains five somites [43,82]. Therefore, as far as this portion is concerned, the vertebrate cranium—like the vertebral column—is segmented, as suggested by transcendental morphologists [2,107,108] (also see [109]).

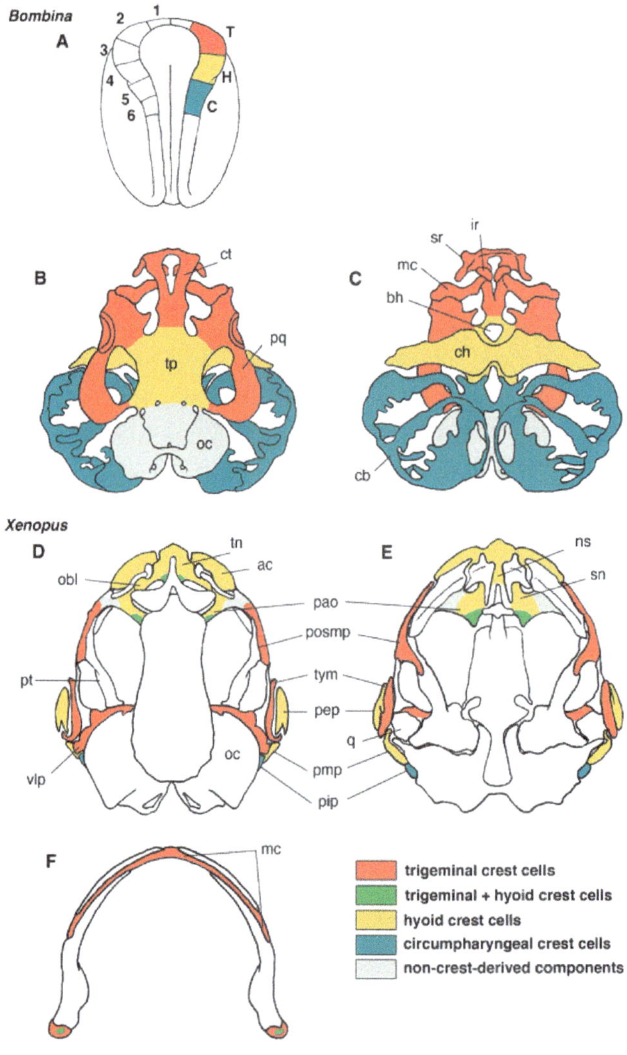

Figure 6 Neural crest mapping of the anuran cranium. **(A- C)** Mapping data in *Bombina orientalis* based on DiI injection onto the neural

fold of the neurula **(A)**. From an experiment performed by Olsson and Hanken (1996) [110]. Origins and differentiation of three crest cell streams are colored in the right neural fold **(A)**, and dorsal **(B)** and ventral **(C)** views of larval chondrocranium. Trigeminal crest cells are colored red, hyoid crest cells yellow, and circumpharyngeal crest cells blue. Numbers on the left neural fold indicate sites of injections. Note that the trabecular plate (tp in **B**), generally derived from the premandibular crest cells, is mapped on the hyoid crest in *Bombina*. **(D- F)** Fate-mapping of adult*Xenopus* cranium. Dorsal **(D)**, ventral **(E)** views. Hyoid crest cells are distributed extensively in the sphenethmoidal region of the cranium. **(F)** Dorsal view of the lower jaw. Note that a part of the articular (proximal end of the Meckel's cartilage) contains hyoid crest cells. Abbreviations: ac, alary cartilage; bh, basihyal; C, origin of circumpharyngeal crest cells; cb, ceratobranchials; ch, ceratohyal; ct, cornu trabecula; H, origin of hyoid crest cells; ir, infrarostral; mc, Meckel's cartilage; ns, nasal septum; oc, otic capsule; obl, oblique cartilage; pao, planum antorbitale; pep, pars externa plectri; pip, pars interna plectri; pmp, pars media plectri; posmp, posterior maxillary process; pq, palatoquadrate; pt, pterygoid; q, quadrate; sn, solum nasi; sr, suprarostral; T, origin of trigeminal crest cells; tp, trabecular plate; tym, tympanic annulus; vlp, ventrolateral process. Redrawn from [111,112].

The developmental origins of the dermatocranium are more enigmatic, creating an obstacle to the understanding of its evolution, and vice versa (Figures 5, 6 and 7). According to classical theory, transcendental morphologists and others believed that the anteroposterior segmentation of the roof of the dermatocranium merely reflected the pattern of cranial mesodermal segments of hypothetical ancestors (reviewed by [92,113,114]; Figure 7A). However, this conventional assumption, which was captured through morphological comparisons (before evolution was conceptualized), is incompatible with our current understanding of developmental origins. Again—in all vertebrate embryos examined so far—the neural crest contributes to both the visceral part of the calvarium and the neural components of the dermatocranium.

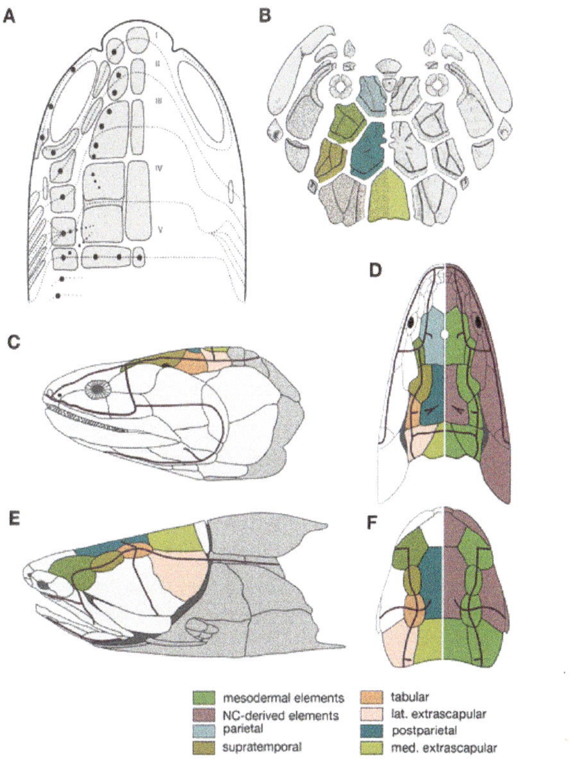

Figure 7 Evolution of dermatocranial elements. **(A)** Traditional scheme of the dermal skull roof, based on the head segmentation scheme of Jollie (1981) [113]. **(B)** Schematized prototype of the arthrodire dermal skull roof as suggested by Heintz (1932) [115]. Homologies between various dermal elements in **B** and **F** are indicated by color. **(C-F)**Dermatocranium of *Eustenopteron* **(C** and **D)** and *Entelognathus* **(E** and **F)**, lateral **(C** and **E)** and dorsal **(D** and **F)**views. Thick red lines represent lateral lines that correspond to patterns of some dermal elements. Presumed homologous dermal elements are shown in the same color in **C** and **E** and the left halves of **D** and **F**. On the right side of **D** and **F**, neural crest- and mesoderm-derived elements are differently colored according to assumptions that the crest–mesoderm interface is primarily found between the frontal and parietal bones (as in the mouse) and that postparietal homologues are consistently derived from the neural crest in sarcopterygians (including tetrapods). **C-F**, redrawn from [59].

The dermal elements of the calvarium are likely patterned according to the lateral line system, and thus the homology of these elements is, in aquatic forms, based on the homology of lateral lines (see [59,114] and references therein; Figure 7C–F). Although the patterns of dermal bones and lateral lines are coupled developmentally, it is unclear whether the lateral line induces the dermal bones, or vice versa (see [116]). Presumably the typical dermal bones found in fishes (including placoderms) became secondarily sunken exoskeletal elements concomitant with the shift in developmental interactions to induce membranous ossification in a deeper layer of the dermis, as found in amniotes. Questions remain regarding homologies (evolutionary continuities) of the dermal elements (reviewed by [8]), as well as their early evolution. The pattern of the dermal skull roof perhaps was first established in placoderms [59] (Figure 7B–F; also see [115]), in which the topographic relationship between dermal bones and lateral lines seen in modern vertebrates is recognizable, at least in part. From lines of circumstantial evidence regarding neural crest contribution and its putative relationship with lateral lines, it is unlikely that the dermal skull roof elements represent segmental organization of the vertebrate head. The lateral lines are not induced as primordia with any segmental prepatterning (for the developmental pattern of the placodes, see [117] and references therein); therefore, the dermal skull roof elements may form independently of any segmental prepattern.

By constructing chick–quail chimeras, Noden found that the rostral part of the dermal skull roof is derived from the neural crest, whereas the posterior arises from the mesoderm [80,81,118,119] (Figure 5A). The boundary between these two cell lineages lies in the frontal bone (for the homology of the avian frontal bone, see [8]). Similar results from a similar experiment were obtained by Le Lièvre (1978) [120]. However, Couly et al. (1993) [82] showed that the entire dermis, as well as the dermatocranial elements, is exclusively of neural crest origin (Figure 5B). To date, systematic fate mapping of the avian craniofacial structures has not been completed; the explanation underlying these inconsistent results remains unclear, but may involve contamination by non-crest tissues or incomplete postsurgical wound healing (summarized by [8]).

Regardless, the views of Couly et al. (1993) [82] once prevailed among zoologists and carried the expectation that the entire exoskeleton of vertebrates—head and trunk—would be of neural crest origin (reviewed by [121]). Another finding that appeared to strengthen this assumption was that the differentiation repertoire of the neural crest is not entirely predetermined differentially along the anteroposterior axis (head versus trunk); heterotopically transplanted trunk neural crest can exhibit skeletogenic potency in the head environment of the embryo [122] (also see [123] for a similar experiment; also see [124]). It was thus speculated that the trunk neural crest is normally suppressed from differentiating into the exoskeleton in animals that have lost most of the postcranial exoskeleton, which, however, can be reactivated under specific circumstances. In fact, all exoskeletal elements in vertebrates, including the dermal skull roof, teleost scales, lepidotrichia, and the extensive head shield in some fossil lineages such as osteostracans and placoderms, were expected to originate from the neural crest [17]—despite the lack of any supporting evidence for this notion. This overly simplified prediction was further extended to postulate the involvement of the neural crest in the turtle shell, which had often been interpreted erroneously as an exoskeletal element (see [12]; see above). Here, the mesoderm-crest duality was related to an in–out topography of endo/exoskeletal parts in the neurocranium, not along the dorsoventral axis.

New embryonic technologies have apparently dispelled the above unsubstantiated assumptions. Shimada et al. (2013) [77], for example, performed transplantations of somites and neural crest in medaka embryos and convincingly showed that the trunk scales of this fish originate from the mesoderm, not the neural crest. Analyses of transgenic lines of zebrafish by several other groups yielded similar results [75,125]. However, several groups suspect that the neural crest contributes to the exoskeleton of the trunk, for example, to the lepidotrichia of the caudal and dorsal fins in zebrafishes [73] and the turtle plastron [124,126]. Furthermore, a recent study tracing the lineages of transgenic cells revealed that trunk neural crest cells do not generate a skeletogenic tissue (that is, ectomesenchyme) [76] although they have skeletogenic potential in the developing head [122]. These lines of evidence, in combination with the fossil evidence from placoderms [52,53] (see above), suggest that the exoskeleton of the trunk develops from the mesoderm in the ancestral condition in the jawed vertebrates and that accretions of the enameloid and dentine tissues (i.e., odontogenic component) to the trunk exoskeleton occurred in many lineages, distinct from what had previously been hypothesized (e.g., [127]).

Consequently, the interface between the neural crest- and mesoderm-derived parts of the exoskeleton again appears to be somewhere in the skull roof, and different results regarding its specific location have been obtained via different experimental methods in embryos of different taxa (reviewed by [8]; Figure 5A, B). Our current understanding regarding the origin of vertebrate skull roof is therefore confused.

Transgenic technology was used to label crest-derived ectomesenchyme and its derivatives in mice (Figure 5C) [30,70,72,83]. In this model, the *Wnt-1* promoter was used to drive *Cre* to activate a reporter gene as a marker for all neural crest cells. This methodology resulted in labeling of the nasal, frontal, and interparietal regions in addition to the more ventrally located dermal elements, and the signal distribution was complementary to the pattern obtained by using *Mesp1-Cre*/R26R to label mesodermal cell lineages (see [72]). This result resembles those of Noden (1978, 1982, 1983, 1984) [80,81,118,128] and Le Lièvre (1978) [120] in avian embryos (Figure 5A; Evans and Noden, 2006 [119], subsequently confirmed these previous results by labeling mesoderm through retroviral infection). Furthermore, these current and previous findings coincide perfectly if we admit misidentification of the boundary between the frontal and parietal regions in mammals and avians: the supraoccipital region is the dorsal portion of a mesodermal element serially homologous with the vertebrae, and the interparietal region may not be present in avians (for the homology and evolution of the interparietal region, see [129] and references therein).

One consistent aspect in this conundrum is that every argument has been based on the firm assumption that evolutionarily conserved bony elements should arise from fixed (homologous) cell lineages in development. This assumption is, of course, profoundly linked to the cell-autonomous and precommitted potency of the neural crest cells in morphological skeletal patterning (see [118,130-133]), which is not per se completely correct [128,134]. Accordingly, the comparative morphological understanding cannot easily be formulated into a simple developmental scheme [8]; in particular, developmental understanding of the neural crest–mesodermal boundaries in the dermatocranial roof is conspicuously unsure compared with that for the cranial base. Several evolutionary scenarios, not always mutually exclusive, may explain the situation regarding the origins of the dermatocranial roof:

1. Morphological homologies of bony elements and the cell lineages that give rise to these elements are regulated at different, decoupled levels, and the bony elements can be conserved through evolution independent from the cell lineages, which are apt to change more rapidly.

2. The ancestral developmental pattern and cell-lineage origins of the dermatocranial elements were established in various fossil taxa, which are reflected in some modern taxa, and are secondarily modified in others, possibly because of the loss or fusion of ancestral elements or the addition of new elements.

3. The dermatocranium (excluding the supraoccipital region) primarily was derived from the cranial neural crest ancestrally, and new mesodermal elements intercalated secondarily to accommodate adaptation to the expansion of the cranial vault in different ways in each animal lineage, obliterating homologies between some bones (as suggested in Figure 7, the parietal bone represents a newly inserted mesodermal element).

4. The dermatocranium (excluding the supraoccipital bone) was primarily derived from the mesoderm ancestrally, and new crest-derived elements were intercalated secondarily to accommodate adaptation to the expansion of the cranial vault in different ways in each animal lineage, thus obliterating homologies of bones.

5. The pattern of dermal elements belongs to most variable parts of the vertebrate body, and developmental constraints assure homologies of dermal elements only within limited levels of taxa (orders, superfamilies, etc.; see [135-137]; reviewed by [113]).

6. Mesodermal dermal elements were associated primarily with various lateral lines in ancestral forms, and other elements were all derived from the neural

crest (Figure 5D and F).

7. The lateral line-induced dermal elements in ancestors have been lost, and the tetrapod dermatocranium, predominantly derived from the neural crest, has been newly reorganized in each animal lineage in its unique way.

None of the above scenarios has been assessed experimentally to date, nor have discrepancies among experimental embryologic data been reconciled. According to the third scenario, the parietal would have to be regarded as a synapomorphy in crown gnathostomes, which however, may be refuted by the fact that the majority of placoderms possess this bone [59]. The situation may be even more confusing than that presented. If the apparent inconsistency in the mesoderm–neural crest boundary could be explained, it may turn out to be attributable to a misnaming of bony elements; this could be resolved by morphological and developmental reexamination of homologous relationships [111]. Unfortunately, however, this confusion may be destined to be insurmountable. Transgenic and chimeric approaches have revealed that the cranium of the frog violates generally accepted rules of development—that is, the developmental origins of the visceral arch and craniofacial skeletons are not found in a canonical set of crest cell streams that are divided into mandibular, hyoid, and branchial arch streams; instead, morphologically homologous dermal elements are derived from inconsistent cell lineages in frog embryos (Figure 6D-F) [111,138]. Therefore, the skeletal development of the frog demonstrates the decoupling of embryonic patterns, cell lineages, and adult morphology in a very radical manner. It is conceivable that, especially in animals that go through metamorphosis, insertion of larval stages causes topographical shifts of the neural crest-derived chondrogenic cells that go on to form adult skeletons (although this does not explain the hyoid crest-origin of the prechordal cranium in amphibians as reported by Olsson and Hanken (1996) [110]). The same may be the case in the development of the dermal skull roof; the morphological patterns and homologies may reside in the local environment of the embryos, such that they become specified during a later phase of development. This potential influence of the local environment recalls the study of Schneider (1999) [139], in which cranial neural crest from the quail embryo was ectopically grafted within mesenchymal populations destined to form the skull wall in the chicken embryo. In resulting chimeras, these grafted cells gave rise to a skeletal element, which in birds is normally derived from the mesoderm. There are many more examples that demonstrate the importance of local tissue interaction in the specification of bony elements [128,140] (also see a review by [92]) by showing potential shifts of cell lineages and populations to generate morphologically conserved skeletal patterns during evolution. A similar situation is seen in the apparent discrepancy of the neural crest contributions to the pectoral girdle bones between amniotes [30,141] and anamniotes [73,142]. It is generally accepted that, within the mesodermal cell population, the developmental basis providing the skeletal identities of the digits shifted between non-homologous primordia in the evolution from dinosaurs to birds (frame-shift hypothesis) [143-145]. No accounts contradict the possibility that skeletal identities similarly shift between neural crest and mesodermal cell populations.

Perspectives—Beyond the Complexity

The vertebrate skull initially attracted the attention of zoologists because of its complex and elegant morphology, but its complexity clearly exceeds all expectations. Theories regarding skeletogenesis and skeletal anatomy and its evolution have been—and still are—fraught with confusion, which never seems to be resolved easily. This situation cannot be ascribed only to the misuse of terminology in non-comparable contexts of discussion; it also reflects the complexity of the developmental and evolutionary diversity of the vertebrate skeletal system per se. Nor is the current developmental understanding of skeletogenesis formulated in an orderly way into the pattern of embryos and cell lineages.

The dilemma described here is tightly linked to the confusion regarding the concept of homology. As noted earlier, morphological homology was in the past reduced to its developmental origins in cell lineages and germ layers, as seen in von Baer's germ layer theory (reviewed by [6]). This theory was refuted as being based on inaccurate concepts of histogenesis, including the concept that skeletogenic differentiation can take place equally in mesodermal and ectodermal (neural crest) cell lineages. Still, the neural crest – mesoderm distinction, as well as endochondral–membranous ossifications, was expected to coincide with specific morphological components of the skull—a belief that could be viewed as a modified version of the germ layer theory. Alternatively, a similar reductionist argument was once widespread with a vague expectation in the dawn of evolutionary developmental biology; namely, that morphologically homologous structures should be patterned through certain unchanged infrastructures, like function of evolutionarily conserved sets of regulatory genes or gene regulatory networks.

Expectations such as these often come true, as typically exemplified by the isomorphic shifts of vertebral formula and Hox code [146] (also see [147]). In this context, the positional identities of vertebrae along the anteroposterior axis of the vertebral column (such as occipital, cervical, thoracic, lumbar, and sacral in mammals) coincide precisely with the expression domains of Hox genes in the prevertebral anlagen, and under this Hox-code-mediated specification the number of segments can vary during evolution (for variable numbers of vertebrae, see [147]). In this case, morphological homology is reduced to the regulation of homologous Hox genes. Similar situations, in which the homology between structure and gene expression is tightly conserved, include the expression of homeobox genes and primordial segments in the developing vertebrate brain, differentiation of somite-derivatives, and dorsoventral specification of the neural tube (reviewed by [148]). In an extreme reductionist argument that is focused on genes, cell-type identities, which are classified by transcript repertoire (that is, molecular fingerprinting of cell types), are comparable among phyla, even between the vertebrate- and annelid body plans, for example, at the level of single neurons [149].

Unfortunately, relationships among homologies at different hierarchal levels—namely at the levels of morphology, histogenesis, cell lineage and genes—remain murky, as homologous skeletal elements can arise from different or shifted cell lineages throughout evolution by means of different mechanisms of development, thus challenging the criteria for morphological homology (e.g., [5,150,151]; reviewed by

[152]). Inconsistency of this type occurs in various phenomena of organogenesis, in which homologous structures are patterned by the actions of non-homologous regulatory genes in each animal lineage [153,154]. In the evolutionary context, there are at least two significant effects worth considering.

One effect is evolutionary novelty and simultaneous loss of homology: the shift in developmental interactions in time and place result in novel regulation of skeletogenic genes, leading to a skeletal pattern incomparable to that in the ancestor. The other effect is developmental drift: the developmental process and mechanisms would shift without changing the readout of the shifted developmental process, thus maintaining the ancestral morphological pattern in the adult. De Beer (1958) [61] noted the heterochronic factor behind similar phenomena, for example, in the creation of the larval stage in development. One of the most conspicuous examples is found in the columella auris (that is, hyomandibular bone) of certain frogs. In *Xenopus*, the anlage of the columella never appears during the larval stage, but arises during metamorphosis [155,156]. In the mouse, the stapes (the homologue of this cartilage bone) is patterned during embryogenesis in the dorsal part of the second pharyngeal arch and is specified through the upregulation of *Hoxa2* [131] in the ectomesenchyme. In *Xenopus*, homology of this skeletal element appears to be maintained—albeit decoupled from the Hox code— and its differentiation is even suggestive of new involvement of the thyroid hormone in the rewired regulatory network. We have already seen, in frog development, how morphologically homologous cranial elements arise from cell populations or pharyngeal arches not identical to those in other vertebrate groups.

It is true that the morphological homology of skeletal elements cannot be reduced directly to the developmental program, or homology of genes, involved in the generation of homologous structures. However, insofar as the criteria for homology largely rest on the relative positions of organs (reviewed by [6]), developmental patterns may, to some extent, explain the impetus behind the manifestation of the homologous patterns. This explanation is especially plausible given that the relative positions suggest evolutionarily maintained topography of cell populations and tissues, which act as the bases for embryonic interactions to establish the identities of the skeletal anlagen, especially through the upregulation of specific sets of transcription factor-encoding genes. Here we recall the experiment of Schneider (1999) [139] to show that neural crest-derived ectomesenchyme and cephalic mesoderm can be exchanged to generate morphologically normal chondrocranium. This experiment indicates that the developmental factor(s) for the morphological homology resides in the "position" in the embryo, not in the embryonic cell lineages. Consistently, a same set of gene expressions has been detected in endochondral ossifications of mesenchymal condensations both derived from neural crest and mesodermal cells [157]. This implication stands in conspicuous contrast to the fact that species-specific shape appears to evolve in the developmental program associated with specific cell lineages [130,133].

To understand the mechanistic background for the burden of development, we have to understand how selective pressure—especially stabilizing selection—at the phenotypic level (adaptation) acts on the developmental program exerted from the genome. In other words, we must identify parts or elements of the developmental program (for example, gene regulatory networks, modules, sets of regulatory genes and their regulatory elements) that can or cannot change when certain fixed phenotypic patterns are favored. These efforts will uncover the aspects of the developmental program that are resistant to change and those that are apt to change during evolution. In evolution, adaptation and constraint cannot be discriminated a priori [158,159]. The key to discriminating between these two causal relationships behind evolution is provided abductively through historical and experimental analyses of the correlation between phenotype and the developmental program behind it (for example, skeletal elements can be considered as a phenotype of a skeletal system). The patterns that allow minimal shifts have been recognized to result from developmental constraint. The concept of developmental constraint has not yet been explained thoroughly, but taxon-specific conserved patterns of embryogenesis have been recognized as the so-called "phylotype," which tends to appear in the organogenetic stage of development ("phylotypic stage" [160]).

In transcendental morphology, the phylotype (pharyngula in vertebrates) has been viewed as an embodiment of the conceptual archetype, a shared morphology of the embryos of animals belonging to the vertebrates, from which various types of adult morphologies can be derived [63]. This "derivation," however, does not necessarily refer to the phylogenetic evolutionary process, but rather to observers' perceptions of homologous patterns and their developmental changes. Morphologically, it is true that the pharyngula-stage embryo in vertebrates is the stage at which the basic body plan, or a set of homologous anlagen, of this animal group becomes established. In the evo-devo context, the phylotypic stage of vertebrate development is recognized as the stage at which so-called tool-kit genes (typically the Hox code) are expressed most conspicuously during development, thus providing the mechanistic bases to explain the significance of this conserved embryonic pattern [160]. Recent transcriptome analyses have shown that the most similar gene expression profiles coincide with the phylotypic stage [161]. Importantly, as indicated by genomic analyses of turtles, the evolutionarily novel patterns of the skeletal system in vertebrates appear to arise through spatiotemporal developmental shifts after the establishment of the above-noted phylotype [162]. Taking into consideration the shifts in morphological homologies—specifically the developmental patterns and processes involved in patterning of the evolutionarily fixed patterns of craniofacial elements—it seems likely that the cranial pattern is specified late relative to the specification of the phylotype. This delay suggests the presence of another developmental constraint, which is more or less uncoupled from those needed to maintain the phylotype. The search for such taxon-specific constraints, as well as their mechanistic importance, is an intriguing focus for future evo-devo studies. The results likely would further our understanding of the synapomorphies used in the reconstruction of evolutionary history.

DECLARATIONS

Acknowledgments

We thank Ruth Elsey and Neil Shubin for the gift of American alligator embryos, and Dai Koyabu for critical reading of the manuscript and valuable discussions. We also thank the two anonymous referees for comments that improved the manuscript.

Authors' Contributions

TH and SK conceived and wrote the review. Both authors read and approved the final manuscript.

REFERENCES

1. Huxley TH: *Lectures on the elements of comparative anatomy.* London: J. Churchill and Sons; 1864.
2. Goethe JW: *Schädelgrüst aus sechs Wirbelknochen aufgebaut. Zur Morphologie, Band 2, Heft 2.* Stuttgart: J. G. Cotta; 1824.
3. Simpson GG: *Tempo and Mode in Evolution.* New York: Columbia University Press; 1944.
4. Baier DB, Gatesy SM, Jenkins FA: A critical ligamentous mechanism in the evolution of avian flight. *Nature* 2007,445:307–310.
5. Wagner GP: The developmental genetics of homology. *Nat Rev Genet* 2007, 8:473–479.
6. Hall BK: *Evolutionary Developmental Biology.* 2nd edition. London: Chapman & Hall; 1998.
7. Patterson C: Cartilage bones, dermal bones and membrane bones, or the exoskeleton versus the endoskeleton. In*Problems in Vertebrate Evolution.* Edited by Andrews SM, Miles RS, Walker AD. London: Academic Press; 1977:77–121. Linnean Society Symposium Series.
8. Gross JB, Hanken J: Review of fate-mapping studies of osteogenic cranial neural crest in vertebrates. *Dev Biol* 2008,317:389–400.
9. Nesbitt R: *Human osteogeny explained in two lectures.* London: J. Noon; 1736.
10. Arendt E: *De capitis ossei Esocis Lucii structura singulari.* Regiomonti (Königsberg): Typis academicis Hartungianis; 1822.
11. Starck D: *Vergleichende Anatomie der Wirbeltiere, Bd. 2. Das Skeletsystem: Allgemeines, Skeletsubstanzen, Skelet der Wirbeltiere einschl. Lokomotionstypen.* Berlin, Heidelberg, New York: Springer-Verlag; 1979.
12. Hirasawa T, Nagashima H, Kuratani S: The endoskeletal origin of the turtle carapace. *Nat Commun* 2013, 4:2107.
13. Janvier P: Homologies and evolutionary transitions in early vertebrate history. In *Major Transitions in Vertebrate Evolution.* Edited by Anderson JS, Sues HD. Bloomington: Indiana University Press; 2007:57–121.
14. Trinajstic K, Sanchez S, Dupret V, Tafforeau P, Long J, Young G, Senden T, Boisvert C, Power N, Ahlberg PE: Fossil musculature of the most primitive jawed vertebrates. *Science* 2013, 341:160–164.
15. Clack JA: *Gaining Ground: The Origin and Evolution of Tetrapods.* 2nd edition. Bloomington: Indiana University Press; 2012.
16. Bellairs AD, Gans C: A reinterpretation of the amphisbaenian orbitosphenoid. *Nature* 1983, 302:243–244.
17. Smith MM, Hall BK: Development and evolutionary origins of vertebrate skeletogenic and odontogenic tissues. *Biol Rev* 1990, 65:277–373.
18. Romer AS: Pectoral limb musculature and shoulder-girdle structure in fish and tetrapods. *Anat Rec* 1924, 27:119–143.
19. Crompton AW, PD G: On the lower jaw of *Diarthrognathus* and the origin of the mammalian lower jaw. *Proc Zool Soc Lond* 1963, 140:697–749.
20. Crompton AW, Parker P: Evolution of mammalian masticatory apparatus. *Am Sci* 1978, 66:192–201.
21. Claessens LPAM: Dinosaur gastralia; origin, morphology, and function. *J Vert Paleontol* 2004, 24:89–106.
22. Hay OP: On Protostega, the systematic position of Dermochelys, and the Morphologeny of the chelonian carapace and plastron. *Am Nat* 1898, 32:929–948.
23. Voeltzkow A, Döderlein L: Beiträge zur Entwicklungsgeschichte der Reptilien III.: Zur Frage nach der Bildung der Bauchrippen. *Abh Senk Naturforsch Ges* 1901, 26:313–336.
24. Völker H: **Über** das Stamm-, Gliedmaßen-, und Hautskelet von Dermochelys coriacea L. *Zool Jahrb Anat Ont* 1913,33:431–552.
25. Zangerl R: The homology of the shell elements in turtles. *J Morphol* 1939, 65:383–406.
26. Ferguson MWJ: Reproductive biology and embryology of the crocodilians. In *Biology of the Reptilia, Vol 14: Development A. Volume 14: Deveolopment A.* Edited by Gans C, Billett F, Maderson PFA. New York: Academic Press; 1985:329–491.
27. Holmgren N: Studies on the head of fishes. Part I. Development of the skull in sharks and rays. *Acta Zool Stockh*1940, 21:51–267.
28. Hayashi S, Carpenter K, Scheyer TM, Watabe M, Suzuki D: Function and evolution of ankylosaur dermal armor. *Acta Palaeontol Pol* 2010, 55:213–228.
29. Broom R: On the development and morphology of the marsupial shoulder girdle. *Trans Roy Soc Edin* 1899, 39:749–770.
30. Matsuoka T, Ahlberg PE, Kessaris N, Iannarelli P, Dennehy U, Richardson WD, McMahon AP, Koentges G: Neural crest origins of the neck and shoulder. *Nature* 2005, 436:347–355.
31. Organ CL: Thoracic epaxial muscles in living archosaurs and ornithopod dinosaurs. *Anat Rec* 2006, 288A:782–793.
32. Smith HM: Classification of bone. *Turtox News* 1947, 25:234–236.
33. Smirnov DG, Tsytsulina K: The ontogeny of the baculum in *Nyctalus noctula* and *Vespertilio murinus* (Chiroptera : Vespertilionidae). *Acta Chiropterol* 2003, 5:117–123.
34. Novacek MJ, Rougier GW, Wible JR, McKenna MC, Dashzeveg D, Horovitz I: Epipubic bones in eutherian mammals from the late Cretaceous of Mongolia. *Nature* 1997, 389:483–486.
35. Jellison WL: A suggested homolog of the Os penis or baculum of mammals. *J Mammal* 1945, 26:146–147.

36. Sereno PC: The evolution of dinosaurs. *Science* 1999, 284:2137–2147.

37. Zhou ZG, Martin LD: Distribution of the predentary bone in Mesozoic ornithurine birds. *J Syst Palaeontol* 2011, 9:25–31.

38. Hill RV: Integration of morphological data sets for phylogenetic analysis of amniota: The importance of integumentary characters and increased taxonomic sampling. *Syst Biol* 2005, 54:530–547.

39. Hill RV: Comparative anatomy and histology of xenarthran osteoderms. *J Morphol* 2006, 267:1441–1460.

40. Vickaryous MK, Sire JY: The integumentary skeleton of tetrapods: origin, evolution, and development. *J Anat* 2009,214:441–464.

41. Bailleul AM, Hall BK, Horner JR: First evidence of dinosaurian secondary cartilage in the post-hatching skull of*Hypacrosaurus stebingeri* (Dinosauria, Ornithischia). *PLoS ONE* 2012, 7:e36112.

42. Scheyer TM: Skeletal histology of the dermal armor of Placodontia: the occurrence of 'postcranial fibro-cartilaginous bone' and its developmental implications. *J Anat* 2007, 211:737–753.

43. De Beer GR: *The Development of the Vertebrate Skull.* London: Oxford University Press; 1937.

44. Maes C, Kobayashi T, Selig MK, Torrekens S, Roth SI, Mackem S, Carmeliet G, Kronenberg HM: Osteoblast precursors, but not mature osteoblasts, move into developing and fractured bones along with invading blood vessels. *Dev Cell* 2010, 19:329–344.

45. Wang NZ, Donoghue PCJ, Smith MM, Sansom IJ: Histology of the galeaspid dermoskeleton and endoskeleton, and the origin and early evolution of the vertebrate cranial endoskeleton. *J Vert Paleontol* 2005, 25:745–756.

46. Donoghue PCJ, Sansom IJ: Origin and early evolution of vertebrate skeletonization. *Microsc Res Techniq* 2002,59:352–372.

47. Abzhanov A, Rodda SJ, McMahon AP, Tabin CJ: Regulation of skeletogenic differentiation in cranial dermal bone.*Development* 2007, 134:3133–3144.

48. Vickaryous MK, Hall BK: Osteoderm morphology and development in the nine-banded armadillo, *Dasypus novemcinctus* (Mammalia, Xenarthra, Cingulata). *J Morphol* 2006, 267:1273–1283.

49. Vickaryous MK, Hall BK: Development of the dermal skeleton in *Alligator mississippiensis* (Archosauria, Crocodylia) with comments on the homology of osteoderms. *J Morphol* 2008, 269:398–422.

50. Sire JY, Donoghue PCJ, Vickaryous MK: Origin and evolution of the integumentary skeleton in non-tetrapod vertebrates. *J Anat* 2009, 214:409–440.

51. Romer AS, Parsons TS: *The Vertebrate Body.* 5th edition. Philadelphia: W. B. Saunders; 1977.

52. Downs JP, Donoghue PCJ: Skeletal histology of *Bothriolepis canadensis* (Placodermi, Antiarchi) and evolution of the skeleton at the origin of jawed vertebrates. *J Morphol* 2009, 270:1364–1380.

53. Giles S, Rücklin M, Donoghue PCJ: Histology of "placoderm" dermal skeletons: Implications for the nature of the ancestral gnathostome. *J Morphol* 2013, 274:627–644.

54. Haines RW, Mohuiddin A: Metaplastic bone. *J Anat* 1968, 103:527–538.

55. Scheyer TM, Sánchez-Villagra MR: Carapace bone histology in the giant pleurodiran turtle *Stupendemys geographicus* : Phylogeny and function. *Acta Palaeontol Pol* 2007, 52:137–154.

56. Scheyer TM, Brüllmann B, Sánchez-Villagra MR: The ontogeny of the shell in side-necked turtles, with emphasis on the homologies of costal and neural bones. *J Morphol* 2008, 269:1008–1021.

57. Witzmann F: Comparative histology of sculptured dermal bones in basal tetrapods, and the implications for the soft tissue dermis. *Palaeodiversity* 2009, 2:233–270.

58. Eames BF, Allen N, Young J, Kaplan A, Helms JA, Schneider RA: Skeletogenesis in the swell shark *Cephaloscyllium ventriosum* . *J Anat* 2007, 210:542–554.

59. Zhu M, Yu XB, Ahlberg PE, Choo B, Lu J, Qiao T, Qu QM, Zhao WJ, Jia LT, Blom H, Zhu YA: A Silurian placoderm with osteichthyan-like marginal jaw bones. *Nature* 2013, 502:188–193.

60. Platt JB: Ectodermic origin of the cartilage of the head. *Anat Anz* 1893, 8:506–509.

61. De Beer GR: *Embryos and Ancestors.* Oxford: Oxford University Press; 1958.

62. De Beer GR: *Homology, An Unsolved Problem.* Oxford: Oxford University Press; 1971.

63. von Baer KE: *Entwicklungsgeschichte der Thiere: Beobachtung und Reflexion.* Born Träger: Königsberg; 1828.

64. Hall BK, Hörstadius S: *The Neural Crest.* New York: Oxford University Press; 1988.

65. Hall BK: *Bones and Cartilage: Developmental and Evolutionary Skeletal Biology.* London: Elsevier Academic Press; 2005.

66. Newth DR: Experiments on the neural crest of the lamprey embryo. *J Exp Biol* 1951, 28:247–260.

67. Newth DR: On the neural crest of the lamprey embryo. *J Embryol Exp Morph* 1956, 4:358–375.

68. Langille RM, Hall BK: Role of the neural crest in development of the trabeculae and branchial arches in embryonic sea lamprey, *Petromyzon-Marinus* (L). *Development* 1988, 102:301–310.

69. Kuratani S, Murakami Y, Nobusada Y, Kusakabe R, Hirano S: Developmental fate of the mandibular mesoderm in the lamprey, *Lethenteron japonicum* : comparative morphology and development of the gnathostome jaw with special reference to the nature of the trabecula cranii. *J Exp Zool B (MDE)* 2004, 302B:458–468.

70. Jiang XB, Iseki S, Maxson RE, Sucov HM, Morriss-Kay GM: Tissue origins and interactions in the mammalian skull vault. *Dev Biol* 2002, 241:106–116.

71. McBratney-Owen B, Iseki S, Bamforth SD, Olsen BR, Morriss-Kay GM: Development and tissue origins of the mammalian cranial base. *Dev Biol* 2008, 322:121–132.

72. Yoshida T, Vivatbutsiri P, Morriss-Kay G, Saga Y, Iseki S: Cell lineage in mammalian craniofacial mesenchyme. *Mech Develop* 2008, 125:797–808.

73. Kague E, Gallagher M, Burke S, Parsons M, Franz-Odendaal T, Fisher S: Skeletogenic fate of zebrafish cranial and trunk neural crest. *PLoS ONE* 2012, 7:e47394.

74. Le Douarin N: *The Neural Crest.* Cambridge: Cambridge University Press; 1982.

75. Lee RTH, Thiery JP, Carney TJ: Dermal fin rays and scales derive from mesoderm, not neural crest. *Curr Biol* 2013,23:R336–R337.

76. Lee RTH, Knapik EW, Thiery JP, Carney TJ: An exclusively mesodermal origin of fin mesenchyme demonstrates that zebrafish trunk neural crest does not generate ectomesenchyme. *Development* 2013, 140:2923–2932.

77. Shimada A, Kawanishi T, Kaneko T, Yoshihara H, Yano T, Inohaya K, Kinoshita M, Kamei Y, Tamura K, Takeda H: Trunk exoskeleton in teleosts is mesodermal in origin. *Nat Commun* 2013, 4:1639.

78. Noden DM: Interactions and fates of avian craniofacial mesenchyme. *Development* 1988, 103:121–140.

79. Kuratani S: Craniofacial development and the evolution of the vertebrates: the old problems on a new background. *Zool Sci* 2005, 22:1–19.

80. Noden DM: Patterns and organization of craniofacial skeletogenic and myogenic mesenchyme: a perspective. *Prog Clin Biol Res* 1982, 101:167–203.

81. Noden DM: Craniofacial development: new views on old problems. *Anat Rec* 1984, 208:1–13.

82. Couly GF, Coltey PM, Le Douarin NM: The triple origin of skull in higher vertebrates: a study in quail-chick chimeras.*Development* 1993, 117:409–429.

83. Chai Y, Jiang XB, Ito Y, Bringas P, Han J, Rowitch DH, Soriano P, McMahon AP, Sucov HM: Fate of the mammalian cranial neural crest during tooth and mandibular morphogenesis. *Development* 2000, 127:1671–1679.

84. O'Gorman S: Second branchial arch lineages of the middle ear of wild-type and *Hoxa2* mutant mice. *Dev Dyn* 2005,234:124–131.

85. Westoll TS: Ancestry of the tetrapods. *Nature* 1938, 141:127–128.

86. Schultze HP, Arsenault M: The panderichthyid fish *Elpistostege* : a close relative of tetrapods? *Palaeontology* 1985,28:293–309.

87. Gegenbaur C: *Elements of Comparative Anatomy.* London: MacMillan & Co.; 1878.

88. Gaupp E: Die Entwicklung des Kopfskelettes. In *Handbuch der Vergleichenden und Experimentellen Entwicklungslehre der Wirbeltiere, Bd 3(2).* Edited by Hertwig O. Jena: Gustav Fischer; 1906:573–874.

89. Wiedersheim R: *Vergleichende Anatomie der Wirbeltiere: Für Studierende bearbeitet.* Jena: Verlag von Gustav Fischer; 1909.

90. Goodrich ES: *Studies on the Structure and Development of Vertebrates.* London: Macmillan; 1930.

91. Portmann A: *Einführung in die vergleichende Morphologie der Wirbeltiere.* Basel: Schwabe & Co; 1969.

92. Jarvik E: *Basic Structure and Evolution of Vertebrates, Volume 2.* New York: Academic Press; 1980.

93. Moore WJ: *The Mammalian Skull.* London: Cambridge University Press; 1981.

94. Hanken J, Hall BK: *The Skull*, Volume 1–3. Chicago: University of Chicago Press; 1993.

95. Hall BK: *The Neural Crest in Development and Evolution.* New York: Springer Verlag; 1999.

96. Wada N, Nohno T, Kuratani S: Dual origins of the prechordal cranium in the chicken embryo. *Dev Biol* 2011, 356:529–540.

97. Kuratani S, Adachi N, Wada N, Oisi Y, Sugahara F: Developmental and evolutionary significance of the mandibular arch and prechordal/premandibular cranium in vertebrates: revising the heterotopy scenario of gnathostome jaw evolution. *J Anat* 2013, 222:41–55.

98. Kuratani S, Matsuo I, Aizawa S: Developmental patterning and evolution of the mammalian viscerocranium: Genetic insights into comparative morphology. *Dev Dyn* 1997, 209:139–155.

99. Kuratani S: Evolution of the vertebrate jaw from developmental perspectives. *Evol Dev* 2012, 14:76–92.

100. Oisi Y, Ota KG, Kuraku S, Fujimoto S, Kuratani S: Craniofacial development of hagfishes and the evolution of vertebrates. *Nature* 2013, 493:175–180.

101. Oisi Y, Ota KG, Fujimoto S, Kuratani S: Development of the chondrocranium in hagfishes, with special reference to the early evolution of vertebrates. *Zool Sci* 2013, 30:944–961.

102. Dupret V, Sanchez S, Goujet D, Tafforeau P, Ahlberg PE: A primitive placoderm sheds light on the origin of the jawed vertebrate face. *Nature* 2014, 507:500–503.

103. Fürbringer M: **Über** die spino-occipitalen Nerven der Selachier und Holocephalen und ihre vergleichende Morphologie. *Festschr für Carl Gegenbaur* 1897, 3:349–788.

104. Kessel M, Balling R, Gruss P: Variations of cervical vertebrae after expression of a *Hox-1.1* transgene in mice. *Cell*1990, 61:301–308.

105. Lufkin T, Mark M, Hart CP, Dolle P, Lemeur M, Chambon P: Homeotic transformation of the occipital bones of the skull by ectopic expression of a homeobox gene. *Nature* 1992, 359:835–841.

106. Kessel M: Respecification of vertebral identities by retinoic acid. *Development* 1992, 115:487–501.

107. Oken L: *Über die Bedeutung der Schädelknochen.* Bamberg: Göbhardt; 1807.

108. Owen RC: *On the Archetype and Homologies of the Vertebrate Skeleton.* London: J. Van Voorst; 1848.

109. Huxley TH: The Croonian Lecture: on the theory of the vertebrate skull. *Proc Zool Soc Lond* 1858, 9:381–457.

110. Olsson L, Hanken J: Cranial neural crest migration and chondrogenic fate in the oriental fire-bellied toad *Bombina orientalis* : defining the ancestral pattern of head development in anuran amphibians. *J Morphol* 1996, 229:105–120.

111. Gross JB, Hanken J: Segmentation of the vertebrate skull: neural-crest derivation of adult cartilages in the clawed frog, *Xenopus laevis* . *Intg Comp Biol* 2008, 48:681–696.

112. Hanken J, Gross JB: Evolution of cranial development and the role of neural crest: insights from amphibians. *J Anat*2005, 207:437–446.

113. Jollie M: Segment theory and the homologizing of cranial bones. *Am Nat* 1981, 118:785–802.

114. Thomson KS: Segmentation, the adult skull, and the problem of homology. In *The Skull, Vol 2.* Edited by Hanken J, Hall BK. Chicago: University of Chicago Press; 1993:36–68.

115. Heintz A: The structure of *Dinichthys* : a contribution to our knowledge of the arthrodira. In *Bashford Dean Memorial Volume: Archaic Fishes.* Edited by Gudger EW. New York: American Museum of Natural History; 1932:111–241.

116. Wada H, Ghysen A, Satou C, Higashijima S, Kawakami K, Hamaguchi S, Sakaizumi M: Dermal morphogenesis controls lateral line patterning during postembryonic development of teleost fish. *Dev Biol* 2010, 340:583–594.

117. Schlosser G: Making senses: development of vertebrate cranial placodes. *Int Rev Cell Mol Biol* 2010, 283:129–234.

118. Noden DM: The role of the neural crest in patterning of avian cranial skeletal, connective, and muscle tissues. *Dev Biol* 1983, 96:144–165.

119. Evans DJR, Noden DM: Spatial relations between avian craniofacial neural crest and paraxial mesoderm cells. *Dev Dyn* 2006, 235:1310–1325.

120. Le Lièvre CS: Participation of neural crest-derived cells in the genesis of the skull in birds. *J Embryol Exp Morph*1978, 47:17–37.

121. Le Douarin NM, Dupin E: The neural crest in vertebrate evolution. *Curr Opin Genet Dev* 2012, 22:381–389.

122. McGonnell IM, Graham A: Trunk neural crest has skeletogenic potential. *Curr Biol* 2002, 12:767–771.

123. Nakamura H, Ayer-Le Lièvre CS: Mesectodermal capabilities of the trunk neural crest of birds. *J Embryol Exp Morph*1982, 70:1–18.

124. Cebra-Thomas JA, Terrell A, Branyan K, Shah S, Rice R, Gyi L, Yin M, Hu YS, Mangat G, Simonet J, Betters E, Gilbert SF:Late-emigrating trunk neural crest cells in turtle embryos generate an osteogenic ectomesenchyme in the plastron. *Dev Dyn* 2013, 242:1223–1235.

125. Mongera A, Nüsslein-Volhard C: Scales of fish arise from mesoderm. *Curr Biol* 2013, 23:R338–R339.

126. Cebra-Thomas JA, Betters E, Yin M, Plafkin C, Mcdow K, Gilbert SF: Evidence that a late-emerging population of trunk neural crest cells forms the plastron bones in the turtle *Trachemys scripta* . *Evol Dev* 2007, 9:267–277.

127. Reif WE: Evolution of dermal skeleton and dentition in vertebrates: the odontode regulation theory. *Evol Biol* 1982,15:287–368.

128. Noden DM: Control of avian cephalic neural crest cytodifferentiation. I. skeletal and connective tissues. *Dev Biol*1978, 67:296–312.

129. Koyabu D, Maier W, Sánchez-Villagra MR: Paleontological and developmental evidence resolve the homology and dual embryonic origin of a mammalian skull bone, the interparietal. *Proc Natl Acad Sci U S A* 2012, 109:14075–14080.

130. Wagner G: Untersuchungen an *Bombinator-Triton* -Chimaeren. *Roux's Arch Ent mech Org* 1959, 151:136–158.

131. Rijli FM, Mark M, Lakkaraju S, Dierich A, Dollé P, Chambon P: A homeotic transformation is generated in the rostral branchial region of the head by disruption of *Hoxa-2* , which acts as a selector gene. *Cell* 1993, 75:1333–1349.

132. Couly G, Grapin-Botton A, Coltey P, Ruhin B, Le Douarin NM: Determination of the identity of the derivatives of the cephalic neural crest: incompatibility between *Hox* gene expression and lower jaw development. *Development*1998, 125:3445–3459.

133. Schneider RA, Helms JA: The cellular and molecular origins of beak morphology. *Science* 2003, 299:565–568.

134. Couly G, Creuzet S, Bennaceur S, Vincent C, Le Douarin NM: Interactions between Hox-negative cephalic neural crest cells and the foregut endoderm in patterning the facial skeleton in the vertebrate head. *Development* 2002,129:1061–1073.

135. Westoll TS: On the evolution of the Dipnoi. In *Genetics, Paleontology and Evolution.* Edited by Jepson GL, Simpson GG, Mayr E. Princeton: Princeton University Press; 1949:112–184.

136. White EI: A little on lungfishes. *Proc Linn Soc Lond* 1966, 177:1–10.

137. Thomson KS, Campbell KSW: The structure and relationships of the primitive Devonian lungfish: *Dipnorhynchus sussmilchi* (Etheridge). *Bull Peabody Mus Nat Hist Yale* 1971, 38:1–109.

138. Gross JB, Hanken J: Cranial neural crest contributes to the bony skull vault in adult *Xenopus laevis* : insights from cell labeling studies. *J Exp Zool B (MDE)* 2005, 304B:169–176.

139. Schneider RA: Neural crest can form cartilages normally derived from mesoderm during development of the avian head skeleton. *Dev Biol* 1999, 208:441–455.

140. Showing J: Mise en **évidence** du rôle inducteur de l'encephale dans l'ostéogenèse du crâne embryonaire du poulet. *J Embryol Exp Morph* 1993, 75:165–188.

141. McGonnell IM, McKay IJ, Graham A: A population of caudally migrating cranial neural crest cells: functional and evolutionary implications. *Dev Biol* 2001, 236:354–363.

142. Epperlein HH, Khattak S, Knapp D, Tanaka EM, Malashichev YB: Neural crest does not contribute to the neck and shoulder in the Axolotl (*Ambystoma*

mexicanum). *PLoS ONE* 2012, 7:e52244.

143. Wagner GP, Gauthier JA: 1,2,3 = 2,3,4: A solution to the problem of the homology of the digits in the avian hand. *Proc Natl Acad Sci U S A* 1999, 96:5111–5116.

144. Tamura K, Nomura N, Seki R, Yonei-Tamura S, Yokoyama H: Embryological evidence identifies wing digits in birds as digits 1, 2, and 3. *Science* 2011, 331:753–757.

145. Xu X, Mackem S: Tracing the evolution of avian wing digits. *Curr Biol* 2013, 23:R538–R544.

146. Burke AC, Nelson CE, Morgan BA, Tabin C: *Hox* genes and the evolution of vertebrate axial morphology. *Development*1995, 121:333–346.

147. Narita Y, Kuratani S: Evolution of the vertebral formulae in mammals: A perspective on developmental constraints. *J Exp Zool B (MDE)* 2005, 304B:91–106.

148. Carroll SB, Greiner JK, Weatherbee SD: *From DNA to Diversity: Molocular Genetics and the Evolution of Animal Design.*2nd edition. Malden: Wiley; 2005.

149. Arendt D: The evolution of cell types in animals: emerging principles from molecular studies. *Nat Rev Genet* 2008,9:868–882.

150. Ahlberg PE, Koentges G: Homologies and cell populations: a response to Sánchez-Villagra and Maier. *Evol Dev*2006, 8:116–118.

151. Sánchez-Villagra MR, Maier W: Homologies of the mammalian shoulder girdle: a response to Matsuoka et al. (2005).*Evol Dev* 2006, 8:113–115.

152. Scotland RW: Deep homology: A view from systematics. *Bioessays* 2010, 32:438–449.

153. Roth VL: The biological basis of homology. In *Ontogeny and Systematics.* Edited by Humphries CJ. New York: Columbia University Press; 1988:1–26.

154. True JR, Haag ES: Developmental system drift and flexibility in evolutionary trajectories. *Evol Dev* 2001, 3:109–119.

155. Kotthaus A: Die Entwicklung des Primordial-Craniums von *Xenopus laevis* bis zur Metamorphose. *Zeit wiss Zool*1933, 144:510–572.

156. De Sá RO, Swart CC: Development of the suprarostral plate of pipoid frogs. *J Morphol* 1999, 240:143–153.

157. Eames BF, Helms JA: Conserved molecular program regulating cranial and appendicular skeletogenesis. *Dev Dyn*2004, 231:4–13.

158. Gould SJ, Lewontin RC: The spandrels of San Marco and the Panglossian paradigm: a critique of the adaptationist programme. *Proc Roy Soc B* 1979, 205:581–598.

159. Olson ME: The developmental renaissance in adaptationism. *Trends Ecol Evol* 2012, 27:278–287.

160. Duboule D: Temporal colinearity and the phylotypic progression: a basis for the stability of a vertebrate bauplan and the evolution of morphologies through heterochrony. *Development* 1994, :Supplement: 135–142.

161. Irie N, Kuratani S: Comparative transcriptome analysis reveals vertebrate phylotypic period during organogenesis.*Nat Commun* 2011, 2:248.

162. Wang Z, Pascual-Anaya J, Zadissa A, Li WQ, Niimura Y, Huang ZY, Li CY, White S, Xiong ZQ, Fang DM, Wang B, Ming Y, Chen Y, Zheng Y, Kuraku S, Pignatelli M, Herrero J, Beal K, Nozawa M, Li QY, Wang J, Zhang HY, Yu LL, Shigenobu S, Wang JY, Liu JN, Flicek P, Searle S, Wang J, Kuratani S *et al*: The draft genomes of soft-shell turtle and green sea turtle yield insights into the development and evolution of the turtle-specific body plan. *Nat Genet* 2013, 45:701–706.

CHAPTER 7

WHY ARE SOME SPECIES OLDER THAN OTHERS? A LARGE-SCALE STUDY OF VERTEBRATES

Laure Cattin, Johan Schuerch, Nicolas Salamin and Sylvain Dubey

Keywords: Species age Intraspecific diversification Latitude Viviparity Oviparity Colour polymorphism

BACKGROUND

If we wish to predict the resilience of a species to current challenges, such as those caused by climate change, one important source of information may be the past history of that taxon. A species that has persisted through very long periods – and thus, survived across a wide range of environmental conditions – may be more resilient than a recently evolved taxon whose vulnerability has not yet been tested to the same degree. The age of a species can then be considered as a consequence of resilience, which, in turn, will be determined by specific phenotypic traits. All else being equal then, we might thus expect "older" species to be more resilient. An important question is then what are the factors that affect a species' age?

Previous research shows that a species' geographic location is linked to how long it has persisted since being formed (defined as the age of species in [1, 2, 3]). On average, terrestrial vertebrate species are older in the Southern than in the Northern Hemisphere (invertebrates, plants, fungi and marine vertebrates have never been considered). The severity of the Last Glacial Maximum was different in these two Hemispheres: extensive ice sheets were almost entirely absent from the Southern Hemisphere, whereas large portions of North America and Eurasia were covered by glaciers [4, 5]. These differences placed fewer constraints on organismal viability, enabling populations to persist over extensive areas in the Southern Hemisphere. As a result, the lack of glaciation reduced the loss of genetic diversity [1, 6]. For similar reasons, the latitudinal distribution of organisms might also impact the age of species. For example, Weir and Schluter [7] found that within New World birds and mammals, the age of species, as well as the divergence between sister species is older towards the tropics.

Intrinsic factors might also influence the resilience of species. In particular, it has been proposed that the reproductive strategies and neonatal behaviour, phenotypic characteristics, as well as the presence of intraspecific phenotypic variations, can have an effect on species ages (e.g. [8, 9, 10, 11, 12]). In this respect, recent studies have shown that intraspecific colour polymorphism allows species to exploit different habitat types or have broader distributions because of potential differences in behaviour, thermoregulation capacities, or prey–predator interactions (e.g. [13, 14, 15]). In addition, a study strictly focusing on snakes highlighted that the age of colour

Citation: Laure Cattin, Johan Schuerch, Nicolas Salamin and Sylvain Dubey, Why are some species older than others? A large-scale study of vertebrates, DOI 10.1186/s12862-016-0646-8

polymorphic species was older than those of monomorphic taxa [12]. In combination, these results suggest that being colour polymorphic enhances the resilience of species. Concerning the reproduction strategies and precociality of neonates, studies on the evolution of viviparity in reptiles suggest that it is a key innovation compared to oviparity, buffering species against the negative impact of past climate fluctuations or enabling them to better exploit cold areas [10, 16, 17]. However, it may also be a dead-end in some circumstances as the transition to viviparity is often irreversible, meaning that cold adapted viviparous species might be at high risk in a context of global warming [17]. Lastly, species with independent (precocial) neonates would be expected to be more resilient. Under changing conditions, independent neonates can move by themselves to find a better environment. In contrast, dependent (altricial) neonates need to be carried during moves, which burdens the parents and lessens their survival rate.

Any investigation of intrinsic factors associated with species age must also consider the indirect effect of body size. It is well known that body size is correlated with generation time and life history traits, as well as with population density. For example, small-bodied species have been shown to be more resilient than large ones (e.g. [9, 18]). Nonetheless, this factor has been neglected in previous studies looking at the age of species.

In the present study, we aim to understand how extrinsic and intrinsic factors influence the resilience of vertebrate taxa and hence their age. Based on a dataset of 600 species worldwide and using phylogenetic comparative methods, we tested for an effect of the geographic location, body size, reproduction mode (oviparity versus viviparity), newborn dependence behaviour (precocial, altricial), and presence of intraspecific colour polymorphism on the timing of the oldest diversification event within species. Our study differs from previous studies by including taxa that are distributed in all areas of the world and we consider in our model all the factors that have been hypothesized to affect the age of species. It is also the first to consider all classes of vertebrates simultaneously and consider for the first time the phylogenetic effects due to shared ancestry between species. It therefore provides a much more complete and clear picture of the factors shaping the resilience of species.

RESULTS

The phylogenetic tree for the 601 vertebrate species (including the shark species *Squalus acanthias* as an outgroup) was congruent with existing knowledge on the relationships of the taxonomic groups included (Fig. 1). Most of the nodes were well supported and the 100 trees sampled from the posterior distribution of trees that were used to account for phylogenetic uncertainty showed small differences (mainly occurring in nodes close to the tips of the tree). The divergence times obtained by maximum likelihood are within the 95 % interval of the posterior samples obtained with BEAST. The two analyses resulted in very similar divergence times and we therefore use the dates obtained by maximum likelihood for the remaining of the analyses. The dates obtained for the main lineages of vertebrates are also congruent with recent large scale phylogenetic reconstructions [19, 20, 21, 22]. We used phylogenetic GLS to test the factors affecting the age of species, which was used as the response variable in our analyses. The backward model selection retained three single factors and one interaction in the final phylogenetic GLS model (Fig. 2a; Additional file 1).

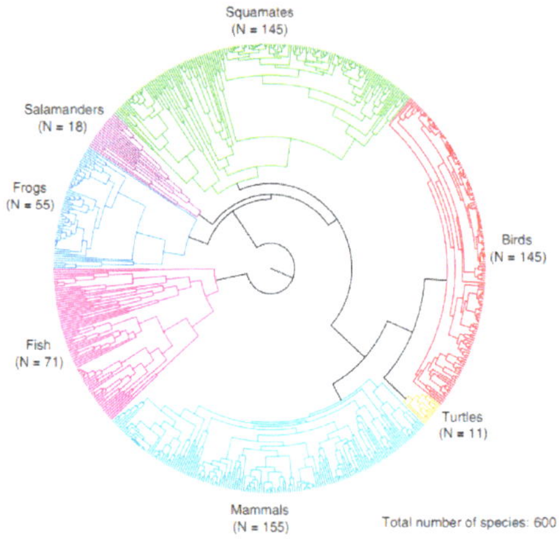

Figure. 1 Best tree obtained from MrBayes and used for the comparative analyses

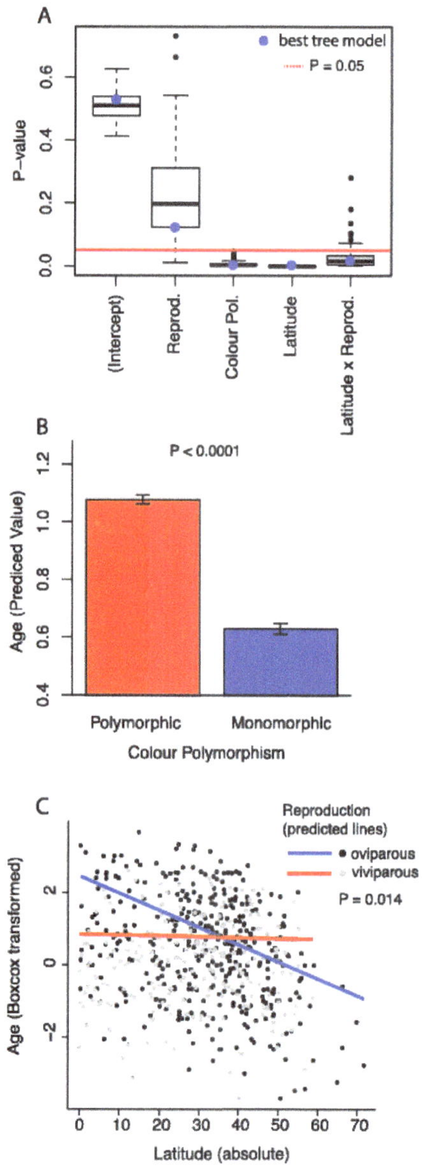

Figure. 2 a Distribution of *p*-values for the different explanatory factors retained in our final model with the age of species as response

variable, for 100 trees sampled from the posterior distribution of trees obtained from MrBayes (best tree: blue dot); Relationships between the age of species and (**b**) the presence or absence of intraspecific colour polymorphism and (**c**) the interaction between the mean latitudinal distribution of species and their reproduction mode (viviparous versus oviparous).

Colour polymorphism and latitude were highly significant (p-value<0.001; Fig. 2a), while mode of reproduction was maintained due to its significant interaction with latitude. Polymorphic species had a much higher resilience than monomorphic ones (fig. 2b) and, on average, the age of the polymorphic species was (when transformed back into the original scale) 1.86 Myr older than the age of monomorphic ones (Fig. 2b; Additional file 2). Latitude was also significant, but its effect was influenced by the mode of reproduction. Oviparous species showed a strong latitudinal trend, with species living far from the Equator being younger than species living in tropical climates (Fig. 2c). In contrast, viviparous species did not show any latitudinal trend in species age (Fig. 2c).

DISCUSSION

Our study, based on a dataset of 600 vertebrate species, allowed us to identify several extrinsic and intrinsic factors that influence significantly the age of vertebrate species. It also clarified results from previous studies that were not based on comparative methods, focused on particular taxa or considered fewer factors (see e.g. [1, 2, 3, 7, 12]). Overall, absolute mean latitude of the species distribution was negatively correlated with species age and colour-polymorphic species were older than monomorphic species. Moreover, the age of oviparous species decreased with latitude, whereas no such pattern was apparent for viviparous ones.

The latitudinal gradient found in this study is consistent with the results of Weir and Schluter [7], who found that within New World birds and mammals, the age of species, as well as the divergence between sister species, is younger towards the poles. Using interspecific divergences, they estimated that the highest recent speciation and extinction rates occurred at high latitudes and declined toward the tropics. This pattern is likely explained by the presence of more stable climatic conditions in the tropics than in temperate areas during Pleistocene climatic fluctuations [5]. According to Weir and Schluter [7], this climatic stability leads to a lower species turnover in the tropics compared to temperate areas, resulting in a lower diversity in the latter. Based on the similarity between our results and those of Weir and Schluter [7] in term of intraspecific latitudinal gradient, this pattern might be generally valid for vertebrate species from both Hemispheres. However, the situation is likely more complex. Indeed, the age of viviparous vertebrate species was not influenced by their mean latitudinal distribution, whereas the age of oviparous species decreased towards the poles. This result suggests a higher resilience of viviparous species to cold climate, and can be attributed to their buffering capacity against inadequate environmental conditions during the embryogenesis. This is particularly true for heliothermic vertebrates, for which a cold year might result in an absence of reproduction if favourable nesting sites are absent. In contrast, viviparous species might be able to better deal with poor environmental conditions by modifying their sun-basking activities (for a review see e.g. [10, 16, 23]). For instance, Pincheira-Donoso et al., [17], found that, in a widespread South American lizard genus (*Liolaemus*), the evolution of viviparity was associated with a radiation into cold climates. This argument is not relevant to birds or placental and marsupial mammals (as opposed to egg-laying monotremes) because their breeding type is constant and they are more widely distributed. However, the argument can be extended to other ectothermic vertebrates (fish and amphibians).

Body colour plays a major role in the evolution of organisms as it is involved, for instance, in thermoregulation, prey–predator interaction, behaviour, protection against UV-light or abrasive agents [24]. Therefore, the presence of intraspecific colour polymorphism may promote the ecological success of species as previously shown for squamates [14, 15], amphibians [25] and owls [26], through their capacity to exploit a larger range of habitat types. In turn, polymorphic species might be more resilient to past and future climatic fluctuations. Our results are also in agreement with the study of Pizzatto & Dubey [12], which found that colour-polymorphic snake species were older than monomorphic species.

Recent studies focusing on temperate vertebrate taxa highlighted that species from the southern Hemisphere are older than those from the northern, and that differences in past climate are likely responsible for this pattern [1, 2, 3]. Interestingly, the Hemisphere of origin is not impacting the age of species in our study, which is likely due to the inclusion of additional intrinsic and extrinsic factors and to the use of phylogenetic comparative methods. Indeed, when the Hemisphere of origin is included in our analyses without accounting for the phylogenetic relationships between taxa, southern species remained significantly older than northern ones (see Additional file 3). The same is true when strictly temperate species of our dataset are included (the use of a comparative method leading to a marginally significant result; Additional file 3). These differences illustrate the importance of considering an array of factors in such analyses. In addition, the phylogenetic relationships of studied taxa might deeply influence such studies, keeping in mind that northern and southern Hemisphere temperate fauna are distantly related genetically, numerous families and genera being solely distributed in one of the two Hemispheres.

CONCLUSIONS

In conclusion, our findings suggest that intrinsic and extrinsic variables might contribute to the resilience of species to past environmental changes and hence are relevant in a context of climate change. Overall and independently of current threats, organisms which exhibit ancient intraspecific diversification events are more likely to deal with future modifications of their environment, those species being colour polymorphic, viviparous, or living at low latitudes.

METHODS

Species Sampling and Traits

We gathered published phylogenetic and phylogeographic studies for vertebrate species using searches within Scopus and ISI Web of Knowledge. We defined the age of a species (in Myr) as the oldest intraspecific diversification event, which is an estimate of the amount of time since extant populations last shared a common ancestor [1, 2, 3, 7]. Species ages were obtained directly from the studies included in our dataset and hence were not estimated in the present study. In addition, only studies including a reasonable number of individuals and covering most of the distribution of the species have been included. We excluded non-monophyletic species and used the Catalogue of Life [27] to eliminate synonyms and ambiguities. Our final sampling contained species from 71 freshwater fish, 73 amphibians, 156 reptiles, 145 birds and 155 mammals (See Fig. 1 & Additional file 4).

We collected for each species their geographical distributions (Additional file 4), as represented by the minimum and maximum latitude taken from IUCN distributional polygon data. We calculated the mean latitude (in absolute values) and latitudinal range with ArcGIS® (ESRI 2006). We further categorized the mean latitude of each species to allocate them to Northern (between 10° and 90° N) or Southern (between 10° and 90° S) Hemispheres. Tropical species (between 10°N and 10°S) as well as species found on both Hemispheres were defined as intermediates. Latitude is considered here as a covariate that can represent several aspects of the life-history that cannot be easily associated with the more than 600 species sampled in our study.

We also collected data on life-history traits from published taxonomies and available databases. As our measure of mean adult size, we used snout-vent length for amphibian, reptile and mammal species, common size for fish species and length for bird species. Both sexes were pooled together and averaged. We also recorded the breeding type for each species and we combined ovoviviparous and viviparous species. The mean litter or clutch size was defined as the mean number of offspring for one female and we only considered the number of offspring in a single clutch/litter when multiple birthing took place in one year. We considered as well the behaviour of newborns and modified the scale of Starck & Ricklefs [28] into four levels to account for the larger taxonomic breadth of our study (Precocial - neonates eat by themselves and move freely; semi-precocial - neonates move freely by themselves but rely on the parents for food; semi-altricial - neonates unable to move freely by themselves, but still have the capacity to cling or act to survive; altricial - neonates are fully dependant on their parents). Finally, we defined a species as colour-polymorphic when it exhibited at least two different colour morphs within the same sex and age class. Thus, at least one of the sexes had to be polymorphic for the species to be considered polymorphic. No distinction was made between geographical colour variation and within-population colour-polymorphism. See Additional file 5 for factors retained in our final analyses.

Phylogenetic Analysis

The phylogenetic tree for the 601 species sampled was based on four mitochondrial genes (cytochrome b, COI, ND2, and ND4) and one nuclear gene (Rag1), but not all species in the dataset were represented for the five genes (TreeBase accession number). The shark *Squalus acanthias* was used as an outgroup. The sequences were downloaded from GenBank and aligned using MUSCLE [29]. The best model for each gene defined using Akaike Information Criterion (AIC) as implemented in the phymltest function from the R package APE [30]. We inferred the Maximum Likelihood tree using the BEST tree search algorithm as implemented in PhyML 2.3 [31] and we performed 100 bootstrap replicates. We also inferred a Bayesian phylogenetic tree using MrBayes 3.2 [32] and we allowed the model of substitution to take different parameter values for each gene present in the concatenated dataset. The MCMC was run for 50 millions generations and we sampled trees and parameter values every 1000 generations. We repeated the analyses twice to assess the convergence of the chain.

We estimated divergence times using 11 fossil calibration points obtained from the literature (Additional file6). Given the size of the phylogenetic tree, a full estimation of the tree topology and the divergence times was not practical. We therefore decided to first use a Maximum Likelihood implementation of a relaxed molecular clock [33] as implemented in the function chronos from the R package APE [30]. We also sampled 100 trees from the posterior distribution of trees obtained by MrBayes and repeated the divergence time analysis to obtain a set of calibrated trees. Second, we used BEAST v. 1.8 [34] to estimate the divergence times using a log-normal relaxed molecular clock while fixing the tree topology to the one obtained by MrBayes. We applied the 11 fossil calibration points using log-normal priors. The means and standard deviations of these prior distributions were selected to include the fossil uncertainties in the 95 % of their density. We used a yule process for the prior on divergence times and ran the MCMC for 10 million generations. We assessed the convergence of the chain using Tracer v. 1.7.5 by looking at the trace of the log-likelihood and the ESS values for each parameter.

Comparative Analysis

We used general least squares (GLS) to understand the effects of intrinsic and extrinsic factors on the age of species. We modified species ages with a Box-Cox transformation (lambda=0.1) and we included the average latitude (in absolute value), Hemisphere, reproduction mode, newborn behavior, body size (Box-Cox transformed, lambda=−0.3) and the presence of colour polymorphism as covariables in the GLS analysis. We started with a full model that included the single factors and all pairwise interactions. We corrected the correlation structure of the GLS by considering the evolutionary relationships between species and we assumed a Brownian correlation structure based on the phylogenetic trees (lambda parameter=1). We performed all analyses using the gls function from the R package NLME [35] and used backward model selection to test the factors. Model validation was performed using graphical tools (as proposed in [36]). We finally tested the effect of phylogenic tree construction on the best model output by repeating the GLS analysis 100 times sampling each time a new phylogenetic tree randomly from their posterior distribution.

DECLARATIONS

Acknowledgements

We thank Richard Shine for helpful comments. This work was financially supported by the Swiss National Science

Foundation (SNSF; n° PZ00P3_136649 and PZ00P3_157244 to SD).

Authors' Contributions

SD, NS, LC, JS contributed with the conceptual development of the work and the writing of the manuscript and carried out the analyses. All authors read and approved the final version of the manuscript.

REFERENCES

1. Dubey S, Shine R. Geographic variation in the age of temperate-zone reptile and amphibian species: southern Hemisphere species are older. Biol Letters. 2011;7:96–7. doi:10.1098/rsbl.2010.0557.
2. Dubey S, Shine R. Are reptile and amphibian species younger in the Northern Hemisphere than the Southern Hemisphere? J Evol Biol. 2012;25:220–6. doi:10.1111/j.1420-9101.2011.02417.x.
3. Tingley R, Dubey S. Disparity in the timing of vertebrate diversification events between the northern and southern hemispheres. BMC Evol Biol. 2012;12:244. doi:10.1186/1471-2148-12-244.
4. Markgraf V, McGlone M, Hope G. Neogene paleoenvironmental and paleoclimatic change in southern temperate ecosystems – a southern perspective. Trends Ecol Evol. 1995;10:143–7. doi:10.1016/S0169-5347(00)89023-0.
5. Hewitt GM. The genetic legacy of the Quaternary ice ages. Nature. 2000;405:907–13. doi:10.1038/35016000.
6. Byrne M. Evidence for multiple refugia at different time scales during Pleistocene climatic oscillations in southern Australia inferred from phylogeography. Quaternary Science Rev. 2008;27:2576–85. doi:10.1016/j.quascirev.2008.08.032.
7. Weir JT, Schluter D. The latitudinal gradient in recent speciation and extinction rates of birds and mammals. Science. 2007;315:1574–6. doi:10.1126/science.1135590.
8. Balon EK. Dynamics of biodiversity and mechanisms of change: A plea for balanced attention to form creation and extinction. Biol Cons. 1993;66:5–8. doi:10.1016/0006-3207(93)90129-O.
9. Isaac JL, Vanderwal J, Johnson CN, Williams SE. Resistance and resilience: quantifying relative extinction risk in a diverse assemblage of Australian tropical rainforest vertebrates. Divers Distrib. 2009;15:280–8. doi:10.1111/j.1472-4642.2008.00531.x.
10. Lynch VJ. Live-birth in vipers (Viperidae) is a key innovation and adaptation to global cooling during the Cenozoic. Evolution. 2009;63:2457–65. doi:10.1111/j.1558-5646.2009.00733.x.
11. Gardner JL, Peters A, Kearney MR, Joseph L, Heinsohn R. Declining body size: a third universal response to warming? Trends Ecol Evol. 2011;26:285–91.
12. Pizzatto L, Dubey S. Colour-polymorphic snake species are older. Biol J Linn Soc. 2012;107:210–8. doi:10.1111/j.1095-8312.2012.01936.x.
13. Shreeve TG. Microhabitat use and hindwing phenotype in *Hipparchia semele* (Lepidoptera, Satyrinae): thermoregulation and background matching. Ecol Entom. 1990;15:201–13. doi:10.1111/j.1365-2311.1990.tb00801.x.
14. Forsman A, Aberg V. Associations of variable coloration with niche breadth and conservation status among Australian reptiles. Ecology. 2008;89:1201–7. doi:10.1890/07-1670.1.
15. Forsman A, Åberg V. Variable coloration is associated with more northerly geographic range limits and larger range sizes in North American lizards and snakes. Evol Ecol Res. 2008;10:1025–36.
16. Shine R. Does viviparity evolve in cold climate reptiles because pregnant females maintain stable (not high) body temperatures? Evolution. 2004;58:1809–18. ISSN: 0014–3820.
17. Pincheira-Donoso D, Tregenza T, Witt MJ, Hodgson DJ. The evolution of viviparity opens opportunities for lizard radiation but drives it into a climatic cul-de-sac. Global Ecol Biogeogr. 2013;22:857–67. doi:10.1111/geb.12052.
18. Damuth J. Population density and body size in mammals. Nature. 1981;290:699–700. doi:10.1038/290699a0.
19. O›leary MA et al. The Placental Mammal Ancestor and the Post-K-Pg Radiation of Placentals. Science. 2013;339:662–7. doi:10.1126/science.1229237.
20. Mulcahy DG et al. Estimating divergence dates and evaluating dating methods using phylogenomic and mitochondrial data in squamate reptiles. Mol Phylogenet Evol. 2012;65:974–91. doi:10.1016/j.ympev.2012.08.018.
21. Zhang P, Zhou H, Chen YQ, Liu YF, Qu LH. Mitogenomic Perspectives on the Origin and Phylogeny of Living Amphibians. Syst Biol. 2005;54:391–400. doi:10.1080/10635150590945278.
22. Zhang GJ et al. Comparative genomics reveals insights into avian genome evolution and adaptation. Science. 2014;346:1311–20. doi:10.1126/science.1251385.
23. Shine R. Life-history evolution in reptiles. Annu Rev Ecol Evol, S. 2005;36:23–46.
24. Dubey S, Roulin A. Evolutionary and biomedical consequences of internal melanins. Pigment Cell Melanoma R. 2014;27:327–38. doi:10.1111/pcmr.12231.
25. Forsman A, Hagman M. Association of coloration mode with population declines and endangerment in Australian frogs. Cons Biol. 2009;23:1535–43. doi:10.1111/j.1523-1739.2009.01244.x.
26. Roulin A, Burri R, Antoniazza S. Owl melanin-based plumage redness is more frequent near than away from the equator: implications on the effect of climate change on biodiversity. Biol J Linn Soc. 2011;102:573–82. doi:10.1111/j.1095-8312.2010.01614.x.
27. Roskov Y, Abucay L, Orrell T, Nicolson D, Flann C, Bailly N, Kirk P, Bourgoin T, DeWalt RE, Decock W, De Wever A. Species 2000 & ITIS Catalogue of Life, 2016 Annual Checklist. 2016. Digital resource at www.catalogueoflife.org/annual-checklist/2016. Species

2000: Naturalis, Leiden, the Netherlands. ISSN 2405-884X.

28. Starck JM, Ricjlefs RE. Avian Growth and Development: Evolution Within the Altricial-precocial Spectrum. Oxford University Press. 1998.

29. Edgar RC. MUSCLE: multiple sequence alignment with high accuracy and high throughput. Nucleic Acids Res. 2004;32:1792–7. doi:10.1093/nar/gkh340.

30. Paradis E, Claude J, Strimmer K. APE: Analyses of Phylogenetics and Evolution in R language. Bioinformatics. 2004;20:289–90. doi:10.1093/bioinformatics/btg412.

31. Guindon S, Gascuel O. A simple, fast, and accurate algorithm to estimate large phylogenies by maximum likelihood. Syst Biol. 2003;52:696–704. doi:10.1080/10635150390235520.

32. Ronquist F et al. MrBayes 3.2: efficient Bayesian phylogenetic inference and model choice across a large model space. Syst Biol. 2012;61:539–42. doi:10.1093/sysbio/sys029.

33. Paradis E. Molecular dating of phylogenies by likelihood methods: A comparison of models and a new information criterion. Mol Phylogenet Evol. 2013;67:436–44. doi:10.1016/j.ympev.2013.02.008.

34. Drummond AJ, Suchard MA, Xie D, Rambaut A. Bayesian phylogenetics with BEAUti and the BEAST 1.7. Mol Biol Evol. 2012;29:1969–73. doi:10.1093/molbev/mss075.

35. Pinheiro J, Bates D, DebRoy S, Sarkar D, the R Development Core Team. Nlme: Linear and Nonlinear Mixed Effects Models. R package version. 2013;3:1–111.

36. Zuur AF, Ieno EN, Walker N, Saveliev AA, Smith GM. In: Gail M, Krickeberg K, Samet JM, Tsiatis A, Wong W, editors. Mixed Effects Models and Extensions in Ecology with R. New York: Springer; 2009.

CHAPTER 8

EFFECT OF TEMPERATURE AND GLIA IN BRAIN SIZE ENLARGEMENT AND ORIGIN OF ALLOMETRIC BODY-BRAIN SIZE SCALING IN VERTEBRATES

Yuguo Yu, Jan Karbowski, Robert NS Sachdev and Jianfeng Feng

Keywords: Brain-body size Allometric scaling Glia Body temperature Brain enlargement Metabolism

BACKGROUND

Brain size is a traditional metric of intelligence. Species with larger brain-body size ratios are often associated with greater intelligence, better tool-making skills, more social interactions, and other enhanced characteristics [1]. Although many correlations between large brains and complex behavioral measures have been posited over the past century, the driving forces and key factors underlying the development of larger brain-body size ratios in mammals and birds are still an open issue [2]-[13]. The evolution of large brains in mammals and birds coincided with the development of endothermy [14] at the end of the Triassic period (~200 million years ago) [15] during which time the first mammals evolved from therapsids (i.e., the mammal-like reptiles) and has continued in primates and humans, which have the largest brain-to-body weight ratios [16], to the present day [17]. In 1891, Snell [18] first described the allometric scaling relationship that emphasizes the dependence of brain size on body size. The most striking observation is that mammals and birds have developed larger brains that are approximately 5–50 times heavier than those of the lower-level vertebrates (e.g., reptiles, fish and amphibians) with similar body mass. It is surprising that mammals and birds developed larger brains, because, brain tissue is extremely demanding in terms of metabolic consumption, because the neuronal electronic signaling that occurs within the brain may have a per-neuron energy cost as high as ten times that of other cells in the body. This difference has attracted much attention to the search for links between brain size, body size and metabolic production. Martin [19] proposed that brain size is linked to maternal metabolic turnover via a 3/4 allometric scaling law [20]. This work indicated that the total amount of energy available from the body to supply the brain may constrain the size of the brain; however, species with larger brains do not have basal metabolic rates significantly higher than those in species with smaller brains [4],[9],[21]-[23]. There is also no significant difference in the percentage of energy supplied by the body to the brain between endotherms and ectotherms [24],[25]. Given these similar metabolic processes and energy supply ratios, why do mammals and birds have larger brains than fish and reptiles of similar weight? How do mammals and birds meet the energy demands of brains that are 5–50 times larger? Although much research has focused on the several-fold differences in the sizes of the brains of primates and those of low-level mammals by considering the improved energy source acquisition styles of primates, such as improvements

Citation: Yuguo Yu, Jan Karbowski, Robert NS Sachdev and Jianfeng Feng, Effect of temperature and glia in brain size enlargement and origin of allometric body-brain size scaling in vertebrates s, DOI: 10.1186/s12862-014-0178-z.

in food hunting style due to the use of tools, the consumption of cooked food, and differences in metabolic allocation, few studies have focused on the proximate causes of the differences in brain size between lower-level animals (e.g., fish, reptiles and amphibians) and higher-level animals (e.g., primates, mammals and birds). What is the role of temperature and metabolism in promoting or constraining brain size? What is the relative contribution of neuron and glial densities in increasing brain size?

In the present study, we re-examined the relationships between temperature, metabolism, neuronal density, energy cost of individual neurons, brain size and body size across of all the major types of vertebrates. Here we specifically focus on the critical effect of temperature and growth of glia cells in the development of large brain-body size ratio in mammals and birds and origin of allometric body-brain size scaling in vertebrates. We then develop a general model to describe the intrinsic relationship among the examined factors.

RESULTS

Body temperature, metabolism, external temperature and brain size

We first re-plotted the relationships between brain size and body size across different species (Figure 1A and Additional file 1: Table S1). In each type of species, the relationship between brain size and body size is highly ordered. Brain size increases with body size, and the relationship can be fit using the following allometric equation [16],[18]:

$$W=CP^{\alpha} \qquad (1)$$

Figure 1. Allometric relationships of the brain and body weights of vertebrate animals on a log-log plot. A. Endothermic animals included 678 mammal species [3],[16],[22],[61] (red open circle), 600 bird species [10] (cyan triangle) and 33 insectivore species [16] (purple cross). Ectothermic animals included 110 fish species [44] (black open circle), 71 reptilian species [11] (black open diamond), and 86 amphibian species [11] (black open triangle). The fitting functions are $W = 0.078P^{0.717}$ (adjusted- $R^2 = 0.95$, $p < 0.01$) for endotherms and $W = 0.014P^{0.578}$ (adjusted- $R^2 = 0.91$, $p < 0.01$) for ectotherms. All these datasets are provided in Additional file 1: Table S1. **B**. The relationship of the brains and bodies of randomly selected fish. Data from 30 polar-water fish species (blue, temperature ~1°C), 70 temperate species (green, 10–20°C), 88 tropical species (red, temperature 20–30°C) and 17 shark species that live in sub-tropical waters (red, core body temperature range 20–30°C [26],[27]) were randomly selected from a fish database (http://fishbase.org) [44]. The fitting functions are $W = 0.0036P^{0.59}$ (adjusted- $R^2 = 0.82$, $p < 0.01$) for polar-water fish, $W = 0.0078P^{0.64}$ (adjusted- $R^2 = 0.92$, $p < 0.01$) for temperate-water fish, and $W = 0.0118P^{0.7}$ (adjusted- $R^2= 0.94$, $p < 0.01$) for tropical-water fish and the sharks. All these datasets are provided in Additional file 2: Table S2. **C**. Brain size as a function of temperature for fixed body masses, i.e., 10 grams and 1000 grams respectively. The open triangles represent the averaged data from 29 fish species with body masses of 5–15 grams, and the open squares represent the average of 33 species with body masses of 500–1500 grams from three temperature conditions. The data for 38°C are from 46 endotherms with body masses of 5–15 grams (solid square) and 52 endotherms with body masses of 500–1500 grams (solid triangle). The fitting equations are $\log_{10}W = 0.205P^{0.0323}$ for P = 1000 g (adjusted- $R^2 = 0.98$, $p < 0.01$ for only fish data) and $\log_{10}W = 0.0095P^{0.0343}$ for 1000 g (adjusted- $R^2 = 0.91$, $p < 0.01$, only for the fish data). **D**. The brain size ratios as a function of body sizes: $Brain_{endotherm}/Brain_{ectotherm}$ (black line), $Brain_{tropic\ fish}/Brain_{polar\ fish}$ (black dashed line) and $Brain_{primate}/Brain_{mammal}$ (black dotted line).

where exponent α is the allometric slope, C is a constant, and W and P are brain and body weight, respectively. Analyses have shown that C = 0.078 and α = 0.689 for endotherms (including mammals, birds, insectivores, primates, dolphins and humans), and C = 0.014 and α = 0.578 for ectotherms (primarily fish, reptiles and amphibians). The parameters C and α vary across species (see Additional file 1: Table S1). A closer examination of the distributions of body mass versus brain mass reveals that the warm-blooded, endothermic animals with relatively constant body temperatures of approximately 8°C (e.g., mammals, birds, primates and dolphins) form one distinct distribution, and cold-blooded ectothermic animals with body temperatures that vary with the environment and average approximately 16°C (i.e., fish, reptiles and amphibians) form a clear second distribution. Warm-blooded species have brains that are 4–43 times larger than those of ectothermic animals. The fact that the most apparent difference between the two groups is temperature suggests that it is a key factor that correlates with brain size.

If temperature is important, we might predict that cold-blooded species living in disparate environmental temperatures might also have different brain sizes. To examine whether the extrinsic, i.e., environmental, temperature also explains the variation in brain size, we used fish brain and body-size data from locations of different temperatures, obtained from a fish database (http://fishbase.org), and plotted the body-brain mass relationships. The fish and environments examined included the following: 1) 35 fish species living in polar conditions in which the average temperature was ~1°C, 2) 70 fish species living in temperate conditions with an average water temperature of ~15°C, 3) 88 species living in tropical conditions with an average water temperature of 25°C, and 4) 17 shark species living in sub-tropical waters that maintain their body temperatures at 20-30°C via steady contractile activity of muscles [26],[27] (Additional file 2: Table S2; these data were randomly selected from the fish database). Surprisingly, we found that the fish brain-to-body mass ratio data fell into 3 distinct groups (Figure 1B). Fish living in warm tropical water had brains that were 2–5 times larger than those of the temperate-water fish and 3–15 times larger than those of the fish living in cold polar water. The fit equation for the fish species living in tropical conditions had an allometric slope of 0.7 and a constant of 0.0118 (red line and symbols), the equation for fish living in the temperate waters had an allometric slope of 0.64 and a constant of 0.0078 constant (green line and symbols), and the fish living in the polar waters had the shallowest allometric slope (0.59) and a constant of 0.0036.

Figure 1C directly quantifies the allometric relationship between the environmental temperature (from 1°C to 25°C) of the fish and their average brain mass for two body masses (10 g and 1,000 g). The fit equations indicate that fish brain mass more than doubles for each 10°C increase in temperature. As a comparison, the average mammalian brain sizes in animals weighing 10 g and 1,000 g are plotted in this figure (with a constant temperature of 38°C). This figure shows a nice fit between body mass and brain temperature for three points, but not for the largest brain sizes and highest temperatures. The data from the larger brain sizes/higher temperatures falls above the fit lines, indicating that temperature alone may not be the sole contributor to the development of larger brains. Other factors may contribute to the development of larger brains in mammals.

Together, these data indicate that temperature may be a critical factor in the marked changes in brain size during evolution in different climate conditions. Specifically, for a given body mass, warmer living conditions should result in larger brains.

In addition to this temperature-driven brain enlargement, the difference in brain sizes of endotherms and ectotherms increases as body size increases (Figure 1D). For small body sizes (below 1 gram), the brains of endotherms are atleast 4 times larger than those of ectotherms (black line in Figure 1D). Similarly, tropical-water fish have brains that are at least 3 times larger than those of cold-water fish (dashed line). In contrast to these effects in animals with small body sizes, for animals with body sizes over 100 kilograms, the brains of endotherms are approximately 20 to 40 times larger than those of ectotherms, and tropical-water fish have brains that are more than 10 to 15 times larger than those of cold-water fish. In comparison, the brains of primates are, relatively consistently, 2–3 times larger than those of other mammals, and this difference is less dependent on body size differences (dotted line in Figure 1D).

The emergence of intrinsically constant warm body temperatures together with an increasing brain size, with increasing body mass may be related to a temperature-dependent regulation of body metabolism. It is possible, that changes in body or environmental temperature affects the metabolic rate and consequently affected the evolution of large brains. This effect may depend on body mass.

The basal metabolic rate of the body (M_{body}) has been found to be governed primarily by two interacting processes [28]: the Boltzmann-Arrhenius factor, which describes the temperature dependence of biochemical processes [29], and an allometric relation that describes how biological rate processes scale with body size for all vertebrates [20],[28],[30]-[42].

$$M_{body} = C_1 P^{\gamma} e^{-\frac{E_v}{kT}} \quad (2)$$

where P is body weight, C_1 is a constant (whose value is different for different species), γ is the allometric scaling factor, T is the absolute temperature, E_v is the activation energy (defined as the minimum amount of energy required to initiate a particular process. It is usually used in the context of chemical reactions, i.e., as the minimum amount of energy that chemical reactants must possess before they can undergo a chemical reaction. For the life metabolic process regulated by the temperature, E_v is expected to be around 0.65 electron Volts (eV) [28]), and k is Boltzmann's constant (8.6×10^{-5} eV/K, here K is the temperature in Kelvin units). Earlier analyses have revealed that $C_1 = 0.0305$ and $\gamma = 0.682$ for endotherms (these data, which are primarily from mammals and birds, were taken from references [42],[43]), measured and scaled at 38°C, and $C_1 = 0.000525$ and $\gamma = 0.81$ for ectotherms when normalized to a temperature of 16°C (see Figure 2A and Additional file 3: Table S3); these findings are consistent with those of recent reports [37],[42],[44]. When the metabolic rates of the endotherms are normalized to a common temperature of 16°C using Eq. (2), the metabolic rates of the endotherms (e.g., mammals and birds) remain several times higher than those of ectotherms (e.g., fish, amphibians and reptiles), particularly for animals with small body sizes (Figure 2B). In large animals, the metabolic rates of the endotherms and ectotherms were similar (Figure 2B and C). These findings suggest that at similar metabolic rates, large endotherms should support brain masses several times greater than those of ectotherms (Figure 2D), given that brain size differences increase with body size among endotherms and ectotherms (Figure 1D).

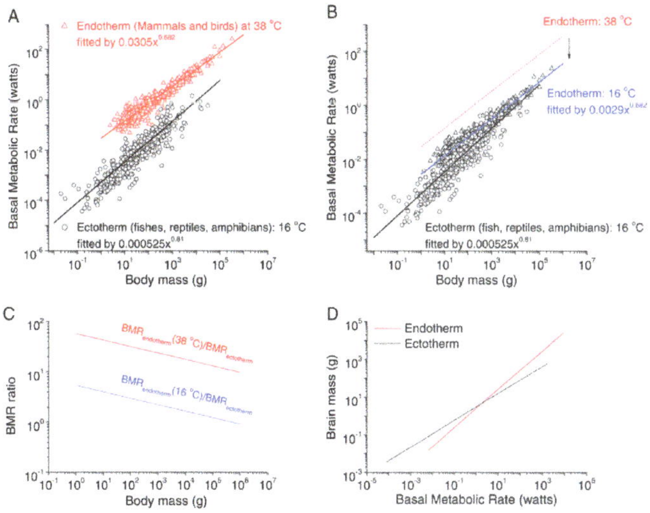

Figure 2. Temperature-regulated basal metabolic rate (BMC) (watts) and brain size (gram). A. Allometric relationships between basal metabolic rate (BMR) (watts) and body mass (gram) for the endotherms (including 525 mammals [43] and 320 birds [37],[42]) (plotted in red triangles) and the ectotherms (including 265 fishes [37],[42],[44], 155 reptiles [37],[42] and 158 amphibians [37],[42]) (plotted in black open circles). The BMR versus body mass data were well fit by the equation $0.0305\times^{0.682}$ (adjusted-$R^2 = 0.933$, $p < 0.01$) for the endotherms and by $0.00525\times^{0.81}$ (adjusted- $R^2 = 0.91$, $p < 0.01$) for the ectotherms. The datasets are given in Additional file 3: Table S3. All the data plotted here for ectotherms were normalized to the temperature of 16°C, and the data for the endotherms were normalized to 38°C based on Eq. (2) (see also the methods in references [28],[37]) to minimize the errors induced by measuring temperatures. B. When the metabolic rate data for the endotherms were normalized to 16°C, they merged into the group of ectothermic metabolic data. However, there some difference between these groups remained, particularly for the small animals. C. The BMR ratios as a function of body size for the endotherms to ectotherms was $BMR_{endotherm}$(38°C)/$BMR_{ectotherm}$ (16°C) (plotted in red line) and $BMR_{endotherm}$(16°C)/$BMR_{ectotherm}$ (16°C) (blue line). D. The brain masses (grams) as a function of body BMRs (watts) for the ectotherms (black line) and endotherms (red line) were based on the fit equations in Figures 1A and 2A.

These analyses suggest that temperature-dependent body metabolism alone cannot explain the differences in brain size among endotherms than among ectotherms. These results also indicate that other factors may play a role in explaining the differences in brain size between warm blooded endothermic animals and cold blooded exothermic animals. It is possible that neural, glial density or the energy demands on these cells are dependent on body temperature.

A general Model Linking Brain and body sizes to Metabolism and Temperature

In this section, we derive a general model to quantify the relationships between brain size, temperature-dependent metabolism, individual neuronal metabolism and body size to account for the origin of Eq. (1). We began by considering the relationship between brain mass and the number of neurons and supporting cells in the brain. Brain mass (W) is determined by the total number of neurons and supporting cells; hence $W = Nw + N_1 w_1$, where N and w are the number and weight of brain neurons, respectively, and N_1 and w_1 are the corresponding quantities for the supporting cells. If we assume that the weight and number of the supporting cells are proportional to those of the neurons, i.e., if $N_1 = k_1 N$, and $w_1 = k_2 w$, then the relationship between brain mass and neurons simplifies to $W = C_2 N$, where $C_2 = (1+k_1 k_2)w$. This theoretical picture in which W is proportional to N is supported by a recent analysis of mammalian and primate brains (see also Figure 3A and Additional file 4: Table S4). Analyses of these data yielded.

$$W=C_2N^{\beta} \quad (3)$$

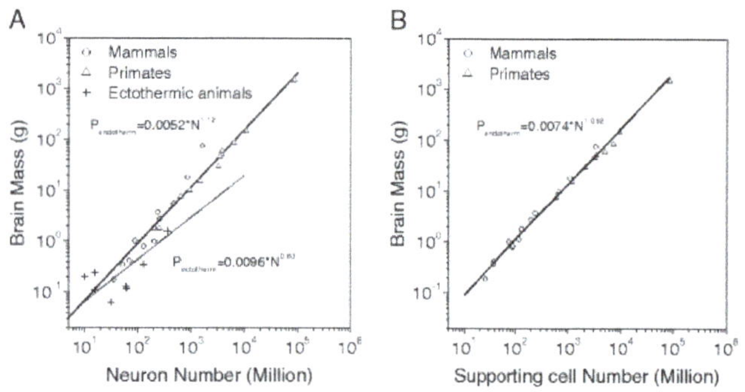

Figure 3. Scaling relationship between the total number of neurons or supporting cells in the brain and brain mass. A Brain mass (in gram) compared with the total number of neurons in the brain for 12 low-level mammal species [47],[62] (solid circle), 11 primates [46],[63] (solid triangle) and 7 ectotherms [44],[48]-[50]. The fit function (in black line) gives $W = 0.0052\ N^{1.12}$ (adjusted- $R^2 = 0.94$, $p < 0.01$) for the endotherms (where $P_{endotherm}$ is the brain mass in grams, N is total number of neurons in the brain in millions, and $P_{ectotherm} = 0.0096\ N^{0.82}$ (adjusted- $R^2 = 0.65$, $p < 0.05$) for the endotherms. **B** Brain mass in grams compared with the total number of non-neuron cells in the brain for 12 low-level mammal species [47],[62] (solid circle), 11 primates [46],[63] (solid triangle) and 7 ectotherms [44],[48]-[50]. The fit functions (in black lines) have $P_{endotherm} = 0.0074\ N^{1.092}$ (adjusted- $R^2 = 0.993$, $p < 0.01$) for the endotherms (where $P_{endotherm}$ is brain mass in grams, and N is the total number of non-neuron cells in millions. The datasets for this figure are given in Additional file 4: Table S4.

where β is close to 1 (here $C_2 = 0.0052$ and $\beta = 1.12$ for both rodents and primates) as reported in recent studies [23],[45]-[47], [63]. The numbers of supporting non-neuronal cells in endotherms are proportional to the numbers of neurons and are also strongly correlated with brain mass (Figure 3B). The supporting cells play an important role in providing mechanical and trophic support to neurons. Additionally, data from several ectothermic animals [44],[48]-[50] indicate that that $C_2 = 0.0096$ and $\beta = 0.83$ (see Additional file 4: Table S4; the numbers of non-neuronal cells are not available for ectotherms).

Neurons are the brain's major energy consumer, and most (~80%) of the brain's energy is consumed by signaling processes within neurons [51],[52]. Brains are energetically costly. Experimental investigations have shown that the brain's metabolic consumption accounts for 1-10% of the total body metabolic rate (M_{body}) in most animal species, although brain mass accounts for no more than 1% of an animal's body weight [24],[25]. Hence, the total energy required to support the brain can be quantified as

$$M_{brain} = \varphi M_{body} \quad (4)$$

where M_{brain} is the brain's basal metabolic rate, M_{body} is the body's basal metabolic rate as defined in Eq. (2), and φ is a variable that varies with species. Re-examination of the data reported by Mink et al. (1981) suggested that φ can be described by the following equation for both ectotherms and endotherms

$$\varphi = C_3 P^{\zeta} \quad (5)$$

where P is the body mass, $C_3 = 0.092$ and $\zeta = -0.102$ for ectotherms, and $C_3 = 0.081$ and $\zeta = -0.092$ for endotherms (see Figure 4A and B and Additional file 5: Table S5 [25]).

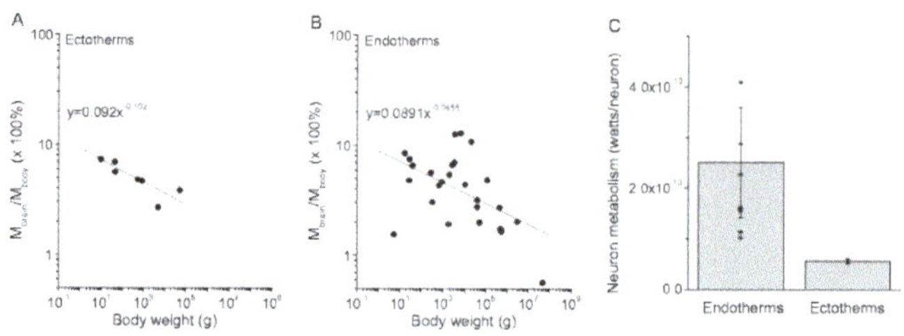

Figure 4. The dependence of brain metabolism on body size and invertebrate species. A) The ratios of brain metabolism to resting body metabolism as a function of body mass (grams) for 7 ectotherms [25] were fit by $y = 0.092\times^{-0.102}$(adjusted-$R^2 = 0.63$, $p < 0.05$), and for 26 endotherms [25] which were fit by $y = 0.0891\times^{-0.0955}$ (adjusted- $R^2 = 0.231$, $p < 0.05$), B). C. The average metabolic cost of an individual neuron for endotherms was $2.51 \pm 1.09 \times 10^{-10}$ watts, and for ectotherms this cost was $0.563 \pm 0.03 \times 10^{-10}$ watts [25]. The datasets used for this figure are given in Additional file 5: Table S5.

The total amount of energy supplied to the brain (M_{brain}) should be equal to the total energy consumption of the brain's neurons. Note that here we make a simplifying assumption, by considering the metabolic costs of neurons only because neurons consume most of the energy while the supporting cells account for 10-20% of total brain energy. If the total number of neurons in the brain is N and the metabolic consumption of each individual neuron is M_{neuron}, we can describe the relationship as follows:

$$N = \frac{M_{brain}}{M_{neuron}} = \frac{\phi M_{body}}{M_{neuron}} \quad (6)$$

where the total number of neurons in the brain (N) contributes directly to the brain's mass (see Eq. 2). By inserting Eqs. (2)-(5) into Eq. (6), we obtained a general model of the relationships between brain mass (W), body mass (P) and temperature (T):

$$W = \lambda P^{\alpha} e^{\frac{-T_V}{T}} \quad (7)$$

where $\lambda = C_2\left(\frac{C_1 C_3}{M_{neuron}}\right)^{\beta}$, $\alpha = \beta\gamma + \beta\xi$, and $T_V = \frac{\beta E_V}{k}$, called effective temperature. Equation (7) links body temperature, body mass, brain mass and neuron metabolic rate (M_{neuron}), provides a general theoretical framework for quantifying changes in brain size during evolution, and provides an origin of the formula of Eq. (1) that was derived from experimental data. The temperature range used for Eq. (7) is 0 ~ 42°C, corresponds to what is normally observed in polar fish to the body temperature of birds. The equations are not valid for temperatures above or below these values, because all the scaling parameters have been derived from animal data acquired within this temperature range.

Equation (7) links brain mass with the metabolic consumption rate of individual neurons (M_{neuron}). Recent studies suggest that the energy cost per neuron (M_{neuron}) in the brain is stable and nearly invariant across mammal species [23]. Calculations based on the available data revealed that M_{neuron} is, on average, 2.51×10^{-10} watts for endotherms and 0.563×10^{-10} watts for ectotherms (see Figure 4C and Additional file 5: Table S5).

We inserted empirically determined parameters (see the analyses in Figures 3 and 4) into Eq. (7) to determine how well this data-based theoretical model predicts actual animal brain size. Figure 5A plots the results of linerar regression analysis between $\log_{10}$ (brain mass) and $\log_{10}$ (brain mass prediction) for endotherms (red line) -- mammals, birds, primates and dolphins, primarily -- at 38°C and endothermic animals at 16°C (black line). The predicted parameters: for endotherms were $\lambda = 0.0749$ and $\alpha = 0.6608$, and the analysis results were a correlation coefficient CC = 0.979, an adjusted-$R^2 = 0.958$, and a mean squared error MSE = 0.042, p-value $p < 1\times10^{-5}$). For endothermic animals at 16°C, the predicted parameters were: $\lambda = 0.0085$ and $\alpha = 0.5876$) and the multiple linear regression analysis results were CC = 0.966, adjusted-R^2 = 0.933, and a MSE = 0.0552 (p-value $< 1\times10^{-5}$). In this analysis body mass P was the regressor variable and temperature was a constant. The model predictions closely matched the real brain data, with data-derived parameters of $C_1 = 0.0305$, $\gamma = 0.682$, $C_2 = 0.0052$, $\beta = 1.12$, $C_3 = 0.081$, and $\zeta = -0.092$ for the endotherm-based data analysis that resulted in the equation $W=0.0749P^{0.6608} e^{\left(\frac{1.12E_V}{k}\left(\frac{1}{T+273.15}-\frac{1}{38+273.15}\right)\right)}$ (where P represents body mass in gram) based on Eq. (7) and $C_1 = 0.000525$, $\gamma = 0.81$, $C_2 = 0.0096$, $\beta = 0.83$, $C_3 = 0.092$ and $\zeta = -0.102$ for ectotherms, which led to the following equation: $W=0.0087 P^{0.5876} e^{\left(\frac{0.83E_V}{k}\left(\frac{1}{T+273.15}-\frac{1}{16+273.15}\right)\right)}$. The difference in brain size of endotherms compared to ectotherm (i.e., $\text{Brain}_{endotherm}/\text{Brain}_{ectotherm}$) also increased from 7.4 to 36-fold with increasing body masses (inset of Figure 5A). If we constrain all parameters in ectotherms, in the same fashion as we have done for endotherms with the exception of the temperature difference, the brain size prediction for the ectotherms is only 7.4-fold lower than that for the endotherms (blue line in Figure 5A and, the inset). Furthermore the scaling parameter of $\alpha = 0.6608$ from Eq. (7) is close to the measured value of 0.689 for endotherms, and $\alpha = 0.5876$ is close to the measured value of 0.578 for ectotherms (Figure 1). Thus, temperature-regulated metabolism can only partially explain the 4-40-fold brain size difference shown in Figure 1A, and other factors, such as increases in non-neuron cell (mostly glial) numbers (as reflected in Figure 4B), may be essential factors in explaining brain size differences. Thus, contributions from metabolism, neural density and temperature together explain most of the differences in brain mass between endotherms and ectotherms shown in Figure 1A.

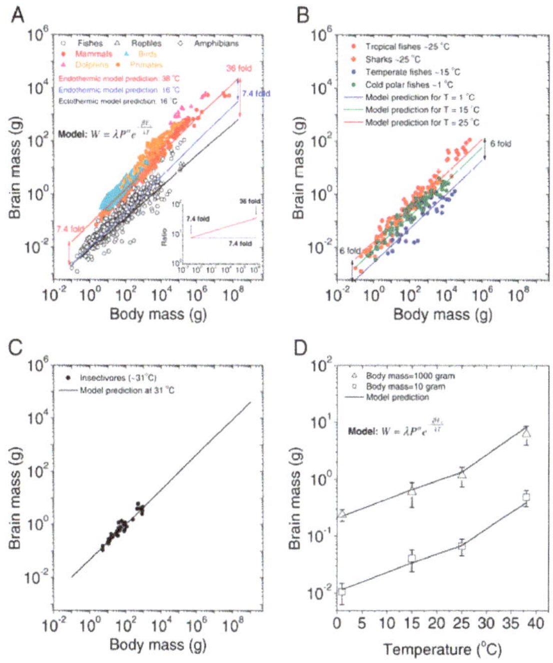

Figure 5. Model prediction for allometric scaling between the brain and body weights of vertebrate animals on a log-log plot. A. For the brain-body mass data listed in Figure 1A, the model predictions based on Eq. (7) for endotherms in 38°C (in red line) and for ectotherms in 16°C (in black line) are displayed. Multiple linear regression analysis (here body mass P is the regressor variable while temperature T is fixed as a constant) for the endothermic prediction yielded adjusted-$R^2 = 0.958$, $p < 10^{-5}$, and for ectothermic prediction, these results were adjusted- $R^2 = 0.933$, $p < 10^{-5}$. Insert: ratios of predicted endothermic brain sizes to ectothermic brain sizes as a function of body mass (red line). For comparison, the ratios of predicted endothermic brain sizes at 38°C to the predicted ectothermic brain sizes at 16°C using all the endothermic parameters are presented (blue line). **B.** For the fish brain-body mass data listed in Figure 1B, the model predictions based on Eq. (7) for fish in 1°C (in blue line), 15°C (in green line) and 25°C (in red line) conditions are shown. Least-squares regression analysis of the model predictions and the actual data revealed that, for the tropical-water fish, N = 105, CC = 0.981, adjusted- $R^2 = 0.962$, $p < 10^{-5}$, for the temperate-water fish N = 70, CC = 0.961, adjusted- $R^2 = 0.923$, $p < 10^{-5}$, and for the polar-water fish N = 35, CC = 0.889, adjusted-$R^2 = 0.79$, $p < 10^{-5}$. **C.** For the brain-body mass data of the insectivores listed in Figure 1A, the model predictions based on

Eq. (7) and the endothermic parameters at the temperature of 31°C (in black line) produced regression analysis results of CC = 0.952, adjusted- $R^2 = 0.91$, $p < 1\times10^{-5}$. **D**. For the fish data in Figure 1C, the model predictions for the relationship between brain size and temperature for the two body sizes of 10 grams and 1000 grams based on Eq. (7) are given in black lines. Multiple linear regression analysis was used here (temperature T is the regressor variable) between $\log_{10}$(actual brain mass) and $\log_{10}$(model prediction).

Note that model predictions for endothermic brain sizes are generally 2–5 times lower than actual brain masses of primates and dolphins, suggesting there other evolutionary factors mat contribute to the larger brain size in these species, as has been indicated in recent reports [12],[13],[53]. The model in the present study does not make any attempt to capture these effects.

The model predictions for tropical-water fish, temperate-water fish, and cold-water fish are shown in Figure 5B, which predicts that brain size at 25°C should be 6-fold higher than that in fish at 1°C and that at 15°C brain size should be 3-fold higher than at 1°C. The model used actual data-derived parameters of $C_1 = 0.00051$, $\gamma = 0.879$, $C_2 = 0.0096$, $\beta = 0.83$, $C_3 = 0.092$ and $\zeta = -0.102$, which for fish results in the equation $W=0.0085\ P^{0.645}\ e^{\left(\frac{0.83E_V}{k}\left(\frac{1}{T+273.15}-\frac{1}{16+273.15}\right)\right)}$ based on Eq. (7). Note, data for fish brain, glial density data are currently not available, so the fitting is imperfect, and only includes the temperature- related differences.

The model prediction for insectivores (with body temperatures of approximately 31°C) is perfectly aligned with the actual data with predicted parameters of $\lambda = 0.0409$ and $\alpha = 0.6608$ (CC = 0.952, adjusted- $R^2 = 0.91$, MSE = 0.0211, $p < 1\times10^{-5}$). Figure 5D quantifies the allometric relationship between the temperatures that fish live at, and their average brain mass for two body masses (10 g and 1,000 g). The equations for the fish were given by W=0.0085 $P^{0.645}e^{\left(\frac{0.83E_V}{k}\left(\frac{1}{T+273.15}-\frac{1}{16+273.15}\right)\right)}$ and for mammals were given by $W = 0.0749\ P^{0.6608}e^{\left(\frac{1.12E_V}{k}\left(\frac{1}{T+273.15}-\frac{1}{38+273.15}\right)\right)}$. At a given body weight, as temperature increases, brain mass of fish increases, but the increase of brain mass with temperature is larger in mammals. This relationship is well captured by the model predictions (see the black line in Figure 5D), with multiple linear regression analysis -- here temperature T is the regressor variable -- between $\log_{10}$ (actual brain mass) and $\log_{10}$(model prediction) for 10 g: CC = 0.998, adjusted- $R^2 = 0.995$, MSE = 0.0041, $p < 0.0023$; and for 1000 g: CC = 0.9999, adjusted- $R^2 = 0.9998$, MSE = 0.0049, $p < 0.0001$.

By applying the general model described by Eq. (7) to all the major types of vertebrates (including endotherms: primates, dolphins, mammals and birds, and ectotherms: fishes, reptiles and amphibians), we found that the actual brain masses and model predictions were aligned well in a log-log plot (Figure 6A). The multiple linear regression results were CC = 0.979, adjusted- $R^2 = 0.958$, MSE = 0.0421, $p < 10^{-5}$ for endotherms and CC = 0.966, adjusted- $R^2 = 0.933$, MSE = 0.0522, $p < 10^{-5}$ for ectotherms. Figure 6B shows the residuals of the log brain sizes i.e., $\log_{10}$(actual brain mass)-$\log_{10}$(model prediction), for both the endotherms and ectotherms as a function of log body size with slopes close to 0, indicating that the model performed well in predicting actual brain mass. Figure 6C shows that actual brain masses of fishes and model predictions were also aligned in a log-log plot (CC = 0.966, adjusted- $R^2 = 0.932$, MSE = 0.0674; $p < 1\times10^{-5}$). The model predictions were underestimates compared to the actual brain masses for the tropical fish with larger body sizes, and were overestimates compared to the actual brain masses for smaller tropical fish. The model underestimated the actual brain masses for fish from cold – i.e. polar -- water (Figure 6C), but for the temperate-water fish the model matched the real data well (Figure 6C). Note that the model is missing parameters, it is worth noting that the model relies on actual data, and that for both tropical-water and polar-water fish there is a dearth of data for neural and supporting cell densities. With adequate samples of all important measures, i.e. the kinds of data we have for mammals and primates, shown in Figure 3, the model predictions would be improved. Figure 6D shows the residuals of the log brain sizes, i.e., $\log_{10}$(actual brain mass)-$\log_{10}$(model prediction), for three groups of fishes as a function of log body sizes. This has a flat relationship with slopes close to 0, indicating that the model performed well in predicting actual fish brain mass.

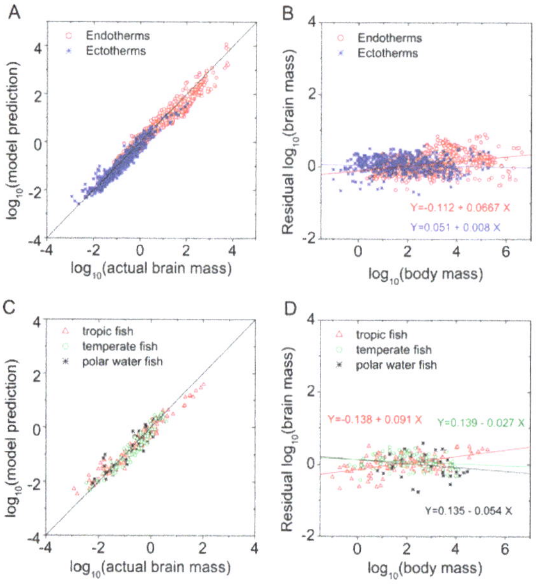

Figure 6. Regression residual analysis of model prediction and actual brain mass. A. Least-squares regression of the $\log_{10}$(model prediction) versus $\log_{10}$(actual brain mass) of the endotherms (including mammals, birds, primates and dolphins) (N = 1335), and ectotherms (including fishes, reptiles and amphibians) (N = 367). **B**. Least-squares regression of the residuals of the $\log_{10}$(brain mass) = $\log_{10}$(actual brain mass)- $\log_{10}$(model prediction) versus the $\log_{10}$(body mass) for the endotherms (red triangles) (N = 1335, CC = 0.411, adjusted- $R^2 = 0.169$, $p < 0.01$) and the ectotherms (blue plus) (N = 367, CC = 0.0488, adjusted- $R^2 = 0.0024$, $p < 0.35$). **C**. Least-squares regression of the $\log_{10}$(brain mass) versus $\log_{10}$(body mass) for the tropical-water fish (N = 105, adjusted- $R^2 = 0.962$, $p < 10^{-5}$), temperate-water fish (N = 70, adjusted- $R^2 = 0.924$, $p < 10^{-10}$), and polar-water fish (N = 35, adjusted- $R^2 = 0.79$, $p < 10^{-5}$). D. Least-squares regression of the residuals of the $\log_{10}$(brain mass) ver-

sus $\log_{10}$(body mass) for the tropical-water fish (N = 105, adjusted- $R^2 = 0.278$, $p < 10^{-5}$), temperate-water fish (N = 70, adjusted- $R^2 = 0.0227$, $p < 10^{-5}$), and polar-water fish (N = 35, adjusted- $R^2 = 0.0302$, $p < 10^{-5}$).

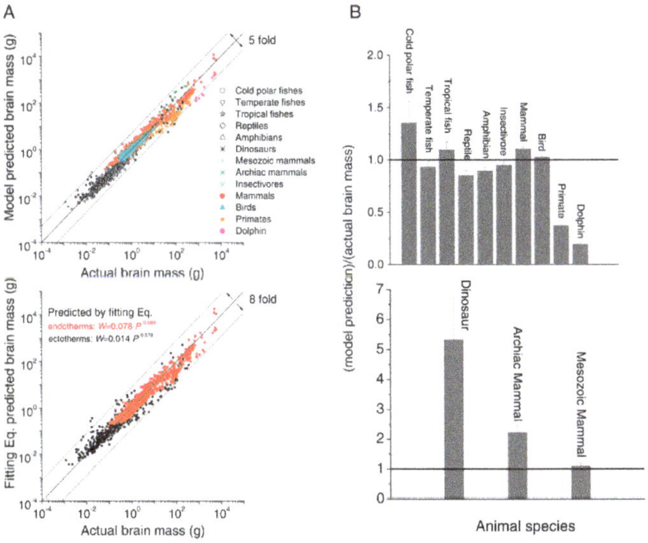

Figure 7. Matching analysis between model predictions and actual brain mass for all the major type of vertebrates. A. The model (based on Eq. (7)) predicted brain mass versus real brain mass for all the major types of vertebrates. The correlation coefficient CC and mean squared error MSE between model prediction and real brain mass are listed here: CC = 0.89 and MSE = 0.1 for cold polar fishes; CC = 0.961 and MSE = 0.05 for temperate fishes; CC = 0.981 and MSE = 0.068 for tropical fishes; CC = 0.958 and MSE = 0.052 for reptiles; CC = 0.943 and MSE = 0.028 for amphibians; CC = 0.803 and MSE = 0.51 for dinosaurs; CC = 0.97 and MSE = 0.0162 for mesozoic mammals; CC = 0.965 and MSE = 0.089 for archaic mammals; CC = 0.952 and MSE = 0.021 for insectivores; CC = 0.987 and MSE = 0.029 for mammals; CC = 0.966 and MSE = 0.007 for birds; CC = 0.956 and MSE = 0.23 for primates; CC = 0.97 and MSE = 0.544 for dolphins. **B**. The ratios defined as (Model predicted brain mass)/(Actual brain mass) for the 13 animal species.

Finally, we used Eq. (7) to predict brain sizes of all the types of vertebrates (also including 7 mesozoic and 18 archiac mammals (with body temperature around 31°C) and 17 dinosaurs (with temperature 16°C)). Top panel in Figure 7A shows that actual brain masses, and the model predictions tracked each other well in a log-log plot (multiple linear regression analysis results: CC = 0.978, adjusted- $R^2 = 0.956$; MSE = 0.0495; $p < 1 \times 10^{-5}$). Compared to the ten- to hundred-fold variance in brain size for the same body mass shown in Figure 1A, the model predictions and actual brain sizes exhibited a rather small variance, and only ~ 5-fold difference (Figure 7A). The predicted brain mass from the fitting equations in Figure 1, and the actual brain mass also tracked each other well, but the variance was a little bit higher, roughly 8 fold than in the analysis used in top panel. The multiple linear regression analysis of these data shows that: CC = 0.973, adjusted- $R^2 = 0.946$; MSE = 0.0544; $p < 1 \times 10^{-5}$. Note that the adjusted- $R^2 = 0.946$ is slightly less than the value 0.956 in top panel using Eq. (7). This suggests that Eq. (7) fits the data better than do the fitting equations from Figure 1A because that both body mass and temperature play important roles in determining the brain mass.

For most of animal species (i.e., temperate and tropic fishes, reptiles, amphibians, insectivores, mammals and birds), model predictions matched real brain data well, with CC > 0.95, MSE = 0.07, $p < 10^{-5}$) with ratios of (model prediction)/(actual brain mass) close to 1 (Figure 7B) while there are larger mean square errors for primates (MSE = 0.23, CC = 0.956) and dolphins (MSE = 0.544, CC = 0.97), and dinosaurs (MSE = 0.51, CC = 0.803) with ratios of (model prediction)/(actual brain mass) either much less or much more than 1 (Figure 7B). The regression analysis gives CC = 0.965, adjusted- $R^2 = 0.932$, MSE = 0.089, $p < 1 \times 10^{-5}$ for archiac mammals, and CC = 0.97, adjusted- $R^2 = 0.94$, MSE = 0.0162, $p < 1 \times 10^{-3}$ for mesozoic mammals.

The predictions of the model could be improved by more accurate measurements of the body temperature, body metabolism, neuronal metabolism and neuronal/glial densities of each group of species.

DISCUSSION AND CONCLUSION

There is increased interest in examining the key factors facilitating enlargement of the brain during evolution. Given that the energy cost of the brain's neurons is more than ten times that of the cells of the body, the factors that regulate the brain's energy supply and energy cost should be given thorough consideration. In this study, we re-examined data on brain size, body size and metabolic rates for all the major types of vertebrates, developed a theoretical framework based on examination of the real data, and identified several critical factors that may trigger the dramatic enlargement of the brains of birds and mammals as compared with fish, reptiles and amphibians. The most critical factor was temperature. When the endothermic animals evolved from ectothermic animals, two unique features were bound together in endotherms: larger brains and higher body temperatures (on average, 20°C higher than those of ectothermic animals). The warm temperature was critical in speeding the biochemical processes and making marked metabolic differences. Our analysis revealed that in small animals the basal metabolic rate of endotherms is 60 times greater than that of ectotherms, and in large animals it is ~ 10 times greate (Figure 2C). However, the brain sizes of small endotherms are only approximately 4 times greater than that of small ectotherms but the brains of large endotherms are more than 40 times greater than the brains of large ectotherms. Clearly, in large animals, the same basal metabolic rate of endothermic animals should support larger brain tissue in ectothermic animals, than it actually does (Figure 2D), suggesting that temperature was only one of the key factors in mediating the enlargement of the brains of endotherms. Note a similar analysis published recently points out that temperature correlates with brain size [54].

Other factors included in our analysis are brain metabolism and number of supporting glial neurons. Analysis of the brain metabolism data showed that that the average metabolic cost of individual neurons in ectothermic brains is ~ 5.63×10^{-11} watts, which is nearly 5 times lower than the average metabolic cost of endothermic brains (Figure 4C). The lower energy cost of individual neurons could be attributed to the effect of colder temperature on metabolism and the occurrence of fewer synapses in ectothermic brains [55], or the lower energy cost of individual neurons could be associated to the smaller number of glia in ectotherms [46],[47],[56]. Glia are the major non-neuronal supporting cells that maintain homeostasis, form myelin, and provide mechanical and tropic support to neurons. In the vertebrate CNS, most glia originate from portions of the developing neural tube—the exception being olfactory ensheathing glia [57]—and the energy cost of glia is generally only a few percent of that of neurons [56],[58]. Recent data suggest that the evolution of the nervous system was accompanied by increases in the number and size of glia [46],[47],[56]. In vertebrates, the glia-to-neuron ratio in the cortex increases with the size of the brain [56],[58]. Figure 3 and Eq. (3) show that brain mass increases much faster in endothermic brains than in ectothermic brains as the number of neurons increase, which is also suggestive of increases in the numbers of non-neuronal supporting cells. Glia contribute to the faster propagation of nerve signals and long-range communications, particularly via the evolution of myelinating glia in developed endotherms [59]. These energetically inexpensive glial cells surround neurons and occupy a relatively large portion of the brain [60], while also providing energy support to neurons and aiding efficient communication between cortical areas [59],[60]. Thus, these cells may be one of the critical factors that promoted brain enlargement.

The evolution of the large brains of endotherms is associated with multiple factors. To quantify the contribution of each factor to the brains of mammals and birds that are tens of times larger than those of fish, reptiles and amphibians, we built up a general theoretical model to examine the origin of body-brain size allometric scaling relationship by linking the key factors of temperature-dependent metabolism, neuron energetics, neuronal density and body size. This basic framework suggests that temperature and the development of supporting non-neuronal cells may have been the critical factors that correlated with, or might even be causal, for the enlargement of the brains of endotherms (Figure 5A). This framework also predicts that the brains of cold-blooded animals living in warmer conditions should be several times larger than those of cold-blooded animals that live in colder conditions. This prediction was confirmed by the analysis of the brain sizes of fish living in various temperatures (Figures 1B and 5B). Indeed, owing to the absence of measurements of neuronal density (and glia-to-neuron ratios) for the brains of fish that live in different temperatures, the model prediction in Figure 5B did not capture the differences in the slopes between the tropical and polar fish that are shown in Figure 1B and D. However, a prediction from the endothermic brain would be that tropical fishes have larger numbers of glia in their brains than do the fish that live in cold conditions. These predictions can be tested by experimental investigations.

METHODS

Ethics

No new animal/human data was generated in this study. All the data used in this study are from references and online database.

Animal data sources

The brain and body size datasets for the endothermic and ectothermic animals (Figure 1A) are from references [3],[16],[22],[61] and are listed in Additional file 1: Table S1. The body and brain size datasets for fish that live in different temperatures and were used to create Figure 1B were collected from the database at www.fishbase.org[44] and are listed in Additional file 2: Table S2. The metabolism-to-body scaling relationships of the endothermic and ectothermic animals were collected from published papers [37],[42]-[44] that are listed in Additional file 3: Table S3 and are plotted in Figure 2. The data regarding the allometric scaling of the brain mass and total neuronal numbers for the endothermic and ectothermic animals were collected from published papers [47],[62] and are listed in Additional file 4: Table S4 and plotted in Figure 3. The datasets for the ratios of brain metabolism to resting body metabolism are listed in Additional file 5: Table S5, are plotted in Figure 4 and were collected from published papers [25].

DECLARATIONS

Acknowledgements

We thank Harry J. Jerison for sharing the brain-body mass data. We express our deep gratitude to Fahmeed Hyder, Gordon M Shepherd, Michael Hines, and David Waxman for valuable discussions and comments. This project was funded by the National Natural Science Foundation of China (31271170) and was also supported by the program for the Professor of Special Appointment (Eastern Scholar SHH1140004) at Shanghai Institutions of Higher Learning

REFERENCES

1. Roth G, Dicke U: Evolution of the brain and intelligence. Trends Cogn Sci. 2005, 9: 250-257. 10.1016/j.tics.2005.03.005. doi:10.1016/j.tics.2005.03.005
2. Jarvis ED, Güntürkün O, Bruce L, Csillag A, Karten H, Kuenzel W, Medina L, Paxinos G, Perkel DJ, Shimizu T, Striedter G, Wild JM, Ball GF, Dugas-Ford J, Durand SE, Hough GE, Husband S, Kubikova L, Lee DW, Mello CV, Powers A, Siang C, Smulders TV, Wada K, White SA, Yamamoto K, Yu J, Reiner A, Butler AB, Avian Brain Nomenclature Consortium: Avian brains and a new understanding of vertebrate brain evolution.Nat Rev Neurosci6:151–159. doi:10.1038/nrn1606 (2005).,
3. Striedter GF: Principles of Brain Evolution. Sinauer Associates; 2005.
4. Isler K, van Schaik CP: Metabolic costs of brain size evolution. Biol Lett. 2006, 2: 557-560. 10.1098/rsbl.2006.0538. doi:10.1098/rsbl.2006.0538
5. Connor RC: Dolphin social intelligence: complex alliance relationships in bottlenose dolphins and a consideration of selective environments for extreme brain size evolution in mammals. Philos Trans R Soc Lond B Biol Sci. 2007, 362: 587-602. 10.1098/rstb.2006.1997. doi:10.1098/rstb.2006.1997
6. Dunbar RI, Shultz S: Evolution in the social brain. Science317:1344–1347. doi:10.1126/science.1145463 (2007).,
7. Niven JE, Laughlin SB: Energy limitation as a selective pressure on the evolution of sensory systems. J Exp

Biol. 2008, 211: 1792-1804. 10.1242/jeb.017574. doi:10.1242/jeb.017574

8. Dunbar RI: The social brain hypothesis and its implications for social evolution.Ann Hum Biol36:562–572. doi:10.1080/03014460902960289912879712 (2009).,
9. Isler K, van Schaik CP: The Expensive Brain: a framework for explaining evolutionary changes in brain size. J Hum Evol. 2009, 57: 392-400. 10.1016/j.jhevol.2009.04.009. doi:10.1016/j.jhevol.2009.04.009
10. Sol D, Garcia N, Iwaniuk A, Davis K, Meade A, Boyle WA, Székely T: Evolutionary Divergence in Brain Size between Migratory and Resident Birds. PLoS One. 2010, 5: e9617-10.1371/journal.pone.0009617. doi:10.1371/journal.pone.0009617
11. Amiel JJ, Tingley R, Shine R: Smart moves: effects of relative brain size on establishment success of invasive amphibians and reptiles. PLoS One. 2011, 6: e18277-10.1371/journal.pone.0018277. doi:10.1371/journal.pone.0018277
12. Navarrete A, van Schaik CP, Isler K: Energetics and the evolution of human brain size. Nature. 2011, 480: 91-93. 10.1038/nature10629. doi:10.1038/nature10629
13. Matsuzawa T: Evolution of the brain and social behavior in chimpanzees.Curr Opin Neurobiol23:443–449. doi:10.1016/j.conb.2013.01.012S0959-4388(13)00028-7 (2013).,
14. Ruben J: The Evolution of Endothermy in Mammals and Birds: From Physiology to Fossils. Annu Rev Physiol57:69–95. doi:10.1146/annurev.ph.57.030195.000441 (1995).,
15. Marshall LG, Butler RF, Drake RE, Curtis GH: Calibration of the beginning of the age of mammals in patagonia. Science. 1981, 212: 43-45. 10.1126/science.212.4490.43. doi:10.1126/science.212.4490.43
16. Jerison HJ: Evolution of the Brain and Intelligence Academic Press; 1973.
17. Kaas JH: The Evolution of Brains from Early Mammals to Humans.Wiley Interdiscip Rev Cogn Sci4:33–45. doi:10.1002/wcs.1206 (2013).,
18. Snell O: Hexenprozesse und Geistesstorung. 1891, Psychiatrische Untersuchungen, Munchen, Lehman size and basal metabolic rate in terrestrial vertebrates. Nature. 1981, 293: 57-60. 10.1038/293057a0.
19. West GB, Brown JH, Enquist BJ: A general model for the origin of allometric scaling laws in biology. Science. 1997, 276: 122-126. 10.1126/science.276.5309.122.
20. Barrickman NL, Lin MJ: Encephalization, expensive tissues, and energetics: An examination of the relative costs of brain size in strepsirrhines. Am J Phys Anthropol. 2010, 143: 579-590. 10.1002/ajpa.21354. doi:10.1002/ajpa.21354
21. Weisbecker V, Goswami A: Brain Size, Life History, and Metabolism at the Marsupial/Placental Dichotomy Proceedings of the National Academy of Sciences; 2010. doi:10.1073/pnas.0906486107.
22. Herculano-Houzel S: Scaling of brain metabolism with a fixed energy budget per neuron: implications for neuronal activity, plasticity and evolution. PLoS One. 2011, 6: e17514-10.1371/journal.pone.0017514. doi:10.1371/journal.pone.0017514
23. Hofman MA: Energy Metabolism, Brain Size and Longevity in Mammals. Q Rev Biol. 1983, 58: 495-512. 10.1086/413544.
24. Mink JW, Blumenschine RJ, Adams DB: Ratio of central nervous system to body metabolism in vertebrates: its constancy and functional basis. Am J Physiol. 1981, 241: R203-R212.
25. Anderson SD, Goldman KJ: Temperature Measurements from Salmon Sharks, Lamna ditropis, in Alaskan Waters. Copeia. 2001, 2001: 794-796. 10.1643/0045-8511(2001)001[0794:TMFSSL]2.0.CO;2.
26. Bernal D, Donley JM, Shadwick RE, Syme DA: Mammal-like muscles power swimming in a cold-water shark.Nature437:1349–1352. (2005)., [http://www.nature.com/nature/journal/v437/n7063/suppinfo/nature04007_S1.html]
27. Gillooly JF, Brown JH, West GB, Savage VM, Charnov EL: Effects of size and temperature on metabolic rate. Science. 2001, 293: 2248-2251. 10.1126/science.1061967. doi:10.1126/science.1061967
28. Kondepudi DK, Prigogine I: Modern Thermodynamics: From Heat Engines to Dissipative Structures. John Wiley & Sons; 1998.
29. Schmidt-Nielsen K: Scaling, Why is Animal Size so Important? Cambridge University Press; 1984.
30. Seymour RS: Biophysics and physiology of temperature regulation in thermogenic flowers. Biosci Rep. 2001, 21: 223-236. 10.1023/A:1013608627084.
31. White CR, Seymour RS: Mammalian basal metabolic rate is proportional to body mass2/3. Proc Natl Acad Sci U S A. 2003, 100: 4046-4049. 10.1073/pnas.0436428100. doi:10.1073/pnas.0436428100
32. Brown JH, Gillooly JF, Allen AP, Savage VM, West GB: Toward a metabolic theory of ecology. Ecology. 2004, 85: 1771-1789. 10.1890/03-9000. doi:10.1890/03-9000
33. Glazier DS: Beyond the '3/4-power law': variation in the intra- and interspecific scaling of metabolic rate in animals. Biol Rev Camb Philos Soc. 2005, 80: 611-662. 10.1017/S1464793105006834. doi:10.1017/S1464793105006834
34. White CR, Seymour RS: Allometric scaling of mammalian metabolism. J Exp Biol. 2005, 208: 1611-1619. 10.1242/jeb.01501. doi:10.1242/jeb.01501
35. Clarke A: Temperature and the metabolic theory of ecology. Functional Ecology. 2006, 20: 405-412. 10.1111/j.1365-2435.2006.01109.x. doi:10.1111/j.1365-2435.2006.01109.x
36. White CR, Phillips NF, Seymour RS: The scaling and temperature dependence of vertebrate metabolism. Biol Lett. 2006, 2: 125-127. 10.1098/rsbl.2005.0378. doi:10.1098/rsbl.2005.0378
37. Karbowski J: Global and regional brain metabolic scaling and its functional consequences. BMC Biol. 2007, 5: 18-10.1186/1741-7007-5-18. doi:10.1186/1741-7007-5-18

38. Savage VM, West GB: A quantitative, theoretical framework for understanding mammalian sleep. Proc Natl Acad Sci U S A. 2007, 104: 1051-1056. 10.1073/pnas.0610080104. doi:10.1073/pnas.0610080104
39. White CR, Cassey P, Blackburn TM: Allometric exponents do not support a universal metabolic allometry. Ecology. 2007, 88: 315-323. 10.1890/05-1883.
40. Glazier DS: A unifying explanation for diverse metabolic scaling in animals and plants. Biol Rev Camb Philos Soc. 2010, 85: 111-138. 10.1111/j.1469-185X.2009.00095.x. doi:10.1111/j.1469-185X.2009.00095.x
41. Makarieva AM, Gorshkov VG, Li BL, Chown SL, Reich PB, Gavrilov VM: Mean mass-specific metabolic rates are strikingly similar across life's major domains: Evidence for life's metabolic optimum. Proc Natl Acad Sci U S A. 2008, 105: 16994-16999. 10.1073/pnas.0802148105. doi:10.1073/pnas.0802148105
42. Clarke A, Rothery P, Isaac NJ: Scaling of basal metabolic rate with body mass and temperature in mammals.J Anim Ecol79:610–619. doi:10.1111/j.1365-2656.2010.01672.x (2010).,
43. Froese RADPE: FishBase. World Wide Web Electronic Publication; 2012. , version., [www.fishbase.org]
44. Tower DB: Structural and functional organization of mammalian cerebral cortex; the correlation of neurone density with brain size; cortical neurone density in the fin whale (Balaenoptera physalus L.) with a note on the cortical neurone density in the Indian elephant. J Comp Neurol. 1954, 101: 19-51. 10.1002/cne.901010103.
45. Herculano-Houzel S, Collins CE, Wong P, Kaas JH: Cellular scaling rules for primate brains. Proc Natl Acad Sci U S A. 2007, 104: 3562-3567. 10.1073/pnas.0611396104. doi:10.1073/pnas.0611396104
46. Herculano-Houzel S, Mota B, Lent R: Cellular scaling rules for rodent brains. Proc Natl Acad Sci U S A. 2006, 103: 12138-12143. 10.1073/pnas.0604911103. doi:10.1073/pnas.0604911103
47. Kemali M, Braitenberg V: Atlas of the Frog's Brain. Springer; 1969.
48. Packard A, Albergoni V: Relative growth, nucleic acid content and cell numbers of the brain in Octopus vulgaris (Lamarck). J Exp Biol. 1970, 52: 539-552.
49. Zupanc GK, Horschke I: Proliferation zones in the brain of adult gymnotiform fish: a quantitative mapping study. J Comp Neurol. 1995, 353: 213-233. 10.1002/cne.903530205. doi:10.1002/cne.903530205
50. Howarth C, Gleeson P, Attwell D: Updated energy budgets for neural computation in the neocortex and cerebellum. J Cereb Blood Flow Metab. 2012, 32: 1222-1232. 10.1038/jcbfm.2012.35. doi:10.1038/jcbfm.2012.35
51. Hyder F, Rothman DL, Bennett MR: Cortical energy demands of signaling and nonsignaling components in brain are conserved across mammalian species and activity levels. In Proceedings of the National Academy of Sciences of the United States of America. doi:10.1073/pnas.1214912110 (2013).
52. Roth G, Dicke U: Evolution of the brain and intelligence in primates.195:413–430. doi:10.1016/B978-0-444-53860-4.00020-9 (2012).,
53. Gillooly JF, McCoy MW: Brain size varies with temperature in vertebrates.Peer J2:e301. doi:10.7717/peerj.301301 (2014).,
54. Braitenberg V, Schüz A: Cortex: Statistics and Geometry of Neuronal Connectivity. Springer; 1998.
55. Hartline DK: The evolutionary origins of glia. Glia. 2011, 59: 1215-1236. 10.1002/glia.21149. doi:10.1002/glia.21149
56. Ramon-Cueto A, Avila J: Olfactory ensheathing glia: properties and function.Brain Res Bull46:175–187 (1998).,
57. Oberheim NA, Wang X, Goldman S, Nedergaard M: Astrocytic complexity distinguishes the human brain.Trends Neurosci29:547–553. doi:10.1016/j.tins.2006.08.004 (2006).,
58. Hartline DK, Colman DR: Rapid conduction and the evolution of giant axons and myelinated fibers.Curr Biol17:R29–35. doi:10.1016/j.cub.2006.11.042 (2007).,
59. Azevedo FA, Carvalho LR, Grinberg LT, Farfel JM, Ferretti RE, Leite RE, Jacob Filho W, Lent R, Herculano-Houzel S: Equal numbers of neuronal and nonneuronal cells make the human brain an isometrically scaled-up primate brain. J Comp Neurol. 2009, 513: 532-541. 10.1002/cne.21974. doi:10.1002/cne.21974
60. Boddy AM, McGowen MR, Sherwood CC, Grossman LI, Goodman M, Wildman DE: Comparative analysis of encephalization in mammals reveals relaxed constraints on anthropoid primate and cetacean brain scaling. J Evol Biol. 2012, 25: 981-994. 10.1111/j.1420-9101.2012.02491.x. doi:10.1111/j.1420-9101.2012.02491.x
61. Sarko DK, Catania KC, Leitch DB, Kaas JH, Herculano-Houzel S: Cellular scaling rules of insectivore brains. Front Neuro. 2009, 3: 8-doi:10.3389/neuro.05.008.2009
62. Gabi M, Collins CE, Wong P, Torres LB, Kaas JH, Herculano-Houzel S: Cellular scaling rules for the brains of an extended number of primate species. Brain Behav Evol. 2010, 76: 32-44. 10.1159/000319872. doi:10.1159/000319872

CHAPTER 9 VERTEBRATE INNATE IMMUNE SYSTEM

Adriana M Montaño, Fumi Tsujino, Naoyuki Takahata , Yoko Satta

BACKGROUND

Innate immunity is the ancient defense system of multicellular organisms against microbial infection. The basis of this first line of defense resides in the recognition of unique motifs or components conserved in microorganisms, and absent in the host. The innate immune system uses sets of pattern recognition receptors to recognize such foreign or non-self motifs. Proteins in the immune system can be located intracellularly, on the cell surface, or secreted into the bloodstream, ready to signal the presence of an intruder in every compartment. In systems lacking the adaptive arm of immunity, the pattern recognition concept serves well to explain the general triggering of the system as well as providing receptors for the limited specificity shown by innate immunity [1, 2, 3, 4].

Peptidoglycan (PGN) is the major structural component of the cell wall of almost all bacterial species. PGN is a large, repetitive macromolecule that forms the rigid cell wall of bacteria. PGN recognition is mediated by the PGRP (PGN recognition protein) family of receptors [5, 6]. PGRPs are a family of innate immunity pattern recognition molecules that were first discovered in silkworms [7]. There are four loci for each PGRPs in humans [8, 9] (PGRP-S, PGRP-L, PGRP-Iα and PGRP-Iβ), while thirteen loci in Drosophila [10, 11], which encode approximately 17 PGRP proteins through alternative splicing, and seven loci in Anopheles [12]. Several other genomes also show relatively large number of PGRPs in invertebrates, but only up to five in vertebrates (Additional file 1). In invertebrates, the functional divergence of each PGRP molecule is well investigated: Some possess an amidase activity that hydrolyzes the amide bond between the N-acetylmuramic acid and the L-alanine of peptidoglycan, others activate Toll, or Imd pathways to induce an expression of anti-bacteria peptides, induce prophenoloxidase cascade, or directly cause phagocytosis and lysis [13, 14, 15, 16, 17, 18]. On the other hand, functions of vertebrate, or mainly mammalian PGRPs, are not fully understood [19]. While PGRP-L has the amidase activity where its role could be to detoxify PGN fragments present in blood and modulate the immune response as insect PGRPs, the PGRP-S, PGRP-Iα and PGRP-Iβ have bacteriostatic and/or bactericidal function [6, 20, 21, 22].

Vertebrates have the acquired immunity system in addition to the innate immunity system, while in insects only the latter is a self-defense system. It is of interest whether possessing acquired immunity has effects on the evolution of

Citation: Adriana M Montaño, Fumi Tsujino, Naoyuki Takahata and Yoko Satta, vertebrate innate immune system, DOI: 10.1186/1471-2148-11-79.

molecules involved in innate immunity. Here, we investigate the birth and death processes of PGRPs by systematically analyzing PGRP genes from a set of diverse eukaryotes, and discuss the role of selection and diversification of this gene family.

RESULTS

Modes of PGRP evolution

To detect lineage-specific expansions 40 sequences of the PGRP family from 21 vertebrate species and 42 sequences from 6 invertebrate species were studied (Additional file 1 and Additional file 2). Both vertebrate and invertebrate PGRPs have a highly conserved C-terminal region of the PGRP domain with three sub-domains (I, II and III). The sub-domains are determined by sequence conservation and not by their function. The PGRP domain shows a sequence similarity (~35%) with bacteriophage T7 lysozyme, which also has amidase activity, indicating that T7 lysozyme would be the origin of PGRP domains [23]. This ancient origin of PGRP domains is also supported by the similarity of the 3D structure between PGRP-L and T7 lysozyme molecules. However, the orthologous relationship of vertebrate and insect PGRP domains has not been ensured [8, 19] due to the limited number of amino acid sites compared and the great extent of sequence divergence among them. Contrary to the conserved C-terminal PGRP domains, the N-terminal region shows no particular similarities among different PGRPs in invertebrates, and partial similarities among PGRP-S, PGRP-Iα and PGRP-Iβ in vertebrates. Therefore, we used only the PGRP domains for the alignment and tree construction. Due to difficulties in identifying orthologous relationship, we performed phylogenetic analyses of PGRP genes in vertebrates and invertebrates separately by using the neighbor-joining (NJ) and minimum evolution (ME) methods. There was no conflict in the topologies obtained with these methods.

In vertebrate PGRPs, the phylogeny shows five clustering groups, four of which corresponds to four loci found in humans; PGRP-L, PGRP-S, PGRP-Iα and PGRP-Iβ. On the other hand, the fifth locus, named PGRP-F, is found only in fish. Including PGRP-F, there are four rounds of gene duplication and a single round of domain duplication, which produced the present-day vertebrate PGRPs (Figure 1, Additional file 3). The first round of gene duplication happened in the stem lineage leading to all jawed vertebrates. This duplication produced PGRP-F and the proto-PGRP that is an ancestor of PGRPs in other jawed vertebrates. In the second round, gene duplication occurred just after the first round and produced proto-PGRP-L and proto-PGRP-S. In addition to these two rounds, there is at least, an additional duplication in proto-PGRP-L in the lineage leading to fish PGRP-L. On the other hand, no descendant of proto-PGRP-S was detected in fish genomes. PGRP-S and proto-PGRP-I were produced after one round of duplication in proto-PGRP-S descendant in the stem lineage leading to tetrapods. The presence of PGRP-S in amphibians suggests the loss of proto-PGRP-I in this lineage. Just after this duplication and before the divergence of therian mammals, the proto-PGRP-I possesses two PGRP domains [24] due to the domain duplication 252~336 million years ago. After the divergence of opossums from placental mammals, the last round of gene duplication occurred 126~168 MYA producing PGRP-Iα and PGRP-Iβ. This observation indicates that PGRP-Iα and PGRP-Iβ are placental mammal-specific genes.

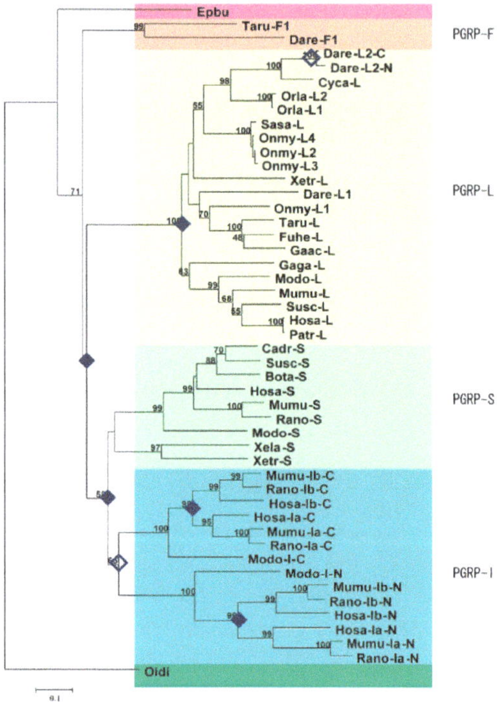

Figure 1. Neighbor-joining tree of vertebrate PGRP amino acid sequences. Filled and open diamonds indicate duplication of loci and domains, respectively. The analyzed sequences contain 145 amino acid sites. Numbers at the nodes represent the bootstrap support for the branch based on 1000 replications. The evolutionary distances were computed using the Poisson correction method and are in the units of the number of amino acid substitutions per site. All positions containing gaps and missing data were eliminated from the dataset. Notation of species names are indicated as follow: Bota (Bos taurus), Cadr (Camelus dromedarius), Cyca (Cyprinus carpio), Dare (Danio renio), Epbu (Eptatretus burgeri), Fuhe (Fundulus heteroclitus), Gaac (Gasterosteus aculeatus), Gaga (Gallus gallus), Hosa (Homo sapiens), Modo (Monodelphis domestica), Mumu (Mus musculus), Oidi (Oikopleura dioica), Onmy (Oncorhynchus mykiss), Orla (Oryzias latipes), Patr (Pan troglodytes), Rano (Rattus norvegicus), Sasa (Salmo salar), Susc (Sus scrofa), Taru (Takifugu rubripes), Xela (Xenopus laevis) and Xetr (Xenopus tropicalis).

The gene structure of vertebrate PGRPs (Figure 2) supports the above scenario, which explains the emergence of the vertebrate PGRPs. The PGRP-S contains two introns, one of which shares the position with both PGRP-Iand PGRP-L, while the other only with PGRP-I. The position of this intron is preserved in the duplicated PGRP domains of PGRP-I. Further, the N-terminal regions of PGRP-Iα and -Iβ genes show some sequence similarity with the PGRP-S N-terminal amino acid sequence. These observations indicate that the PGRP-L diverged first, PGRP-I is originated from PGRP-S, and the second PGRP domain in PGRP-I has been produced by domain duplication in PGRP-S. In addition to the main events, which originated the PGRP family commonly found in mammals, we also observed a recent domain duplication event in zebrafish PGRP-L where the domains exhibit homology of 99% (Figure 1).

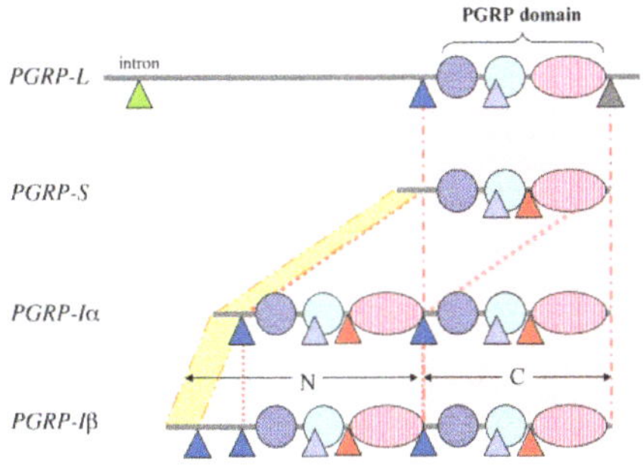

Figure 2. Gene structure of four human PGRP genes. Vertical lines indicate corresponding regions between different genes. A triangle shows the position of the introns and the same color indicates that the position was shared.

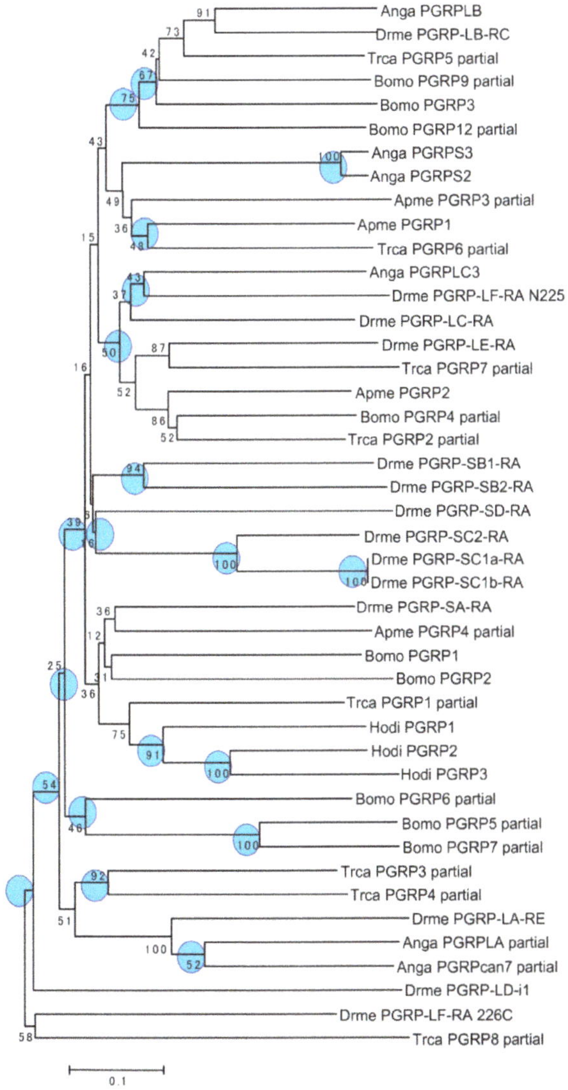

Figure 3. Neighbor-joining tree of invertebrate PGRP amino acid sequences. Circles on nodes indicate orthologous pairs of genes. The analysed sequences contain 207 amino acid sites. The evolutionary distances were computed using the Poisson correction method and are in the units of the number of amino acid substitutions per site. All positions containing gaps and missing data were eliminated from the dataset. Notation of species names are indicated as follows: Anga (Anopheles gambiae), Apme (Apis mellifera), Bomo (Bombyx mori), Drme (Drosophila melanogaster), Hodi (Holotrichia diomphalia), and Trca (Tribolium castaneum).

In contrast to this relatively small number of gene duplications and gene losses in vertebrate PGRPs, the birth and death process shows a different pattern in invertebrate PGRPs. The number of PGRP loci in the invertebrate genome ranges from four in A. mellifera to 14 in B. mori. Using 44 different sequences retrieved from databases (Additional file 2), we reconstructed the phylogenetic tree of PGRP domains for insects. In contrast to vertebrates, invertebrate PGRP genes are not clearly classified into orthologous groups. As representatives of the class Insecta, we used four genomes, which correspond to four different orders (Diptera, Hemiptera, Coleoptera, and Lepidoptera). The divergence time of these orders is similar to that of vertebrates. Gene duplication and loss are rather frequent and taxon-specific sets of PGRPs are evident. For example, in Drosophila only six of the thirteen are found in different orders, and seven seem to be Drosophilaspecific. The phylogenetic tree reveals that at least 14 rounds of gene duplication and two of gene losses are required to produce the extant repertoire of PGRPs in Drosophila genome (Figure 3, Additional file 4). A similar pattern of species-specific gene duplication and gene loss is observed in other insects, too.

The mentioned observation could be confirmed by using other methods that can predict the number of gene gains and losses. We verified our observations by using the program NOTUNG and EvolMAP [25, 26]. The results with NOTUNG showed a similar tendency of the number of gene gains predicted (Additional file 5 and Additional file 6), however the number of gene losses seems to be overestimated in vertebrates, especially in fish. In NOTUNG, absence of a gene in a particular taxon means gene loss. Thus the number of losses in fish became enormously large. EvolMAP, on the other hand, predicts that gene gains are 8.6 times more frequent than gene losses in invertebrates over all branches. This suggests an expansion of PGRPs in invertebrates. For vertebrates, EvolMAP analysis shows no evidence of expansion or contraction for this gene family (Additional file 7 and Additional file 8). Overall, for the genes and species analyzed here, we find that the number of gains detected in invertebrates is twice the number of gains in vertebrates. Thus we could confirm the large number of gene gains and losses in invertebrates when compared to vertebrates.

Ancestral PGRP Genes

Due to the limited number of sites compared and long divergence time of sequences, we could not elucidate the relationship among the ancestors of vertebrate and invertebrate PGRPs from the phylogenetic tree of vertebrate and invertebrate PGRPs, including T7 lysozyme. Thus whether the origin of vertebrate PGRPs is monophyletic or paraphyletic to invertebrate PGRPs remains to be an open problem. However, our analysis clearly shows that for vertebrate PGRPs, the first major divergence took place between PGRP-L and PGRP-S. Therefore, in the following we focus on the vertebrate PGRPs to infer the functions of the ancestral PGRPgenes.

To elucidate the function of ancestral PGRP molecules in vertebrates, the amino acid sequences of proto- PGRP-L and PGRP-S molecules were estimated by the maximum-likelihood (ML) method with the JTT substitution matrix [27]. It is known that seven amino acids are responsible for PGRP function [1, 8]. Four amino acid residues (H17, Y46, H122, and C130) are essential for the amidase activity, whereas three (H36, W41, and K128) are important for Zn^{2+} ligand-binding in the bacteriophage T7 lysozyme. Since all seven amino acids are conserved in both the proto- PGRP-L and PGRP-S sequences, the ancestor of both proto-PGRPs is likely to possess the amidase activity (Additional file 9). While the present-day PGRP-L has reserved its original function of amidase activity [20], PGRP-S has lost it and instead obtained the bacteriostatic function [21, 22]. On the other hand, the invertebrate PGRPs possessing the amidase activity are paraphyletic to each other. This suggests independent gain or loss of the amidase activity in invertebrate PGRP at an early stage of the evolution.

Selection acting on PGRPs and Evolutionary Rates

Next important question is whether some kind of selection process has acted on each amino acid site of vertebrate PGRP genes that will lead to their functional divergence after gene duplication. We identified positively or negatively selected sites in vertebrate PGRPs using Single Likelihood Ancestor Counting (SLAC) analysis as described in Methods [28]. Only the site 138 in PGRP-L is positively selected among all the vertebrate PGRPs. This site, which is involved in substrate binding, shows a high degree of amino acid variation in PGRP-L of different species.

Average values of the ratios of non-synonymous to synonymous substitutions of PGRP-S, L and I are 0.16, 0.13 and 0.17, respectively, and the overall value for PGRPs is 0.20. This indicates strong functional constraint, suggesting that amino acid sequences of these domains are well conserved among vertebrates.

Although there is a report where a few positively selected sites were observed in the PGRP domain of Drosophila PGRP-LC [29], we could not apply the above SLAC analysis to invertebrate PGRPs, because the clustering pattern and phylogenetic relationship among sequences are not ensured and the results of SLAC strongly depend on the tree topology (data not shown).

We further examined parallel and convergent evolution at the amino acid level to infer the operation of natural selection. We aligned 39 vertebrate PGRP sequences for each locus and deduced the ancestral amino acids [27] at all internal nodes of the phylogenetic tree (Figure 1), in order to estimate the presence of parallel and convergent substitutions, which may have been driven by the functional importance of sites. Subsequently the probability of a parallel or convergent substitution by chance was estimated as described in Methods. Our analysis revealed that thirteen sites have experienced parallel and twenty-three sites convergent substitutions, of which occurrence is statistically significant ($p \leq 0.05$) (Table 1, Figure 4). Comparison of these sites based on the tertiary structure of Drosophila PGRP-LB [23], Drosophila PGRP-SA [30], and Human PGRP-Iα [24] showed that all these residues are located on α helices (sites 38, 41, 43, 45 and 52 in α1; 105, 108, 109, 113, 114, 117, 120, 121 and 123 in α 2; and 145 in α 3) and on β-sheets (site 2 on β1; 12 and 15 on β2; 89 and 96 on β6). Sites 2, 5, 12 and 15 are located on the PGRP specific fragment (Table 1).

Table 1. Tests of parallel and convergent evolution of PGRPs.

Species	n	P	Site positions
A. Parallel changes			
Pig-S & Cow-S	3	0.003	2, 78, 113
Frog-S2 & Rat-S	2	0.02*	34, 77
Rainbow trout-L1 & stickleback-L	3	0.006	45, 52, 120
Rat-Ia-C & Rat-S	1	0.03	77
Fugu-L & Pig-L	1	0.04*	89
Chicken-L & Salmon-L	1	0.049	108
Rainbow trout-L1 & Zebrafish-L1	3	0.03	45, 100, 114
Stickleback-L & Rat-Ia-C	1	0.048	125
B. Convergent changes			
Human-Ib-N & Rat-Ib-C	1	0.029	5
Cow-S & Frog-L	2	0.0071	12, 125
Pig-S & Frog-S1	2	0.002	15, 123
Fugu-L & Hagfish-S	2	0.007	21, 78
Frog-S2 & Rat-S	1	0.02*	100
Mouse-L & Killifish-L	1	0.02	38
Mouse-L & Rat-Ia-N	1	0.05	38
Rat-Ia-N & Killifish-L	1	0.035	38
Carp-L & Hagfish-S	2	0.007	41, 140
Mouse-Ia-N & Pig-S	1	0.016	43
Mouse-S & Mouse-Ib-C	1	0.014	52
Rat-Ib-N & Pig-S	1	0.02	54
Rainbow trout-L1 & Mouse-Ia-N	1	0.02	77
Fugu-L & Pig-L	1	0.04*	78
Pig-L & Human-Ia-C	1	0.02	96
Rat-S & Rainbow trout-L1	1	0.05	100
Mouse-L & Zebrafish-L1	1	0.048	101
Rat-S & Cow-S	1	0.015	101
Human-S & Salmon-L	1	0.008	108
Pig-S & Salmon-L	1	0.008	108
Human-S & Chicken-L	2	0.007	105, 108
Frog-S1 & Mouse-Ia-C	1	0.038	109
Mouse-L & Mouse-Ia-N	1	0.02	117
Human-Ia-C & Rat-S	1	0.03	121
Carp-L & Fugu-L	1	0.021	123
Zebrafish-L2-N & Medaka-L1	1	0.009	145

n = Number of sites where parallel or convergent evolution occurred.

P = Observed probability of parallel or convergent change

* = Observed probability of parallel and convergent change Internodal branches are omitted.

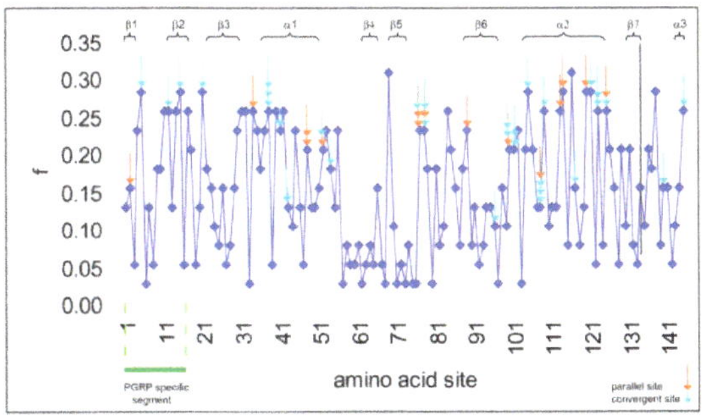

Figure 4. Profile of the variation of amino acid sites, and occurrence of convergent and parallel substitutions. The PGRP specific segment absent in T7 lysozyme is indicated. Orange arrows correspond to parallel substitutions, and blue arrows indicate convergent substitutions. The secondary structure assignment is depicted above the profile. f is defined as the ratio of the number of different amino acids at a specified site to the total number of sequences compared.

Among the six parallel sites and ten convergent sites, which occurred with a significance level of 1%, sites 45, 78, and 123 may be potential adaptive sites. Site 45 has three independent parallel changes of S (Ser) to A (Ala) only in fish PGRP-L (three spine stickleback, rainbow trout and zebrafish). Serine is the only amino acid residue present in this site in PGRP-L except for these fishes. Moreover, exclusively non-polar character of this site indicates that it may have an important role in the function of these proteins especially because this residue is involved in substrate binding. Site 78 has suffered one parallel and one convergent substitution. The chemical profile of this site in mammal PGRP-S is exclusively hydrophobic. The variability which results from the change of I (Ile) to the polar T (Thr) in pig and cow may suggest an effect on the structure since this amino acid is located in the hydrophobic core. At the site 123 there is a convergent change of A (Ala) or Y (Tyr) to V (Val) in pig and frog PGRP-S, respectively located in the hydrophobic core. Both substitutions create a possible adaptive site in PGRP-S gene.

In vertebrates, Tajima›s relative test did not show rate heterogeneity in PGRP-I genes, but showed it in PGRP-Sand PGRP-L genes [31] (Figure 5). We examined the genetic distances (poisson and gamma distances), their relationship between the divergence times, and found acceleration in the substitution rate in recent evolutionary stages.

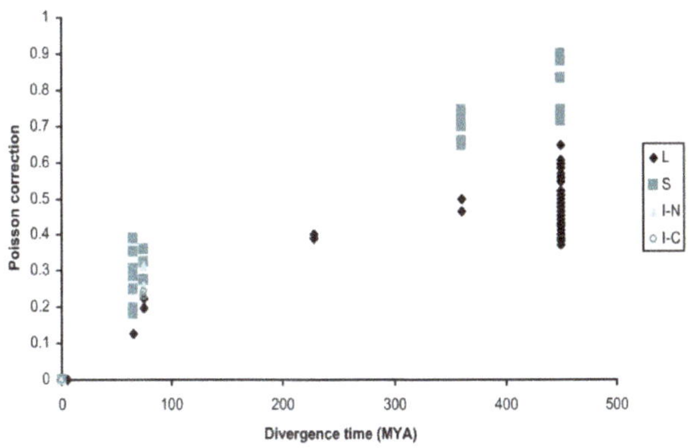

Figure 5. Acceleration of evolutionary rate in the latest stages of vertebrate evolution. Poisson distances of 19 taxa are plotted against species divergence time. The divergence times depicted in the abscissa correspond to: human-artiodactyls, human-mouse, human-chicken, human-frog; and human-fish with 65 MYA, 80 MYA, 228 MYA, 360 MYA and 450 MYA, respectively [38]. MYA: Million years ago.

We studied in invertebrates the genetic distances and the relationship between the divergence times and found that PGRP-LC, PGRP-LE, PGRP-LB, PGRP-LF and PGRP-LA have an evolutionary rate that is not constant (Figure 6).

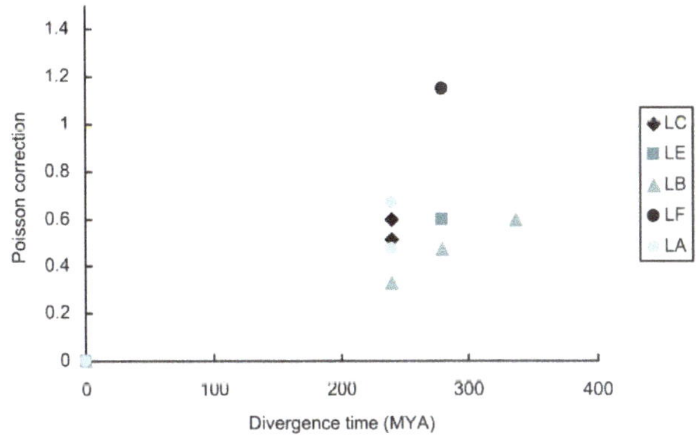

Figure 6. Evolutionary rate in invertebrates. Poisson distances of six taxa are plotted against species divergence time. The divergence times depicted in the abscissa correspond to: 241 MYA (Diptera - Hemiptera), 280 MYA (Diptera - Coleoptera), 336.1 MYA (Diptera - Lepidoptera). MYA: Million years ago.

DISCUSSION

Evolutionary characteristics of PGRPs in vertebrates and invertebrates

When we compared PGRP evolution in vertebrates and invertebrates, we observed several differences, which are characteristic of each mode of evolution. First of all, despite the similar divergence time of 450 myr for the most recent common ancestor in each vertebrate and invertebrate (insect) species, the phylogenetic trees clearly show different patterns. In invertebrates each PGRP shows a relatively longer branch than those in vertebrates, suggesting a relatively ancient origin of each PGRP in invertebrates. In addition, in invertebrates the clustering pattern was rarely orthologous among PGRP genes, while in vertebrates orthologous relationship was clearly seen. This observation shows that higher rates of birth and death processes are seen in invertebrates than in vertebrates. Although the repertoire of PGRPs in each species may depend on some ecological and biological conditions, a less frequent birth and death process in vertebrates could reflect the presence of acquired immune system.

Consequences of Natural Selection on PGRPs

In vertebrate PGRP proteins we observed changes as consequence of parallel and convergent amino acid substitutions, with significance greater than the random chance expectation. The changes may be either due to conservation of chemical property of amino acids at the site or due to modification of their properties. Convergent or parallel substitutions can provide evidence for the action of natural selection for keeping the function or structure [32]. PGRP proteins play an important role in innate immunity, which requires updated and immediate responses, because pathogens may change frequently. As a consequence, a high turnover rate is expected to happen. Actually, in invertebrates, the frequent turnover of PGRP repertoire was observed. On the other hand, several motifs indispensable for peptidoglycan recognition should be conserved through evolution of PGRPs in both vertebrates and invertebrates. In addition, we have observed that the amino acid residues that are located on the hydrophobic groove have high degree of conservation and do not show any parallel or convergent amino acid substitution.

Evolution of the PGRP family in vertebrates and invertebrates and functional implications

This study provides for the first time a description of the origin and mode of evolution on vertebrate PGRPs, compared with invertebrates, namely insects, PGRPs.

PGRPs are proposed to be a family of genes that evolved by birth and death process with different rates in vertebrates and invertebrates. In the model of the birth and death process [33, 34], some of the duplicated genes diverge functionally, but others become pseudogenes due to deleterious mutations or are deleted from the genome. The end result of this mode of evolution is a multi-gene family with a mixture of divergent groups of genes and highly homologous genes. We have observed that PGRPs have experienced several rounds of gene duplications and some duplicated genes have been deleted from the genome. This lineage specific birth and death process has been observed both in vertebrates and invertebrates.

The PGRP proteins are involved in innate immunity, which responds to protect the organisms from invading pathogens. Therefore, several motifs in PGRP domains are, of course, indispensable for pathogen recognition and have been conserved through the vertebrate and invertebrate evolution. However, since vertebrates possess acquired immunity, the significance of innate immunity might be more relaxed than in insects whose immune systems depend solely on innate immunity. This difference in the evolutionary patterns could be related to the plasticity of the receptors to detect a broad spectrum of microbial pathogens and it is clearly reflected in the birth and death process of the PGRP molecules in vertebrates and invertebrates.

CONCLUSIONS

PGRP gene family reveals an example of genetic and functional variation of which roles in the immune systems are understood through an analysis of comparative genomics. Especially the analysis reveals that the mode of PGRP evolution was characterized by birth and death process. Vertebrates and invertebrates show striking differences in the evolutionary tempo and mode of PGRP genes. Broad repertoire of pathogen recognition proteins is advantageous in invertebrates, due to the absence of adaptive immunity, in contrast to the moderate repertoire in vertebrates. This reveals that the mode of evolution of a system strongly depends on other systems, which interact with the former both directly or indirectly.

METHODS

Sequence Data

The sequences were retrieved from the genomic and EST NCBI database http://www.ncbi.nlm.nih.gov, the Ensembl database http://www.ensembl.org and the TIGR database http://www.tigr.org/tdb/tgi/ using TBLASTN and PSI-BLAST http://www.ncbi.nlm.nih.gov/BLAST/. Search was performed using each exon of human PGRP-S, PGRP-L, PGRP-Iα, and PGRP-Iβ as a probe. Takifugu rubripes PGRP-L full cDNA and genomic sequences were predicted using Genscan and confirmed by sequence analysis. Takifugu rubripes liver tissue was kindly provided by Dr. Shugo Watabe of the University of Tokyo, Japan. The Eptatretus burgeri PGRP cDNA sequence was kindly provided by Dr. Masanori Kasahara of The Graduate University for Advanced Studies (Sokendai), Hayama, Japan.

GenBank accession numbers

The nomenclature used in this study and the accession numbers are listed on Additional Files 1 and 2.

Rapid amplification of cDNA ends (RACE)

Based on the partial sequence information of chicken PGRP-L retrieved from databases, we reconstructed the missing 3' region of the transcript using the BD SMART™ RACE cDNA Amplification Kit (Clontech, USA) according to manufacturer's instructions.

Data analyses

Sequences were aligned using the CLUSTALW version 1.83 computer program with its default parameter setting [35] and manually adjusted using the GeneDoc program version 2.6.002 [36]. Phylogenetical analyses were done using the neighbor-joining (NJ) and minimum evolution (ME) methods in MEGA version 4 [37]. The NJ and ME trees were based on the number of differences, and reliability was assessed by bootstrap values with 1000 replications. The reconciliation between species tree and gene tree along with the confirmation of the gene loss/duplication scenario were determined by using Notung 2.6 [25] and EvolMAP software [26]. To detect positive selection at single amino acid sites the Data Monkey software program was used with its default parameter setting http://www.datamonkey.org/. Poisson and gamma genetic distances were determined by using MEGA version 4 [37].

Test of convergence

The ancestral sequences were determined using the program ANCESTOR [27], and the significance of the convergent and parallel sites was estimated using the program CONVERG2 [32].

ACKNOWLEDGEMENTS

AMM is grateful to the Japan Society for the Promotion of Science (JSPS) for support. This work was supported in part by a Grant-in-Aid for Scientific Research on Priority Areas (13143202) from The Ministry of Education, Culture, Sports, Science, and Technology of Japan. We thank V. Byrappa for helpful discussions on Fugu PGRPs and Dr O. Sakarya for assistance with EvolMAP software. We thank Dr. A. Noguchi and Dr. Ş. K. Özdemir for critical reviewing of the manuscript. This paper is dedicated to Dr. Fumi Tsujino, who was deceased before completing the manuscript, on 27 October, 2006.

REFERENCES

1. Steiner H: Peptidoglycan recognition proteins: on and off switches for innate immunity. Immunol Rev. 2004, 198: 83-96. 10.1111/j.0105-2896.2004.0120.x.
2. Dziarski R, Gupta D: The peptidoglycan recognition proteins (PGRPs). Genome Biol. 2006, 7: 232-10.1186/gb-2006-7-8-232.
3. Girardin SE, Philpott DJ: The role of peptidoglycan recognition in innate immunity. Eur J Immunol. 2004, 34: 1777-1782. 10.1002/eji.200425095.
4. Janeway CA, Medzhitov R: Innate immune recognition. Annu Rev Immunol. 2002, 20: 197-216. 10.1146/annurev.immunol.20.083001.084359.
5. Aggrawal K, Silverman N: Peptidoglycan recognition in Drosophila. Biochem Soc Trans. 2007, 35: 1496-1500. 10.1042/BST0351496.
6. Chaput C, Boneca IG: Peptidoglycan detection by mammals and flies. Microbes Infect. 2007, 9: 637-647. 10.1016/j.micinf.2007.01.022.
7. Yoshida H, Kinoshita K, Ashida M: Purification of a peptidoglycan recognition protein from hemolymph of the silkworm, Bombyx mori. J Biol Chem. 1996, 271: 13854-13860. 10.1074/jbc.271.23.13854.
8. Kang D, Liu G, Lundström A, Gelius E, Steiner H: A peptidoglycan recognition protein in innate immunity conserved from insects to humans. Proc Natl Acad Sci. 1998, 95: 10078-10082. 10.1073/pnas.95.17.10078.
9. Liu C, Xu Z, Gupta D, Dziarski R: Peptidoglycan recognition proteins: a novel family of four human innate immunity pattern recognition molecules. J Biol Chem. 2003, 276: 34686-34694. 10.1074/jbc.M105566200.
10. Werner T, Liu G, Kang D, Ekengren S, Steiner H, Hultmark D: A family of peptidoglycan recognition proteins in the fruit fly Drosophila melanogaster. Proc Natl Acad Sci. 2000, 97: 13772-13777. 10.1073/pnas.97.25.13772.
11. Werner T, Borge-Renberg K, Mellroth P, Steiner H, Hultmark D: Functional diversity of the Drosophila PGRP-LC gene cluster in the response to lipopolysaccharide and peptidoglycan. J Biol Chem. 2003, 278: 26319-26322. 10.1074/jbc.C300184200.
12. Christophides GK, Zdobnov E, Barillas-Mury C, Birney E, Blandin S, Blass C, Brey PT, Collins FH, Danielli A, Dimopoulos G, et al: Immunity-related genes and gene families in Anopheles gambiae. Science. 2002, 298: 159-165. 10.1126/science.1077136.
13. Takehana A, Yano T, Mita S, Kotani A, Oshima Y, Kurata S: Peptidoglycan recognition protein (PGRP)-LE and PGRP-LC act synergistically in Drosophila immunity. EMBO J. 2004, 23: 4690-4700. 10.1038/sj.emboj.7600466.
14. Bischoff V, Vignal C, Boneca IG, Michel T, Hoffmann JA, Royet J: Function of the Drosophila pattern-recognition receptor PGRP-SD in the detection of Gram-positive bacteria. Nat Immunol. 2004, 5: 1175-1180. 10.1038/ni1123.
15. Choe KM, Werner T, Stoven S, Hultmark D, Anderson KV: Requirement for a peptidoglycan recognition protein (PGRP) in Relish activation and antibacterial immune responses in Drosophila. Science. 2002, 296: 359-362. 10.1126/science.1070216.
16. Leulier F, Parquet C, Pili-Floury S, Ryu JH, Caroff M, Lee WJ, Mengin-Lecreulx D, Lemaitre B: The Drosophila immune system detects bacteria through specific peptidoglycan recognition. Nat Immunol. 2003, 4: 478-484. 10.1038/ni922.
17. Mellroth P, Karlsson J, Steiner H: A scavenger function for a Drosophila peptidoglycan recognition protein. J Biol Chem. 2003, 278: 7059-7064. 10.1074/jbc.M208900200.
18. Chang CI, Pili-Floury S, Hervé M, Parquet C, Chelliah Y, Lemaitre B, Mengin-Lecreulx D, Deisenhofer J: A Drosophila pattern recognition receptor contains a peptidoglycan docking groove and unusual L, D-carboxypeptidase activity. PLoS Biol. 2004, 2: E277-10.1371/journal.pbio.0020277.
19. Dziarski R, Gupta D: Mammalian PGRPs: novel antibacterial proteins. Cell Microbiol. 2006, 8: 1059-1069. 10.1111/j.1462-5822.2006.00726.x.
20. Gelius E, Persson C, Karlsson J, Steiner H: A mammalian peptidoglycan recognition protein with N-acetylmuramoyl-L-alanine amidase activity. Biochem Biophys Res Commun. 2003, 306: 988-994. 10.1016/S0006-291X(03)01096-9.
21. Dziarski R, Platt KA, Gelius E, Steiner H, Gupta D: Defect in neutrophil killing and increased susceptibility to infection with non-pathogenic gram-positive bacteria in peptidoglycan recognition protein-S (PGRP-S)-deficient mice. Blood. 2003, 102: 689-697. 10.1182/blood-2002-12-3853.
22. Lu X, Wang M, Qi J, Wang H, Li X, Gupta D, Dziarski R: Peptidoglycan recognition proteins are a new class of human bactericidal proteins. J Biol Chem. 2006, 281: 5895-5907. 10.1074/jbc.M511631200.
23. Kim MS, Byun M, Oh BH: Crystal structure of peptidoglycan recognition protein LB from Drosophila melanogaster. Nature Immunology. 2003, 4: 787-793. 10.1038/ni952.
24. Guan R, Malchiodi EL, Wang Q, Schuck P, Mariuzza RA: Crystal structure of the C-terminal peptidoglycan-binding domain of human peptidoglycan recognition protein Iα. J Biol Chem. 2004, 279: 31873-31882. 10.1074/jbc.M404920200.
25. Durand D, Halldorsson BV, Vernot B: A Hybrid Micro-Macroevolutionary Approach to Gene Tree Reconstruction. J Comput Biol. 2006, 13: 320-335. 10.1089/cmb.2006.13.320.
26. Sakarya O, Kosik KS, Oakley TH: Reconstructing ancestral genome content based on symmetrical best alignments and Dollo parsimony. Bioinformatics. 2008, 24: 606-612. 10.1093/bioinformatics/btn005.

27. Zhang J, Nei M: Accuracies of ancestral amino acid sequences inferred by the parsimony, likelihood, and distance methods. J Mol Evol. 1997, 44 (Suppl 1): S139-S146. 10.1007/PL00000067.

28. Pond SL, Frost SD: Not so different after all: a comparison of methods for detecting amino acid sites under selection. Mol Biol Evol. 2005, 22: 1208-1222. 10.1093/molbev/msi105.

29. Sackton TB, Lazzaro BP, Schlenke TA, Evans JD, Hultmark D, Clark AG: Dynamic evolution of the innate immune system in Drosophila. Nat Genet. 2007, 39: 1461-1468. 10.1038/ng.2007.60.

30. Reiser JB, Teyton L, Wilson IA: Crystal structure of the Drosophila peptidoglycan recognition protein (PGRP)-SA at 1.56 Å resolution. J Mol Biol. 2004, 340: 909-917. 10.1016/j.jmb.2004.04.077.

31. Tajima F: Simple methods for testing the molecular clock hypothesis. Genetics. 1993, 135: 599-607.

32. Zhang J, Kumar S: Detection of convergent and parallel evolution at the amino acid sequence level. Mol Biol Evol. 1997, 14: 527-536.

33. Nei M, Rooney AP: Concerted and birth-and-death evolution of multigene families. Annu Rev Genet. 2005, 39: 121-152. 10.1146/annurev.genet.39.073003.112240.

34. Nei M, Gu X, Sitnikova T: Evolution by the birth-and-death process in multigene families of the vertebrate immune system. Proc Natl Acad Sci USA. 1997, 94: 7799-7806. 10.1073/pnas.94.15.7799.

35. Thompson JD, Higgins DG, Gibson TJ: CLUSTAL W: Improving the sensitivity of progressive multiple sequence alignment through sequence weighting, positions-specific gap penalties and weight matrix choice. Nucleic Acids Res. 1994, 22: 4673-4680. 10.1093/nar/22.22.4673.

36. Nicholas KB, Nicholas HB, Deerfield DWII: GeneDoc: Analysis and visualization of genetic variation. EMBNEW.NEWS. 1997, 4: 14.

37. Kumar S, Tamura K, Jakobsen IB, Nei M: MEGA2: Molecular evolutionary genetics analysis software. Bioinformatics. 2001, 17: 1244-1245. 10.1093/bioinformatics/17.12.1244.

38. Blair Hedges S, Kumar S: Genomic clocks and evolutionary timescales. Trends Genet. 2003, 19: 200-206. 10.1016/S0168-9525(03)00053-2.

CHAPTER 10

ORIGIN OF VERTEBRATE SOMITES

Takayuki Onai, Toshihiro Aramaki, Hidehiko Inomata, Tamami Hirai and Shigeru Kuratani

Keywords: Amphioxus Somites Notch signalling Origin of vertebrates Enterocoel theory

INTRODUCTION

Segmented structures composed of repetitive units, called somites, that arise transiently during embryogenesis are a key feature of the vertebrate body plan. The somites lie laterally to the notochord, and a spinal nerve forms a segmental unit assigned to somitic derivatives in the trunk [1, 2]. During development, somites differentiate into myotomes and skeletal elements that form the basic supporting structure in adults. From an evolutionary perspective, somites are conserved between vertebrates and amphioxus but are secondarily lost in tunicates [3, 4, 5]. Amphioxus (*Branchiostoma floridae*), also known as cephalochordate or lancelet, has a notochord, neural tube, and pharynx, but lacks a neural crest and placodes. In amphioxus, somites extend into the anterior end, unlike in vertebrates, and amphioxus is thus considered not to possess a homolog of the vertebrate unsegmented head mesoderm; however, this contention remains controversial (Additional file 1: Figure S1) [6, 7]. A recent genome analysis indicated that amphioxus retains most of the developmental genes, such as Hox clusters, present in vertebrates and is thus the best proxy to address the origin of the vertebrate body plan [8].

Although somites are common in chordates, there are some differences in developmental sequences between amphioxus and vertebrates. For example, in amphioxus, the rostral somites develop from the dorsal mesoderm during early embryogenesis. The mesoderm is a single layer located in the dorsal roof of the archenteron (endodermal/mesodermal structure) at the early neurula stage (Fig. 1a). By the mid-neurula stage, a prospective somite swells dorsolaterally and the bottom narrows (Fig. 1b). At the late neurula stage, all the rostral somites pinch off from the archenteron roof simultaneously and line up laterally to the notochord (Fig. 1c). Beginning from the late neurula stage, the caudal somites bud off one by one directly from the lateral epithelial cells in the tail bud by schizocoely (a coelom is formed by separating mesenchymal cells) [9]. Whereas, in vertebrates, somites are formed in pairs in the anterior end of the PSM through gradual epithelialization [10]. These developmental differences raise the question of the ancestral origin of somitogenesis.

Citation: Takayuki Onai, Toshihiro Aramaki, Hidehiko Inomata, Tamami Hirai and Shigeru Kuratani, origin of vertebrate somites, DOI: 10.1186/s40851-015-0033-0.

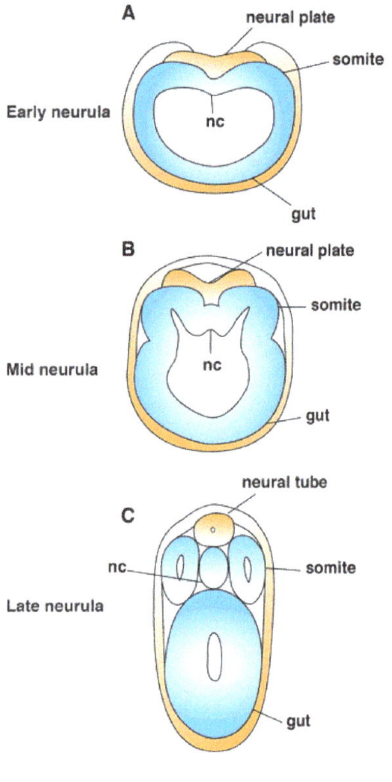

Figure 1. Development of amphioxus rostral somites. a At the early neurula stage, the dorsal roof of the anterior archenteron begins to expand dorsolateral to the ectoderm. b At the mid-neurula stage, the rostral somites swell and form a U-shape that remains part of the dorsal roof of archenteron. c At the late neurula stage, the somites pinch off from the archenteron roof. nc notochord

Somites do not occur in any group other than chordates. However, a previous study indicated that the evolutionary origin of somites could be found in cnidarians, diploblastic animals that are thought to be ancestral to the bilaterians [11]. According to Sedgwick's enterocoel theory, chordate somites are derived from the alimentary pouches of coelenterates (ctenophores and cnidarians) [11]. Nonetheless, it is still debated whether coelomic cavities are ancestral to all bilaterians (e.g., the basal bilaterian Acoela does not have coelomic cavities), as there is no persuasive evidence to support this hypothesis [12, 13, 14]. The enterocoel theory promotes the view that the rostral somites are more ancestral than the caudal somites in amphioxus and the vertebrate somites. However, without grounding in molecular genetic comparisons, the somite evolutionary scenario remains enigmatic.

Recent molecular studies have revealed the developmental mechanisms underlying somitogenesis. In vertebrates, somites are formed by several developmental sequences. Starting from gastrulation, the PSM is internalized around the blastopore and becomes part of the tail bud at the posterior end of the body. A future pair of somites is gradually epithelialized in the anterior PSM, and a boundary between each somite is determined. One pair of somites is formed every 120 min in mice and every 25 min in zebrafish, indicating the interspecies differences in somitogenetic rhythms among vertebrates [15]. Molecular oscillators control this rhythmic somitogenesis. Of these, Notch signalling is a key factor in the anterior PSM, or "determination front", and is essential for the boundary formation of future segments [10]. At the determination front, the lunatic fringe (*Lfng*) glycosyltransferase, a downstream modulator of Notch signalling, has been reported to be essential for generating a boundary by modifying the Notch receptor in mouse embryos [15]. Additionally, in zebrafish, *Lfng* is expressed in the tail bud where, unlike in mouse and chick embryos, expression does not oscillate, suggesting a diverse role for somitogenesis, which may have been secondarily lost in the anole lizard, as *Lfng* is not expressed in the PSM in lizard embryos [16, 17]. Upstream of *Lfng*, mesoderm posterior (*Mesp2*), a member of the basic helix-loop-helix transcriptional factor family, regulates Notch signalling input [18].

In amphioxus, the dorsal mesoderm includes somites and the notochord the initial somite boundary formation is recognized in the dorsal mesoderm at the mid-gastrula stage. A pair of stripes of Bf*Delta* (the homolog of DLL1) expression is detected in the first somites [19]. Additionally, several homologs of vertebrate segmentation genes, such as *Hairy* and *Uncx4.1*, are expressed in somites from the gastrula to the larval stages [9]. However, no direct evidence of periodic expression of cyclic genes in the tail bud has been reported. Although the amphioxus genome harbours a *Mesp* homolog, it is not likely to be expressed during embryogenesis [9]. Bf*Fringe* is expressed in the neural tube and the endodermal gut, but not in the somites, during the neurula stages [20]. Evidence suggests that the periodic expression of cyclic genes and the establishment of the determination front by the *Mesp* and *Lfng* genes are vertebrate-specific genetic networks.

In this study, we investigated the molecular mechanisms underlying somitogenesis in amphioxus embryos with a focus on Notch signalling. Since current evidence suggests that vertebrate somites evolved from somites of an amphioxus-like ancestral animal, we performed gene expression and functional analyses using amphioxus and *Xenopus* embryos.

MATERIALS AND METHODS

Collection of amphioxus embryos

Amphioxus adults were collected in Old Tampa Bay, Florida, USA during the summer breeding season in 2011. The animals were subjected to electric shock in filtrated seawater at night (21:00 to midnight), as described previously [21], to promote spawning. The fertilized eggs were cultured in culture dishes (6–36 h) with filtrated seawater at room temperature.

Xenopus experiments

Following *in vitro* fertilization, the embryos were cultured in 0.1× Barth's solution until DAPT (Tocris, Bristol, UK) treatment. Staging was performed based on the normal table of Nieuwkoop and Faber [22].

DAPT treatments

DAPT was dissolved in DMSO (100 mM). Fifty [23] to 100 micromolar DAPT was added to amphioxus embryos at the late blastula stage at room temperature. The larval morphology of 50–100 μM DAPT-treated embryos was similar.

The same amount of DMSO was applied to the seawater as a control. The DAPT-treated embryos were further cultured until they reached the late gastrula or larval stages for fixation. In *Xenopus*, 200 μM DAPT or the same amount of DMSO for controls was added to 0.1× Barth's solution at the late blastula stage, and the embryos were further cultured until fixation. One hundred micromolar DAPT treatment in *Xenopus* did not lead to any overt phenotypes, as seen in Fig. 6. Embryos were fixed in 4 % paraformaldehyde in 3-(N-morpholino) propanesulfonic acid (0.1 M) solution for amphioxus or in 3-(N-morpholino) propanesulfonic acid (0.1 M)/2 mM EGTA/1 mM magnesium sulphate/3.7 % formaldehyde solution for *Xenopus* embryos. Amphioxus embryos were stored in 70 % EtOH, and *Xenopus* embryos were stored in 100 % MeOH until further analysis.

Whole-mount in situ hybridization (WISH) and F-actin staining

WISH and F-actin staining with BODIPY FL phallacidin (Molecular Probes, Minato, Tokyo, Japan) were performed following previously published protocols [23, 24, 25]. WISH was performed for the following genetic markers: Bf*Delta* [19], Bf*Hairy-b* [26], Bf*Gsc*, Bf*Bra* [21], Bf*Muscle-actin* [25], Bf*Fringe* [20], Bf*Mrf1* [27], Xl*Gsc*[28], Xl*Bra* [29], Xl*Delta2* [30], Xl*MyoD* [31], and Xl*Tbx1* [32]. Zeiss LSM 710 or LSM 780 confocal microscope (Zeiss, Shinjuku, Tokyo, Japan) was used for detecting *F-Actin* signals and fluorescence WISH.

Embedding and sectioning of amphioxus embryos

Following WISH, the embryos were washed in distilled water for 3 min. The embryos were stained with Ponceau S solution (SIGMA, Shinagawa, Tokyo, Japan), incubated for 1 h, and then washed in an EtOH series (75–100 %). After removal of the EtOH, resin (Polysciences, Inc., Warrington, PA, USA) was added to the dishes and incubated for 30 min. Samples were transferred to an embedding mould and treated with fresh resin for 30 min. The mould was then placed in an oven (67 C overnight). The embedded samples were cut with a sharp glass knife, and the sections were carefully transferred to glass slides.

RESULTS

Dorso-ventral boundary formation between the rostral somites and the gut is controlled by Notch signalling

In vertebrates, Lfng suppresses Notch signalling in the anterior PSM and controls segmental boundary determination of future somites [33]. In amphioxus, Bf *Fringe* is expressed in the neural plate but not in the somites or tail bud [20]. However, at the late neurula stage, Bf*Fringe* is expressed in the ventral part of the dorso-ventral boundary between somites and the gut, suggesting that Bf*Fringe* regulates dorso-ventral boundary formation of the archenteron [20]. We examined the expression pattern of Bf*Fringe* at the late gastrula stage and found that it was expressed in the ventral half of the archenteron (Fig. 2a). The dorsal limit of Bf*Fringe* expression was located ventral to Bf *Delta* expression in the presumptive rostral somites (Fig. 2b). These findings suggest that the Fringe/Delta cassette controls dorso-ventral boundary formation between the mesoderm (somites) and the endoderm (gut) in the archenteron.

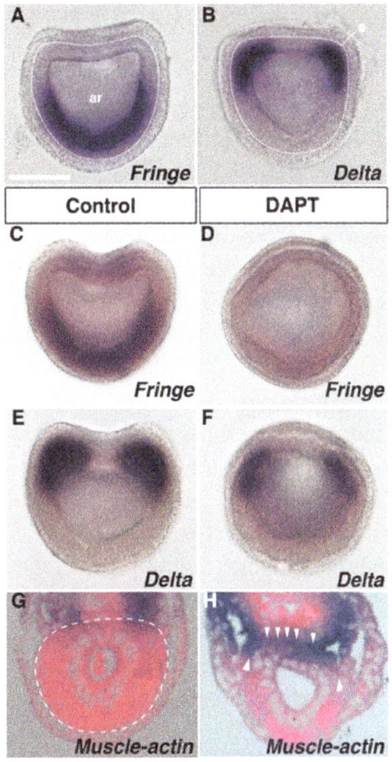

Figure 2. Notch signalling controls the pinching off process of the rostral somites. a–b BfFringe expressed in the anterior endoderm (ventral part of the archenteron) and BfDelta was expressed in the presumptive rostral somites at the late gastrula stage. Blastopore views with the dorsal side up. The white dotted circle indicates the archenteron. ssomite, ar archenteron. Scale bar, 50 μm. c–f Effect of 100 μM DAPT treatment on BfFringe (n = 7, 100 %) or BfDelta(n = 8, 100 %). Anterior views with the dorsal side up. g In the DMSO-treated control larval embryo, the segmental boundary between the somite and the dorsal gut roof was clear (n = 1). h In larval embryos treated with DAPT, the boundary was unclear and ectopic expression of Muscle-actin was observed (n = 1). Transverse sections with the dorsal side up. White arrowheads indicate somite and gut fusion locations. The white dotted circle indicates the border between the gut and somites

In vertebrates, Notch signaling regulates *Lfng* expression in the PSM. For example, DAPT treatment in chick embryos affects the cyclic expression of *Lfng* [34]. To test whether Notch signalling controls the expression of Bf*Fringe* and Bf *Delta* in amphioxus embryos, we treated the embryos from the late blastula stage with 100 μM DAPT. DAPT-treated embryos displayed severe defects in the pinching off process during somitogenesis. At the very late gastrula stage, DMSO-treated control embryos exhibited swelling of the lateral dorsal roof of archenteron, while morphological transformation of the dorsal archenteron roof was not detected in the DAPT-treated embryos (Fig. 2c and d). Furthermore, the expression of Bf*Fringe* was eliminated from the anterior endoderm in DAPT-treated embryos (Fig. 2c and d). In contrast, Bf*Delta* expression was detected in the presumptive somites both in the control and DAPT-treated embryos, but the expression was reduced in the latter (Fig. 2e and f). Later in development, DAPT treatment expanded the localization of expression of Bf*Muscle-actin* (a conserved structural protein in muscles expressed in myotomes) medially into the gut at the larval stage, and the boundary between the somites and the gut became vague (Fig. 2g and h). These results indicate that Notch signalling and the Fringe/Delta cassette are essential for dorso-ventral boundary formation between the somites and the gut.

Notch signalling is essential for rostral somite Formation in amphioxus embryos

The essential role of Notch signalling on formation of the dorso-ventral boundary between the gut and somites suggests that Notch signalling may also play a role in segmental boundary formation between somites in amphioxus embryogenesis. To address this questions, we studied the role of Notch signalling in segmental boundary formation between somites. In amphioxus, the first and second somites express the somite markers Bf*Delta* and Bf*Hairy-b* at the late gastrula stage [19, 26]. Treatment with 100 μM DAPT from the late blastula stage reduced Bf*Delta* and Bf*Hairy-b* expression (Fig. 3a–d), indicating that Notch signalling is essential for rostral somite formation. Treatment with DAPT had little effect on the expression of the axial mesoderm marker Bf*Gsc* and the pan-mesoderm marker Bf*Bra* (Fig. 3e–h). These findings suggest that the role of Notch signalling is to specify the rostral somites rather than to induce the entire dorsal mesoderm. To determine whether the effect of DAPT treatment disrupts segmental boundaries in the rostral somites, we examined later stage phenotypes. At the larval stage, DAPT-treated embryos displayed severe morphological abnormalities; both notochord and somites bent mediolaterally and epidermal cells were much thicker than those of the control embryos (Fig. 4a and b). In control embryos, Bf*Muscle-actin* expression was detected in myotomes that were asymmetric along with the left/right (L/R) body axis (Fig. 4c and e). In DAPT-treated embryos, Bf*Muscle-actin* expression was observed in myotomes; however, the size of each myotome was smaller and the boundaries between myotomes were less clear compared with those of the control embryos. In addition, the asymmetry of somites was disturbed in DAPT-treated embryos (Fig. 4d and f) [35, 36].

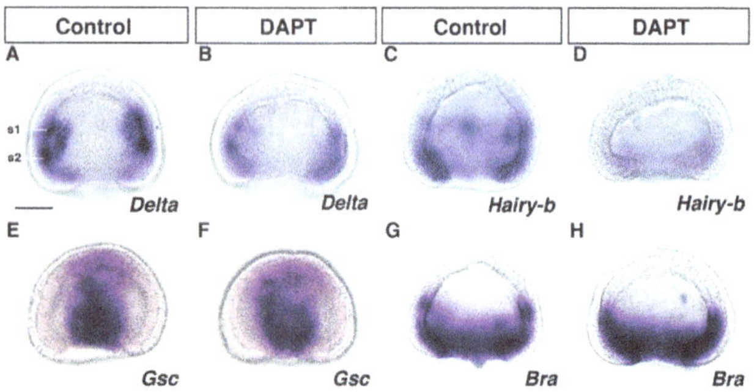

Figure 3. DAPT treatment affects somitogenesis in amphioxus embryos. Effect of treatment with 100 μM DAPT at the late blastula stage on mesodermal gene expression (a, b Delta; c, d, Hairy-b; e, f, Gsc; g, h, Brachyury; n = 10 each, 100 %)at the gastrula stage. Dorsal views with the anterior side up. s, somite. Scale bar, 50 μm (a)

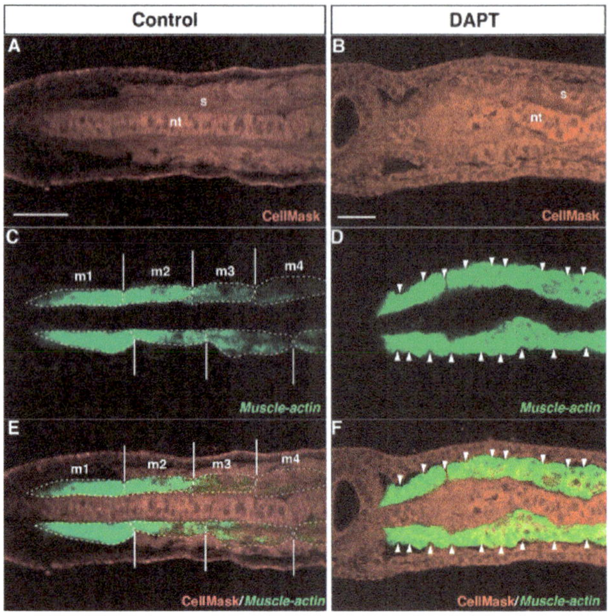

Figure 4. Effect of DAPT treatment on BfMuscle-actin expression. a, c, e DMSO-treated control embryo (n = 5, 100 %). Anterior to the left. Dorsal view. nt, notochord; s, somite; m, myotome. CellMask (Red) labelled the plasma membrane. BfMuscle-actin (Green) was stained by fluorescence in situ hybridization. The white dotted circle indicates a myotome. The white line indicates the segmental border between myotomes. b, d, f DAPT-treated embryo (n = 5, 100 %). Anterior to the left. Dorsal view. The white arrowhead indicates a possible segmental border between myotomes. Scale bars, 20μm

The incomplete formation of the segmental boundary of each somite in the DAPT-treated embryos prompted us to test whether actin filament formation was normally organized in the somites. In the DMSO-treated control embryos, three or four muscle fibres were formed in each somite (Fig. 5a). However, the DAPT-treated embryos did not have actin filaments (no F-actin signal was detected in the somites; Fig. 5b–d). Together, these findings suggest that Notch signalling is essential for both segmental boundary formation of the somites and differentiation of muscle fibres. As Bf*Fringe* is not expressed in the somites, the Notch signalling-mediated regulation of segmental boundary formation between somites involves genetic networks distinct from that underlying dorso-ventral boundary formation between the gut and somites.

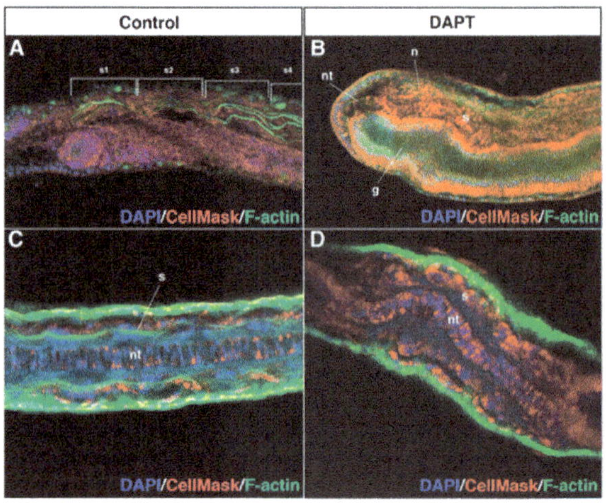

Figure 5. Filament formation of actin is suppressed by DAPT treatment. a DMSO-treated control embryos display filamentous actin in

the rostral somites (n=12, 100 %). b F-actin staining in the somites. Treatment with 100 μM DAPT from the late gastrula stage onwards (n=10, 100 %). The images were taken using an LSM 710 confocal microscope (Zeiss). cDorsal view of DMSO-treated control embryos. d Dorsal view of DAPT-treated embryos. s somite, nt notochord, nneural tube, g, gut. Nuclei were labelled in blue and the plasma membrane was labelled in red

Notch signalling is required for somitogenesis from the gastrula stages in vertebrate embryos

Suppression of Notch signalling resulted in reduced somite marker expression in amphioxus at the late gastrula stage (Fig. 3a–d). In vertebrates, the dorsal mesoderm is regionalized into the head and trunk mesoderm during the gastrula stages, and the somite progenitor cells are generated in the trunk mesoderm [37]. However, it is unclear whether Notch signalling is crucial for the initial differentiation of somites during gastrula stages. To address this, we investigated the role of Notch signalling in somitogenesis during the gastrula stages. In *Xenopus,* Xl*Gsc* and Xl*Bra* were expressed in the head mesoderm and the notochord, respectively, while Xl*Delta-2* was expressed in the region of the trunk somites at the late gastrula stage (Fig. 6a, c, e). Similar to the findings in amphioxus embryos, treatment with DAPT from the late blastula stage did not affect the expression of Xl*Gsc* and Xl*Bra*, while Xl*Delta-2* expression was severely disrupted (Fig. 6a–f). This suggests that, as in amphioxus, Notch signalling regulates the formation of somites, but not mesoderm induction, during the gastrula stages.

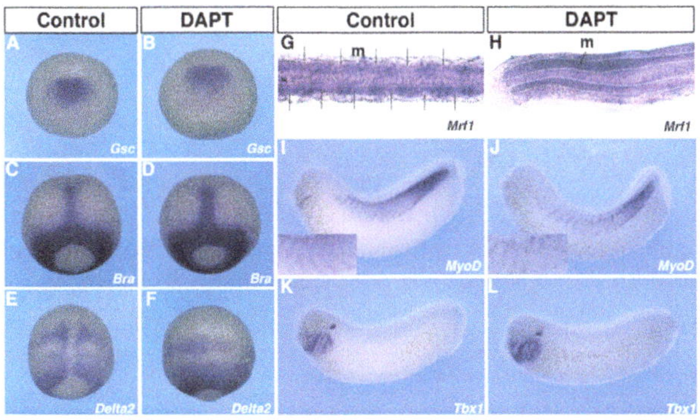

Figure 6. Inhibition of Notch signalling results in loss of segment formation in Xenopus embryos. a–f DAPT treatment did not affect the expression pattern of Gsc (a n=42, 93 %; b n=38, 87 %) and Brachyury (c n=20, 95 %; d n=14, 94 %), whereas it disrupted the paired-stripe expression pattern of Delta-2 (e n=33, 100 %; f n=47, 51 %). a–b Anterior views with dorsal side up. c–f Dorsal views with the anterior side up. g–h DAPT treatment resulted in loss of segmental expression of Mrf1 in amphioxus (g n=12, 100 %; h n=11, 100 %. Dorsal views with anterior to the left), MyoD in Xenopus (i n=23, 100 %; j n=15, 93 %), and Tbx1 (k n=23, 100 %; l n=13, 54 %). i, j The lower left panel displays the magnification of the somite expression of MyoD

Additionally, DAPT-treated *Xenopus* embryos exhibited severe phenotypes at the tail bud stage. Bf*Mrf1* is a homolog of the vertebrate *MyoD* gene, a key marker of early muscle differentiation [27]. Treatment of the embryos with 100 μM DAPT from the late blastula stage onwards resulted in loss of segmental Bf*Mrf1*expression in the somites (Fig. 6g and h). DAPT treatment had a severe effect on the segmental expression of Xl*MyoD* in the somites. In control embryos, Xl*MyoD* expression was observed in each somite in a V-shaped pattern. However, this V-shaped expression was disrupted, and the borders of Xl*MyoD* expression in each somite were indistinct in DAPT-treated embryos (Fig. 6i and j). In contrast, expression of the pharyngeal mesoderm marker Xl*Tbx1* was relatively normal in the DAPT-treated embryos (Fig. 6k and l). Taken together, these results suggest that Notch signalling controls somitogenesis from the gastrula stage in amphioxus and vertebrates.

DISCUSSION

Somitogenesis in amphioxus embryos

Previous molecular studies of amphioxus somitogenesis revealed that although rostral and caudal somites display differences in their developmental processes, both express conserved somite segmentation markers (e.g., *Tbx15/18/22, Delta, Notch, Hey, HairyC, HairyD, IrxA, NeuroD, Ripply, Six1/2*). These are thus considered to involve similar developmental mechanisms despite the observation that some genes, such as Bf*En*, are expressed only in rostral somites [9, 38].

A functional study of Fgf signalling, however, demonstrated that loss of Fgf signalling affects only rostral and not caudal somite segmentation [39]. This suggests that the molecular mechanisms underlying somite segmental boundary formation and AP body axis formation in amphioxus are different. In this study, we demonstrated that Notch signalling is essential for dorso-ventral boundary formation between the somites and the gut in the archenteron. In addition, our data suggest that the Fringe/Delta cassette under Notch signalling is likely important in dorso-ventral boundary formation in the archenteron. Notch signalling plays also an essential role in segmental boundary formation between somites. We have also shown that DAPT treatment results in loss of the F-actin formation in myotomes. Finally, our data indicate that Notch signalling is important for the L/R asymmetry of somites. Current data suggest that Notch and Fgf signalling regulate the segmental boundaries of rostral somites, but it is unclear whether crosstalk between Notch and Fgf signalling exists in rostral somite segmentation.

In amphioxus, caudal somites develop directly from the tail bud. In the control larval embryos, segmental expression of Bf*Muscle-actin* was detected in caudal myotomes (Additional file 1: Figure S2A, C, E). Additionally, caudal myotomes were asymmetric (Additional file 1: Figure S2E). In DAPT-treated embryos, expression of Bf*Muscle-actin* was somewhat continuous, and clear segmental boundaries between somites were nearly undetectable (Additional file 1: Figure S2B, D, F). Furthermore, the asymmetric organization of somites along with the L/R body axis was irregular (Additional file 1: Figure S2F). These results indicate that Notch signalling plays an essential role in caudal somitogenesis. As Fgf signalling does not play a primary role in caudal somite segmentation, Notch signalling is probably the major factor driving segmental border formation in caudal somites.

Gut pouches and the origin of the somites

Based on the results of the current study, we propose that Notch signalling-dependent segmental boundary formation between the gut and somites in an amphioxus-like ancestor may have been co-opted by vertebrates and used in segmental border determination between the propsective somites and the anterior PSM (Fig. 7). According to the enterocoel theory, the dynamic transformation of a part of the endoderm to mesoderm appears to have happened during evolution from a cnidarian-like diploblastic ancestor to triploblastic animals [11, 14, 40]. Recent molecular phylogeny proposed that deuterostomes are in a basal lineage of bilateria [41].

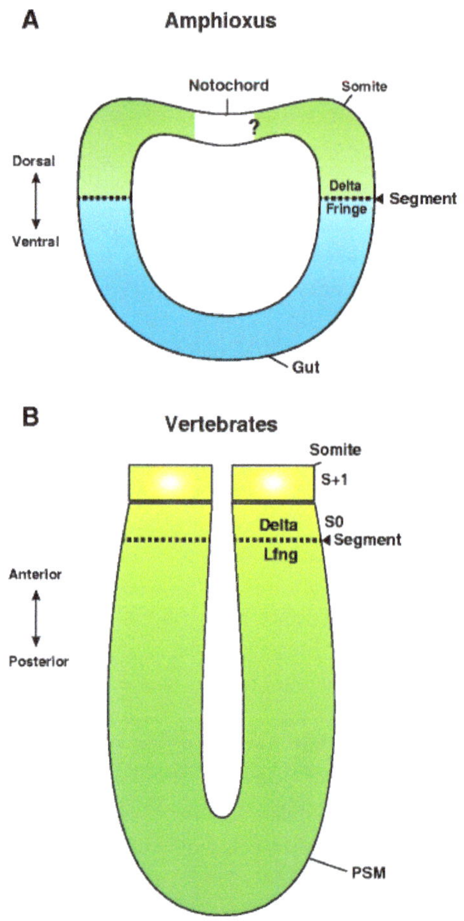

Figure 7. A new evolutionary scenario of the chordate somite formation. a In amphioxus, mesoderm/endoderm boundary formation is organized by Fringe/Delta cassette under Notch signalling. b The Notch signalling-dependent mechanism of segmental boundary formation in (a) was co-opted to the future segments in the anterior PSM of amniotes embryos. The mesoderm is presented in green. The endoderm is in light blue. The dotted line indicates a segmental border

In amphioxus, the mesoderm and endoderm originate from the same single layer (archenteron), and the mesoderm differentiates into the musculature somites and the notochord, from which muscle fibres extend to the neural tube [42, 43]. A recent gene expression profiling study demonstrated that mesodermal genes are expressed in the embryonic endoderm of the sea anemone *Nematostella* (a member of the phylum Cnidaria) [44], suggesting that a new genetic program for mesoderm formation in endoderm evolved in Urbilateria. For the evolution of mesoderm in the archenteron, axial patterning events that occur during early embryogenesis could be important. In deuterostomes, dorsal/ventral (D/V) axial determination is regulated by members of the TGF-β super family, including *Nodal* and *Bmps* [21]. Nodal and Bmp antagonize each other to generate D/V polarity. Conserved Nodal/Bmp antagonism in deuterostomes is implicated in a fundamental genetic mechanism specifying and maintaining D/V polarity in mesoderm development [21]. Future studies will address how the mesoderm specification gene networks evolved in the archenteron by reorganizing the relationship between Nodal signalling and key mesodermal genes (e.g. *Gsc, Bra,* and *Twist*) and generating the mesoderm/endoderm boundary in the archenteron by Notch signalling. The genome of the ctenophore (*Mnemiopsis leidyi*) has many Notch signalling components, suggesting an ancient role for Notch signalling in boundary formation [45].

In hemichordates, a sister group of chordates, the longitudinal muscles run just underneath the dorsal nerve cord [46]. In some hemichordate species (e.g., *S. kowalevskii*), the mesoderm is derived from the endoderm by enterocoely (the mesodermal coelom forms by outpocketing of the gut) [46, 47]. Currently, it is unclear whether the genetic programs that underlie enterocoelic mesoderm formation are shared between hemichordates and amphioxus.

CONCLUSIONS

Our results reveal that Notch signalling and the Fringe/Delta cassette regulates dorsal-ventral boundary formation in the archenteron to segregate somites from the gut in the basal chordate amphioxus. These findings increase our understanding of how the vertebrate body plan evolved by recapitulating ancestral developmental programs (e.g. segmental boundary formation in amphioxus archenteron) and adapting them to a novel developmental event (e.g. segmental border formation in vertebrate anterior PSM) as a heterotopic shift.

DECLARATIONS

Acknowledgements

The authors thank Drs. Holland L.Z. and Holland N.D. for their help with amphioxus sampling and discussion. This project was supported by a KAKENHI Grant-in-Aid for Young Scientists (B) from the Japan Society for the Promotion of Science (grant number 24770222).

REFERENCES

1. Goethe JW. Das Schädelgrüt aus sechs Wirbelknochen aufgebaut. Zur Naturwissenschaft überhaupt, besonders zur Morphologie II 2. 1790.
2. Oken L. Über die Bedeutung der Schädelknochen. Bamberg: Göbhardt; 1807.
3. Goodrich ES. On the development of the segments of the head in *Scyllium*. QJ Microsc Sci. 1918;63:1–30.
4. Onai T, Irie N, Kuratani S. The evolutionary origin of the vertebrate body plan: the problem of head segmentation. Annu Rev Genomics Human Genet. 2014;15:443–59. doi:10.1146/annurev-genom-091212-153404.
5. Holland LZ, Onai T. Early development of cephalochordate (amphioxus). WIREs Dev Biol. 2011;1:167–83.
6. Kuratani S. Is the vertebrate head segmented?–evolutionary and developmental considerations. Integr Comp Biol. 2008;48(5):647–57.

7. Holland LZ, Holland ND, Gilland E. Amphioxus and the evolution of head segmentation. Integr Comp Biol. 2008;48(5):630–46.
8. Putnam NH, Butts T, Ferrier DE, Furlong RF, Hellsten U, Kawashima T, et al. The amphioxus genome and the evolution of the chordate karyotype. Nature. 2008;453(7198):1064–71.
9. Beaster-Jones L, Kaltenbach SL, Koop D, Yuan S, Chastain R, Holland LZ. Expression of somite segmentation genes in amphioxus: a clock without a wavefront? Dev Genes Evol. 2008;218(11–12):599–611.
10. Pourquie O. Vertebrate segmentation: from cyclic gene networks to scoliosis. Cell. 2011;145(5):650–63.
11. Sedgwick A. On the origin of metameric segmentation and some other morphological questions. Q J Microsc Sci. 1884;24:43–82 + pl. II-III.
12. Hejnol A, Martindale MQ. Acoel development supports a simple planula-like urbilaterian. Phil Trans R Soc B. 2008;363(1496):1493–501.
13. Remane A. Zur Metamerie, Metaerismen und Metamerisation bei Wirbeltieren. Zool Anz. 1963;170:489–502.
14. Masterman AT. On the diplochorda. Q J Microsc Sci. 1898;40:281–366.
15. Hubaud A, Pourquie O. Signalling dynamics in vertebrate segmentation. Nat Rev Mol Cell Biol. 2014;15(11):709–21.
16. Holley SA, Nusslein-Volhard C. Somitogenesis in zebrafish. Curr Top Dev Biol. 2000;47:247–77.
17. Eckalbar WL, Lasku E, Infante CR, Elsey RM, Markov GJ, Allen AN, et al. Somitogenesis in the anole lizard and alligator reveals evolutionary convergence and divergence in the amniote segmentation clock. Dev Biol. 2012;363(1):308–19.
18. Morimoto M, Takahashi Y, Endo M, Saga Y. The Mesp2 transcription factor establishes segmental borders by suppressing Notch activity. Nature. 2005;435(7040):354–9.
19. Rasmussen SL, Holland LZ, Schubert M, Beaster-Jones L, Holland ND. Amphioxus AmphiDelta: evolution of Delta protein structure, segmentation, and neurogenesis. Genesis. 2007;45(3):113–22.
20. Mazet F, Shimeld SM. Characterisation of an amphioxus Fringe gene and the evolution of the vertebrate segmentation clock. Dev Genes Evol. 2003;213(10):505–9.
21. Onai T, Yu JK, Blitz IL, Cho KW, Holland LZ. Opposing Nodal/Vg1 and BMP signals mediate axial patterning in embryos of the basal chordate amphioxus. Dev Biol. 2010;344(1):377–89. doi:10.1016/j.ydbio.2010.05.016.
22. Nieuwkoop PD, Faber J. Normal table of Xenopus laevis. Amsterdam: North-Holland Publishing Co; 1956.
23. Lu TM, Luo YJ, Yu JK. BMP and Delta/Notch signaling control the development of amphioxus epidermal sensory neurons: insights into the evolution of the peripheral sensory system. Development. 2012;139(11):2020–30.
24. Onai T, Sasai N, Matsui M, Sasai Y. *Xenopus* XsalF: anterior neuroectodermal specification by attenuating cellular responsiveness to Wnt signaling. Dev Cell. 2004;7(1):95–106.
25. Onai T, Takai A, Setiamarga DH, Holland LZ. Essential role of Dkk3 for head formation by inhibiting Wnt/beta-catenin and Nodal/Vg1 signaling pathways in the basal chordate amphioxus. Evol Dev. 2012;14(4):338–50.
26. Minguillon C, Jimenez-Delgado S, Panopoulou G, Garcia-Fernandez J. The amphioxus Hairy family: differential fate after duplication. Development. 2003;130(24):5903–14.
27. Schubert M, Meulemans D, Bronner-Fraser M, Holland LZ, Holland ND. Differential mesodermal expression of two amphioxus MyoD family members (AmphiMRF1 and AmphiMRF2). Gene Expr Patterns. 2003;3(2):199–202.
28. Cho KW, Blumberg B, Steinbeisser H, De Robertis EM. Molecular nature of Spemann's organizer: the role of the *Xenopus*homeobox gene goosecoid. Cell. 1991;67(6):1111–20.
29. Smith JC, Price BM, Green JB, Weigel D, Herrmann BG. Expression of a *Xenopus* homolog of Brachyury (T) is an immediate-early response to mesoderm induction. Cell. 1991;67(1):79–87.
30. Peres JN, McNulty CL, Durston AJ. Interaction between X-Delta-2 and Hox genes regulates segmentation and patterning of the anteroposterior axis. Mech Dev. 2006;123(4):321–33.
31. Hopwood ND, Pluck A, Gurdon JB. MyoD expression in the forming somites is an early response to mesoderm induction in *Xenopus* embryos. EMBO J. 1989;8(11):3409–17.
32. Ataliotis P, Ivins S, Mohun TJ, Scambler PJ. XTbx1 is a transcriptional activator involved in head and pharyngeal arch development in *Xenopus laevis*. Dev Dyn. 2005;232(4):979–91.
33. Zhang N, Gridley T. Defects in somite formation in lunatic fringe-deficient mice. Nature. 1998;394(6691):374–7.
34. Dale JK, Maroto M, Dequeant ML, Malapert P, McGrew M, Pourquie O. Periodic notch inhibition by lunatic fringe underlies the chick segmentation clock. Nature. 2003;421(6920):275–8.
35. Kusakabe R, Satoh N, Holland LZ, Kusakabe T. Genomic organization and evolution of actin genes in the amphioxus *Branchiostoma belcheri* and *Branchiostoma floridae*. Gene. 1999;227(1):1–10.
36. Kusakabe R, Kusakabe T, Satoh N, Holland ND, Holland LZ. Differential gene expression and intracellular mRNA localization of amphioxus actin isoforms throughout development: implications for conserved mechanisms of chordate development. Dev Genes Evol. 1997;207:203–15.
37. Winklbauer R, Damm EW. Internalizing the vegetal cell mass before and during amphibian gastrulation: vegetal rotation and related movements. WIREs Dev Biol. 2012;1:301–6.
38. Holland LZ, Kene M, Williams NA, Holland ND. Sequence and embryonic expression of the amphioxus

engrailed gene (AmphiEn): the metameric pattern of transcription resembles that of its segment-polarity homolog in *Drosophila*. Development. 1997;124(9):1723–32.

39. Bertrand S, Camasses A, Somorjai I, Belgacem MR, Chabrol O, Escande ML, et al. Amphioxus FGF signaling predicts the acquisition of vertebrate morphological traits. Proc Natl Acad Sci U S A. 2011;108(22):9160–5.

40. Haeckel E. The gastraea-theory, the phylogenetic classification of the animal kingdom and the homology of the germ-lamellae. Q J Microsc Sci. 1874;14:142–65. 223–47.

41. Dunn CW, Hejnol A, Matus DQ, Pang K, Browne WE, Smith SA, et al. Broad phylogenomic sampling improves resolution of the animal tree of life. Nature. 2008;452(7188):745–9.

42. Bone Q. The origin of the chordates. J Linn Soc Lond. 1960;44:252–69.

43. Wicht H, Lacalli TC. The nervous system of amphioxus: structure, development, and evolutionary significance. Can J Zool. 2005;83:122–50.

44. Martindale MQ, Pang K, Finnerty JR. Investigating the origins of triploblasty: 'mesodermal' gene expression in a diploblastic animal, the sea anemone *Nematostella vectensis* (phylum, Cnidaria; class, Anthozoa). Development. 2004;131(10):2463–74.

45. Ryan JF, Pang K, Schnitzler CE, Nguyen AD, Moreland RT, Simmons DK, et al. The genome of the ctenophore Mnemiopsis leidyi and its implications for cell type evolution. Science. 2013;342(6164):1242592.

46. Kaul-Strehlow S, Stach T. A detailed description of the development of the hemichordate *Saccoglossus kowalevskii* using SEM, TEM, histology and 3D-reconstructions. Front Zool. 2013;10(1):53.

47. Green SA, Norris RP, Terasaki M, Lowe CJ. FGF signaling induces mesoderm in the hemichordate *Saccoglossus kowalevskii*. Development. 2013;140(5):1024–33.

CHAPTER 11 BIRTH AND DEATH OF GENE OVERLAPS IN VERTEBRATES

Izabela Makałowska, Chiao-Feng Lin and Krisitina Hernandez

BACKGROUND

3.2 billion base pairs of the human genome harbor about 23,000 protein coding genes. With the average size of a gene equal to 48 kb, they cover approximately one third of the genome. It seems that there's enough space in the genome for each gene to be separated by a large distance. Yet, between five and fourteen per cent of genes in the vertebrate genomes do overlap [1]. The unexpected abundance of complementary pairs of sense/antisense transcripts poses major challenges to achieve a comprehensive understanding of a gene structure and expression at the genomic level. Studies of individual overlapping gene pairs in eukaryotes have shown that they regulate gene expression by different mechanisms such as genomic imprinting [2], RNA interference and translational regulation [3], transcriptional interference [4], alternative splicing [5], and X-inactivation [6]. Many genes involved in overlaps are known to be involved in disease development, e.g. *CACP*gene is responsible for camptodactyly-arthropathy-coxa vara-pericarditis syndrome [7] or are responsible for some important morphological features, e.g. *SLC24A5* is partially responsible for skin coloration [8].

Several mechanisms have been proposed to explain gene overlap origination. For instance, Keese and Gibbs [9] suggested that overlapping genes arise as a result of overprinting a process of generating new genes from preexisting nucleotide sequences. This process supposedly took place after divergence of mammals from birds and overlapping genes represent young, phylogeneticaly restricted genes encoding proteins with diverse functions, and are therefore specialized to the present life-style of the organism in which they are found. Shintani et al. [10] suggested that the overlap between genes *ACAT2* (acetyl-Coenzyme A acetyltransferase 2) and *TCP1* (t-complex 1) arose during the transition from therapsid reptiles to mammals in one of two ways. In one scenario, one of genes was translocated and the rearrangement has been accompanied by the loss of a part of the 3› UTR, including the polyadenylation signal, from one gene. By chance the 3› UTR of the new neighbor on the opposite strand contained all the signals necessary for transcription termination so that the translocated gene could continue to function. Alternatively, the two genes become neighbors through the rearrangement but at first did not overlap. Later, one of the genes lost its original polyadenylation signal, but was able to use a signal that happened to be present on the non-coding strand of the other gene. The *ACAT2-TCP1* overlap evolution was placed, similarily as in Keese and

Citation: Izabela Makałowska, Chiao-Feng Lin and Krisitina Hernandez, Birth and death of gene overlaps in vertebrates, DOI: 10.1186/1471-2148-7-193.

Gibbs hypothesis, after the divergence of mammals from birds. Dahary et al. [11] place the origin of most vertebrate overlaps much earlier. They found that human antisense genes have largely conserved linkage in torafugu which may imply that big fraction of human overlapping genes represents vertebrates› ancestral overlaps. However, our previous study of human and mouse overlapping genes showed that even between closely related species overlaps are not that well conserved [12]. Out of 255 cases in which both members of the human overlapping gene pair had mouse orthologs, only 95 were overlapping in both species. In addition, significant fraction of these 95 gene pairs show different overlap patterns in the two genomes. Lack of the overlap conservation was also observed in other studies [13, 14, 15].

Here we present results of the comparative analysis of seven vertebrate genomes: human, chimpanzee, mouse, rat, chicken, fugu, and zebrafish. This comparative study shows that on one hand, many of the vertebrate gene overlaps are not conserved and are lineage specific. On the other hand, this work reveals new, not published before, cases of genes overlap conservation in vertebrates. We also show new mechanisms of overlapping genes evolution and demonstrate, for the first time, that evolutionary events could not only lead to the new gene overlaps origin but also to the loss of an ancient overlap. Therefore lack of strong overlaps conservation between even closely related species may result from the origin of a new, lineage specific overlap as well as from the loss of overlaps in many lineages. Findings about evolutionary changes in the gene structure and organization are very important in our quest toward understanding genomes and genes expression. Changes in the gene structures may lead to modifications in the gene expression and expression correlation between involved genes, which may further explain some differences between species such as discrepancies in the orthologous genes expression patterns [16].

RESULTS

Conservation of overlaps in vertebrate genomes

Fraction of the overlapping genes in various vertebrates differs significantly (Table 1). In tetrapoda over 10% of all genes are involved in some type of overlap, while in fish only 5–7%. The exception is the rat genome where only 4.87% of all genes are overlapping with another gene. However, this is most likely due to the annotation incompleteness and not a specific feature of the rat genome. Many rat genes do not have UTRs annotated and as we learned from this and other studies [12, 17, 18] the majority of gene overlaps are in the UTR regions. Incomplete annotations are likely to be responsible for some discrepancies between human and chimpanzee and could also be true for fish genomes. Another possibility is that many overlaps evolved after *Actinopterigii*diverged from *Sacropterigii* and most overlaps observed in these lineages arose independently. Differences in overlapping genes frequencies between human and other species were tested using chi-square test (Table 1). In all cases but mouse the chi-square is higher than critical value at $\alpha = 0.0005$ and therefore differences are statistically significant. For mouse the difference is significant at $\alpha = 0.05$.

Table 1. Overlapping genes in vertebrate genomes

	Number of genes analyzed	Number and fraction (in parentheses) of genes involved in overlaps	Chi-square test value (when compared to human)	Number of overlaps	Nested genes	Exon/exon overlaps (NATs) *		Intron/exon overlaps*
						Total	CDS involved	
Human	22,291	2,978 (13.4)	NA	1,766	972	634	417	160
Chimpanzee	21,506	2,219 (10.3)	73.0888	1,276	665	479	317	132
Mouse	25,383	3,456 (13.6)	5.0750	2,053	1,071	819	565	163
Rat	22,159	1,080 (4.9)	895.1585	607	458	102	100	47
Chicken	17,709	1,960 (11.1)	32.2585	1,135	474	511	471	150
Fugu	20,796	993 (4.8)	880.5199	556	174	290	290	92
Zebrafish	23,524	1,625 (6.9)	472.4534	1,026	767	98	85	161

* excluding nested genes

Number of genes involved in overlaps is smaller than the number of overlaps multiplied by two since some genes are involved in more than one overlap. Multiple genes may be nested in one host gene as well as a gene may be overlapping other genes on both ends as reported previously [12]. Differences in overlapping genes frequencies between human and other species were tested using chi-square test. In all cases but mouse the chi-square is higher than critical value at $\alpha = 0.0005$ and therefore differences are statistically significant. For mouse the difference is significant at $\alpha = 0.05$. However, no definite conclusions can be drawn by this test results since some differences may result from the annotation problems but not real differences in the overlapping genes fraction.

Table 2 shows results of the analysis of overlap conservation, which indicate that sets of overlapping genes differ between species. This demonstrates that many overlaps are species or lineage specific and only a small fraction of them are shared among vertebrates. Although some conserved overlaps are not observed due to missing data, we can assume that this would affect both, species specific and conserved overlaps in similar way and therefore the proportions and the general picture are not affected. Similar disproportion in the sense-antisense transcripts abundance was also shown by Zhang et al. [19].

Table 2. Overlapping genes conserved between species

	Human	Chimpanzee	Mouse	Rat	Chicken	Fugu	Zebrafish
Human	-	1100	274	98	64	23	17
Chimpanzee	477	-	NA	NA	NA	NA	NA
Mouse	146	NA	-	141	76	26	16
Rat	11	NA	48	-	45	19	6
Chicken	9	NA	10	2	-	22	13
Fugu	1	NA	0	0	0	-	13
Zebrafish	5	NA	5	0	0	1	-

Above diagonal shows total number of conserved overlaps, and below diagonal shows numbers of conserved exon/exon overlaps. For chimpanzee the data is provided only in relation to human, gene orthology relation between chimpanzee and other than human species is not established and annotated in the Ensemble database yet.

Patterns of human overlapping genes evolution

Using Ensembl gene homology data [20] we identified homologs of human overlapping genes in other species. Out of 2,978 human genes involved in overlaps, 264 were human specific and had no homologs in any other analyzed genome, including chimpanzee. Interestingly, we couldn't find a rodent homolog for about 25% of human overlapping genes, whereas genome wide comparison shows that 89–90% of rat genes possess a single ortholog in the human genome [21]. Similarly, it was observed that 25% of human genes do not have torafugu orthologs [22], while our study shows that in the case of the overlapping genes 46.17% of human genes lack a torafugu ortholog. These results imply that a lot of genes involved in overlaps are young, lineage specific genes and do not have orthologs in other lineages. This supports the 'overprinting' hypothesis but is in sharp contrast to observation made by Dahary et al. who based on comparison of the human and torafugu genomes concluded that most human overlaps are ancient [11]. However, their conclusion may be an artifact of the applied method, because they analyzed only those human genes that have identifiable orthologs in the torafugu genome.

Based on the presence or absence of an ortholog in species representing two other lineages, i.e. rodents and fish, we divided human overlapping genes pairs into those that have: both orthologs in both lineages (476 pairs); both orthologs in rodents and only one in fish (279 pairs); both orthologs in rodents and none in fish (111 gene pairs); an ortholog of one gene only in both lineages (466 pairs); ortholog of one gene in rodents and none in fish (200 pairs); and no orthologs in neither lineage (92 pairs). Next, we analyzed genomic arrangement in all cases where both orthologs of human overlapping genes were found. According to our results we divided gene pairs into: overlapping (if they also overlap in particular species), neighboring (if they were not overlapping but placed one next to each other without any gene between them), and separated (if they were on different chromosomes, contigs or were separated by other genes).

Considering previously published major events leading to the gene overlaps; genes rearrangements or transposition, extension of genes by adoption of signals or new exons, and new genes origination [9, 10] we developed 15 theoretical scenarios of overlapping genes evolution (Figure 1). Results from the above analysis provide support for every theoretical pattern of overlap evolution. Presented examples, for each scenario, were carefully examined and confirmed by the presence of cDNA and/or EST sequences. Also, sequence analysis was performed to ensure the missing orthologs are truly not present in a given species or lack of overlaps is not resulting from annotations problem. Our requirements here were very conservative and each gene pair where there was disagreement between species from the same lineage was removed from studies.

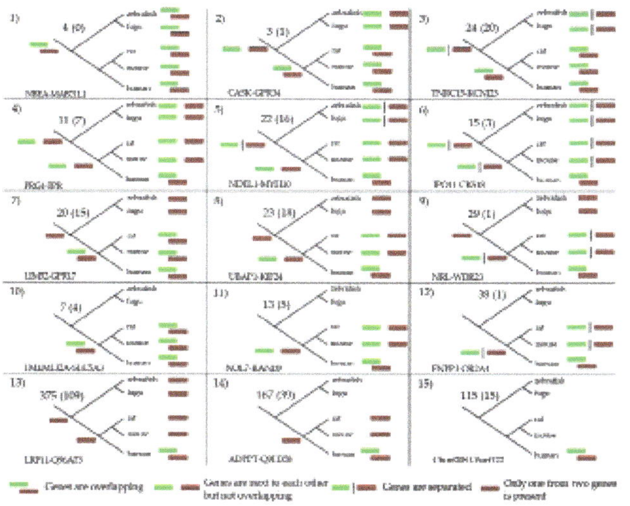

Figure 1. Putative patterns of human overlapping genes evolution together with examples from our data set. Numbers in parenthesis represent cases of exon/exon overlaps in each category. In each case an example from our studies is given. Analysis was done based on the October 2004 Ensembl release. The bar between two genes indicates that the genes are located on different chromosomes.

Mechanisms Leading to Gene Overlaps

Gene overlaps evolve by a variety of mechanisms and not by a single universal mechanism. Essentially, any mechanism that gives rise to a new gene, such as gene duplication or retroposition, may result in a gene overlap. Alternative splicing represents another major source of proteome diversity in mammals and origination of a new splice form may lead to a gene overlap as well.

In the analyzed data, we found cases supporting all the proposed hypotheses of gene overlap origination, i.e. overprinting, a gene translocation, or adoption of a new transcription termination signal (changing a gene structure in more general terms). Among identified human overlapping genes in 115 cases both genes involved in the overlap did not have the ortholog in neither rodent nor fish lineage, in 64 cases one ortholog was present in rodents and none in fish, and only in 68 cases both orthologs were present in rodent and fish genomes. Figure 1 shows fifteen scenarios of overlapping gene evolution. Patterns 7, 13, and 14 (Figure 1) fit exactly the overprinting hypothesis because they show human overlaps where one of gene is an old gene present in all vertebrates and the second gene is a young one not present in fish nor rodents. Dan et al. [13] showed that a recently evolved overlap between MINK and CHRNE genes resulted from mutations in the polyadenlylation signal and acquisition of a new downstream signal within a neighboring locus. Evolutionary scenarios represented by models 2–6, 8, 9, 11, and 12 (Figure 1) indicate involvement of translocation and possible signal adoption in the overlapping genes origin.

Although our models support published hypotheses we should consider much broader range of events which could lead to genes overlaps. Summarizing published hypotheses in a more general way we can say that major events playing a role in overlapping genes evolution are: translocation (or transposition), change in the gene structure (extension of UTR would fall into this category), and development of a new gene or a new splice variant.

Development of a new Splice Variant

Gene overlaps might not be conserved among species due to different gene structures [12]. In addition to adopting a new termination and an extension of the last exon, conversion of the previously unused genetic material in the form of a new splicing variant may lead to the gene overlap. There are two possible scenarios, an additional splice variant arises or the ancestral variant may be replaced by a new one.

Developing a new, additional, splice variant may be considered as a special case of overprinting since the new splice variant represents a new transcript. In fact, the case described by Keese and Gibbs [9] falls into this category because reported the new gene is just a new splice variant of TRalpha (TRHA) gene. Comparative analysis of the genomic region containing TRHA and NR1D1 (nuclear receptor subfamily 1, group D, member 1) genes revealed that the overlap is conserved among placental mammals, who have two splice variants of TRHA. Only one of these, the one which does not overlap with NR1D1, was identified in marsupials and all non-mammalian lineages. Close examination of the genomic region alignments showed that an insertion of new genetic material occurred some time after divergence of placental mammals and this inserted sequence was used for a new splice variant. This finding disagrees with the overprinting hypothesis as a new variant wasn't built from old existing material but rather new genetic information, not present in other genomes. However, we cannot exclude possibility that this genomic fragment wasn't lost in other genomes.

The same mechanism can be attributed to the origination of ITFG3 (integrin alpha FG-GAP repeat containing 3) and RGS11 (regulator of G-protein signalling 11) overlap in primates. ITFG3 has two splice variants in primates and one of them is overlapping at 3' end with RGS11 gene. Figure 2 shows genomic organization of both genes in human; similar organization is observed in the chimpanzee and macaque genomes. In all non-primate species only the non-overlapping (shorter) variant is present. To exclude the possibility that the overlapping splice variant was missed in annotations in non-primate genomes we investigated alignments of human and other genomes in two genomic browsers, Ensemble and UCSC browser. In both cases there was not a good alignment in the region occupied by the primate specific exon. Similarly, search against GenBank databases did not reveal any similarity between proteins encoded by the overlapping exon and any other non-primate protein, genomic or EST sequence. This clearly shows that overlapping splice variant of gene encoding ITFG3 is lineage specific and arose recently after divergence of primates.

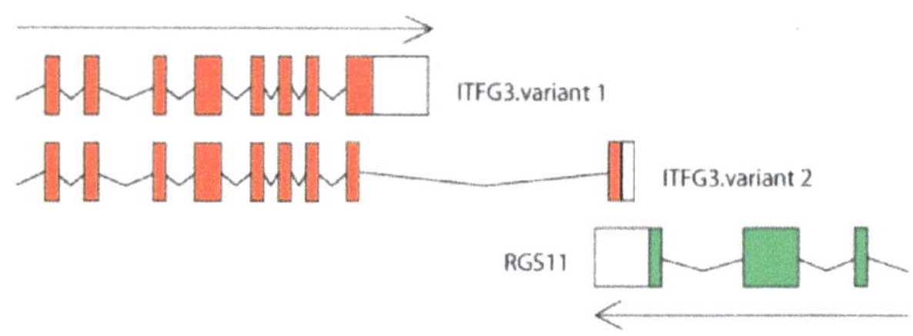

Figure 2. Genomic organization of ITFG3 and RGS11 in primates. In other mammals, although, these genes are neighbors on the same chromosome, variant 2 of ITGG3 is not observed and genes do not overlap. Coding sequences are colored and empty boxes denote 3'UTRs of both genes. 5' end of both genes were trimmed for the presentation purpose.

Another example of primate specific overlap that resulted from a new splice variant is a pair of genes THAP3, THAP domain containing, apoptosis associated protein 3, and DNAJC11, DnaJ (Hsp40) homolog, subfamily C, member 11. Both THAP3 and DNAJC11 homologs were found in a majority of analyzed vertebrate species: human, macaque, chimpanzee, mouse, rat, dog, cow, opossum, and zebrafish. Interestingly, the THAP3 was missing in chicken, frog, and tetraodon. We couldn't identify the gene in these genomes by any standard comparative methods including BLASTn and tBLASTn. However, zebrafish is apparently not the only fish species THAP3 gene. EST sequences DT157701, DT154094, DT175180, DT175179, and DT157700 from *Pimephales promelas* show high similarity to zebrafish THAP3 protein (64–77% identity) and likely represent THAP3 transcript. In primates, THAP3 has two splice variants one of which overlaps with DNAJC11. In all other species only one, shorter variant is present and it is not overlapping with DNAJC11. Comparative analysis of genomic sequences in the region of the overlap shows that there is no conservation in this region and most likely the longer variant is primate specific. We also did not identify any EST sequence which could show the presence of longer (overlapping) variant in non primate vertebrates. This analysis led to conclusion that longer splice variant of THAP3 and THAP3-DNAJC11 overlap are primate specific.

Development of a new gene

The new splice variant origination seems to be one of the most common events leading to the lineage specific overlaps. However, our data strongly suggest that this is not the only case where we observe 'overprinting'. The powerful evidence that origination of new lineage specific genes plays a big role in the evolution of overlapping genes comes from data that many human genes do not have orthologs in other lineages. It is known that the many of human genes are not found in rodent [21], chicken [23] or fish [22] genomes, so our result could just reflect these findings. Interestingly, the fraction of human genes with missing orthologs is higher for overlapping genes than non-overlapping genes. Approximately 10% of human genes are missing in mouse genome while 27% of human overlapping genes are missing in mouse. Similarly, about 20% and 24% of human genes are not found in chicken and torafugu but this fraction is much higher in our studies: 44.9% and 46.17% in chicken and torafugu, respectively.

A confirmation for new gene origination as a source of gene overlap comes from the case of TMEM16C, transmembrane protein 16C, and MUC15, mucin 15; located on 11p14.2-3. Human TMEM16C has 27 exons spanning 331,865 base pairs. Two splice variants of MUC15 are embedded in the TMEM16C gene occupying introns 13 and 14 and overlapping, in the 3'UTR area, with exon 14 (Figure 3). Gene MUC15 is present only in mammalian genomes and there is experimental evidence that it is overlapping with TMEM16C at least in primates, rodents and cow. This gene is not present in any other lineages including chicken, xenopus and zebrafish and alignment of genomic sequences shows no conservation in areas covered by MUC15. However alignments at the protein level show some traces of similarity in chicken in MUC15 exon four of splice variant 2 and in xenopus in the same exon four in part of exon three. We could not detect any similarity in zebrafish as well as in invertebrates. The MUC15 sequence appears to be specific for vertebrates only and there are two possible evolutionary scenarios.

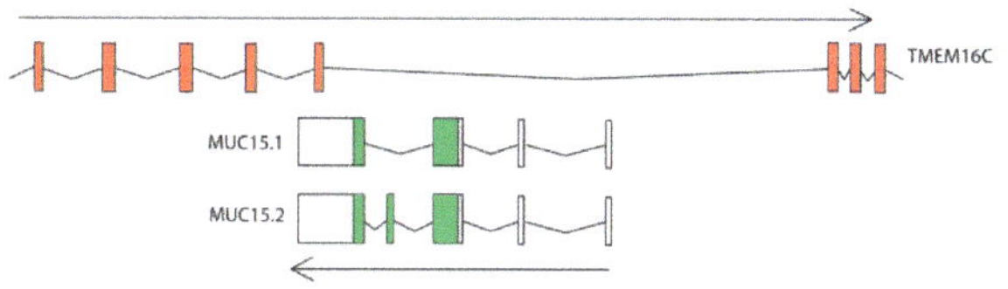

Figure 3. Genomic organization of TMEM16C and MUC15 in mammals. In chicken, Xenopus and zebrafish MUC15 is not observed.

In one, the sequence was present in early chordates and in the process of neutral mutation gained the coding potential which was used to build a new gene in mammalian ancestor. Another possibility is that this gene, was lost in majority of lineages with traces left in few genomes, and was maintained only in mammals. At any rate TMEM16C and MUC15 represent an overlap between an old gene TMEM16C and a newer mammalian specific gene (MUC15).

Changes in the gene structure

In the cases described above the gene overlap evolved through the origin of a new, longer splice variant. In many instances we observed a slightly different situation, a new variant arose and replaced the ancient one, so the number of variants was the same in analyzed lineages; however, they differ in their genomic organization. Examples of ACAT2-TCP1 [10] and MINK-CHRNE [13] overlaps are simple cases of changes in the gene structure where the most 3' exon was extended as a result of adopting the closest polyA signal after the original one was lost.

Example of BLZF1 gene (basic leucine zipper nuclear factor 1), overlapping at 3' end with the gene C1orf114 (open reading frame 114 on human chromosome 1, position 167603818–167663296) shows a more drastic shift in the gene structure. This overlap exists in the human and chimpanzee genomes but the two genes are neighbors, and do not overlap in other mammals including opossum where they are located on chromosome 2 about 6 kb apart. A similar arrangement is observed in chicken (chromosome 1, 2 kb separation) and in Xenopus (the same scaffold, 9 kb separation). Interestingly, in zebrafish these two genes are located on different chromosomes – BLZF1 on chromosome 1 and a homolog of C1orf114 most likely is located on chromosome 8. Although there's no gene annotated in the cognate region, both human protein similarity and zebrafish EST alignments strongly suggest existence of the C1orf114 gene in this region. Figure 4 shows genomic organization of these overlapping genes in human and mouse. Analysis of multiple alignments of the region containing the last exon of BLZF1 in human and corresponding region in other vertebrates revealed that this fragment is not conserved among vertebrate lineages. Clearly there was an insertion in an ancient primate genome; a fragment belonging to the last, primate specific, exon does not align with non-primate genomes. Analysis of this fragment showed that the very 3' end of human BLZF1 is occupied by AluS, an old primate specific retroelement. Similarly, middle part of the last BLZF1 exon in mouse, contains rodent specific element B4A. Apparently independent insertions of repetitive elements occurred in both, primate and rodent, genomes and both are associated with exonification and new splice variant development.

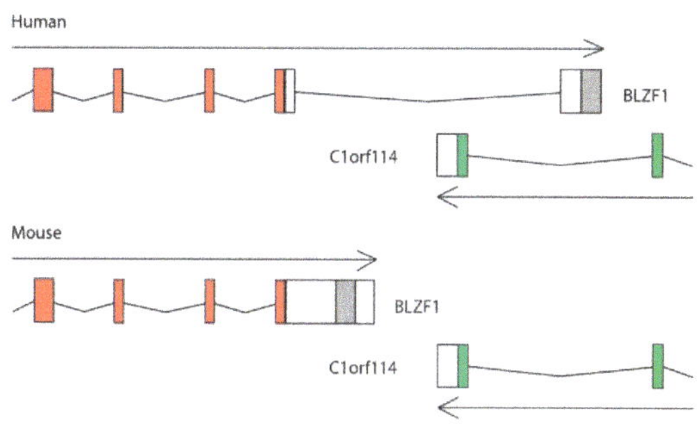

Figure 4. Genomic organization of human and mouse BLZF1 and C1orf114 genes. Shaded areas of 3'UTRs represent repetitive elements location.

Gene Duplication and Retrotransposition

Gene duplication is a common mechanism for the origin of new genes [24, 25]. Retrotransposition is an interesting mechanism that allows a gene to move to a distant location on the same or different chromosome. Retro(pseudo)genes are products of reverse transcription of a spliced (mature) mRNA and they are characterized by lack of introns, presence of polyA track, and flanking direct repeats. Because they are copies of mature mRNAs, they usually lack promoters and cannot be transcribed. However, in some rare instances, after insertion near an existing promoter or exaptation of anonymous sequence as a promoter, they can gain transcriptional activity and create a new functional gene [24, 26].

An example of a new gene overlap due to formation of a new gene origination comes from the ribosomal protein RPS27 retrogene and TSPAN9 (tetraspanin 9, known also as NET-5) gene. RPS27 has two intron-containing paralogs: RPS27 and RPS27L, and both of them gave rise to multiple retrocopies in the human genome. We identified 24 retro(pseudo)genes of RPS27; ten of them are nested in another gene. Although multiple RPS27 retrogenes can be identified in other mammalian genomes, none of the host-nested gene pairs are the same in the human and rodent genomes. The aforementioned retrocopy of RPS27, nested in the human tetraspanin 9 gene, has an intact open reading frame and potentially encodes for 84 amino acid protein 100% identical to the spliced version of the gene on chromosome 1. We also identified two EST sequences, AV763564 and CD386048 that are 99% identical to this gene and show weaker similarity to other RPS27 genes, which may imply that this gene is expressed. However, because of relatively low quality of EST sequences, these results are not conclusive and further analysis would be required to confirm expression of this gene. This retrosequence is present in the human and chimp genomes but missing from orthologous location in macaque and all other vertebrates that we analyzed. This confirms recent origin of the RPS27 retrosequence and makes its expression assessment based on an intact ORF impossible. However, this example demonstrates a potential route to new overlaps in the vertebrate genomes.

Loss of Gene Overlaps

While young gene overlaps can arise from new splice variants, ancient overlaps can disappear due to the loss of the overlapping splice variant. In fact, we did observe such cases in our dataset that showed that analysis of narrow range of vertebrate or mammalian lineages may lead to false conclusions due to incompleteness of the data. The human NDEL1 gene coding for a thiol-activated peptidase and MYH10 gene that codes for myosin heavy chain 10 reside on chromosome 17 and they overlap in primates but not in rodents, chicken or fish. In primates there are three splice variants of NDEL1 and only one is overlapping with MYH10. In mouse and rat only two non overlapping transcripts are present, and in chicken we observe only one variant. The same variant as in chicken is present in zebrafish, however NDEL1 and MYH10 genes are separated. In addition, in zebrafish NDEL1 has two copies representing the same splice variant, one located at chromosome 12, position 35 35,318,946–35,338,461 and another located at chromosome 3 at position 56,442,046–56,469,437. MYH10 in zebrafish is on chromosome 6, position 5,082,544–5,106,092. Based on these observations the most parsimonious explanation would be that one of these genes was translocated after fish divergence and later on the overlapping splice variant arose in primates. However, when we looked at the genomic arrangement of these genes in other species we observed that the dog genome has all three primate's splice variants. Further analysis showed that *Xenopus* has even more, four splice variants, two of them confirmed by cDNA and EST sequences and two predicted. One of the predicted variants is overlapping with MYH10 and protein coded by this variant shows high similarity to protein encoded by overlapping transcript observed in primates and in dog. Analysis of all known vertebrate transcripts and proteins encoded by them suggests that in a tetrapoda ancestor genome there were at least three splice variants of NDEL1. During the evolution one or two of them were lost in most lineages. Only primates retained three variants and all of them are confirmed by EST or cDNA sequences. In dog and *Xenopus* overlapping variant was predicted but it is shorter than the one in primates and up to date there is no experimental confirmation for its expression. However, the analysis of proteins coded by these variants (Figure 5) clearly shows that the overlapping splice variant in *Xenopus* is the same as the one in dogs and primates. In zebrafish both copies of NDEL1 represent variant 1 that seems to be the most common in vertebrate genomes. Variants of NDEL1 most probably evolved at early chordates because only one is present in *Ciona intestinalis*. Sequence divergence did not allow us to establish which, if any, of three variants is shared with *Ciona intestinalis*. Figure 6 shows the phylogenetic tree with all analyzed species. Bars to the right represent splice variants of NDEL1.

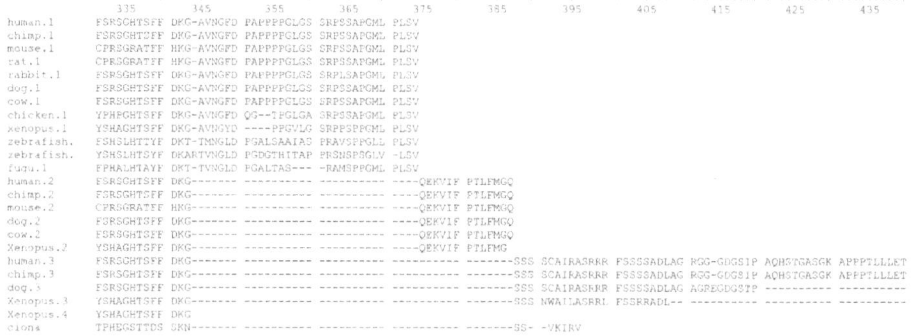

Figure 5. Alignments of vertebrate NDEL1 proteins coded by all splice variants.

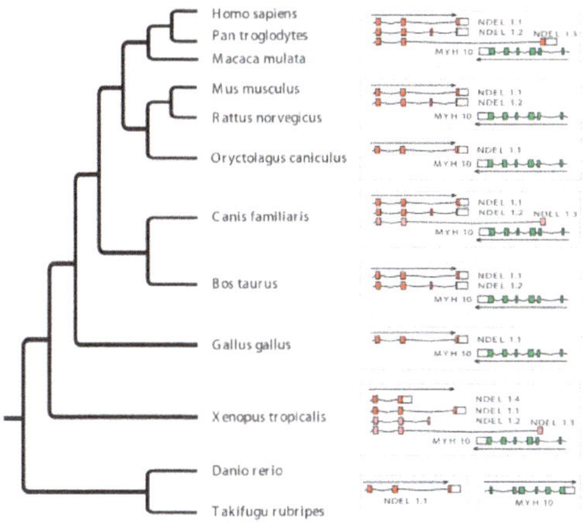

Figure 6. Phylogenetic tree with MYH10 and NDEL1 variants and genomic organization in vertebrates.

The most likely mechanism of the variant loss in some lineages is a weakening of the splicing signal that leads to skipping of an exon. Similar effect can be observed in case when the other exon acquires signal so strong that dominates completely splicing and results in a constitutive exon. Another possibility is that this transcript is using signals from L1 elements which are in its 3' UTR. [27] These elements are in human sequence but are partially or completely lost in other genomes. However, more extensive analyses that would include some wet lab experiments are required in order to determine what the case is here.

Time of Overlapping Genes Evolution

Equally important, to the mechanism leading to overlaps, is the time when particular event occurred. Two widely accepted hypotheses of vertebrate overlapping genes evolution [9, 10] assume that explosion of gene overlaps occurred in early mammals. On the contrary, Dahary et al. [11] and Zhang et al. [19] suggested that most naturally occurring anti-transcripts observed in human represent ancient vertebrate gene overlaps. To answer this question, we looked for evidence of putative gene overlap before and after mammalian radiation. We studied cases where the pattern of gene arrangement differs in fish and rodent, i.e. two genes overlap in rodents but not in fish. As shown in Figure 7, genes arrangement in the chicken genome in some cases is similar to the one in rodents in other to the one in fish. For example genes CIO32 and ASB6 overlap in mammals only but genes RNF123 and GNPPB overlap also in chicken. In another example gene pair UBAP1 and KIF24 is overlapping in human but not overlapping, although located next to each other, in rodents and chicken. Only one of these genes, UBAP1 is present in fish. On the other hand, genes RFESD and SPATA9 overlap in human but not in rodents where they are located next to each other. In chicken, as well as in fish gene SPATA9 is not found. This demonstrates that the events leading to gene overlaps happened before and after mammalian radiation. The development of gene overlaps is a long, continuous process, not a 'big bang' that happened during a specific short period of vertebrate evolution. Also examples discussed in previous sections showed that gene overlaps were evolving at different stages of vertebrate evolution and could arise in early chordates, early mammals, or just recently in primates and other mammalian lineages.

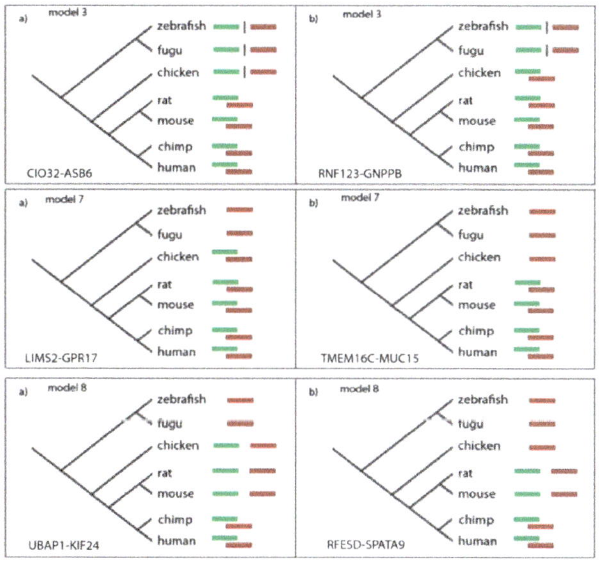

Figure 7. Examples of genomic arrangements of human overlapping genes orthologs in chicken. In left column genes arrangement in chicken resembles this in primates and rodents, in the right column genes in chicken are arranged the same way as in fish.

DISCUSSION

Although large scale studies of overlapping genes have been available since 2002 we still do not understand how these overlaps evolved and what, if any, is the functional meaning of sharing the genomic locus between genes in eukaryotic genomes. Results published so far show evidence for both, relatively new, lineage specific [9, 10, 13] as well as conserved overlaps among vertebrate [11, 19] and even all eukaryotes [28] gene overlaps. However, none of the papers, even those with gene overlaps origin hypotheses, fully explains this evolutionary phenomenon. This study brings us a little closer to that goal and the major conclusion is that there's no single mechanism responsible for the overlap origination. In principle, any mechanism of a new exon or a new gene origination may lead to a gene overlap. In the light of presented results, we can conclude that the major forces in the overlapping genes evolution are transposition and exaptation – a process that gives rise to new genes or new variants from preexisting nucleotide sequences. [29]. UTR extensions in the course of new polyA signal adoption [10, 13] or new splice variant development [9] are perfect examples of exaptation. Another type of exaptation is building a new gene structure by adopting an inserted transposable element. Transposable elements are known to contribute to host gene regulation [30, 31] or structure [32, 33]. Our study on BLZF1 and C1orf114 showed that transposable elements also contribute to origin of a new class of genetic novelties, namely overlapping genes. The analyzed data also provided evidence that new gene origination is truly observed in the process of overlaps evolution supporting even further hypothesis by Keese and Gibbs [9]. 'Overprinting' hypothesis was constructed based on the new splice variant origination. Overlap between TMEM16C and mammalian specific gene MUC15 showed that overlaps may involve pairs of an old ancient gene and a new lineage specific gene. However, we cannot agree that this is true for all vertebrate overlaps as hypothesized. In many of analyzed gene pairs both genes were old and conserved through eukaryotes. Nevertheless, our study shows a number of gene overlaps that are lineage specific and are not conserved among vertebrates which supports our earlier studies of overlaps in human and mouse [12]. This is in a contradiction with the study of Dahary et al. [11] on human and fugu genomes that concludes that most human overlaps are ancient. However, they analyzed only those human genes that have identifiable orthologs in the fugu genome and therefore, young overlaps involving lineage specific genes were excluded from the study by definition. Also, their judgment was based on cases of overlapping human genes that were on average closer to each other in the fugu genome then other genes, and not based on true conserved overlaps. So, some of the gene pairs in fugu although close one to each other, are not necessary overlapping as is clear from our analysis.

In summary, we should emphasize that overlapping genes do not present any special case in regard to mechanisms of evolution. Events like gene translocation or exaptation, driving forces in genome evolution, are also common and major mechanisms in gene overlaps origin. There wasn't also any rapid origin or a 'big bang' of the overlapping genes after the split of bird and mammal lineages as suggested by Keese and Gibbs [9], nor are most of the human overlaps ancient as described by Dahary et al [11]. Birth as well as most likely death of gene overlaps is a continue process that occurred over the entire time of vertebrate evolution, similarly like any other genes arose or die over a long process of the eukaryotic genomes evolution [34].

Our results also imply that origin of overlapping genes is not an issue of saving space and contracting genomes size. Although there are some implications on functional importance of overlapping genes, the present analysis shows that most gene overlaps evolve stochastically, the same way as other genomic features, and without any positive pressure on the overlap presence. If overlaps have some functional meaning it is not a common case and most likely this function evolved by chance as a consequence of new genes arrangement.

This study also demonstrates that in order to fully understand the evolution of overlapping genes one has to study many genomes in minute details. Studies on a limited number of species may lead to false conclusions as shown in the case of NDEL1 and in many other cases we investigated during this study. Many gene pairs were moved from one category to another as a result of detailed examination of annotations and additional analysis. This shows that although human and other genomes are considered to be complete, their annotation is still far from final and in many cases cannot be trusted. Therefore, careful examination of any gene pair by a human expert followed by, in an ideal world, some wet-lab experiments is a key to sound results. We are very well aware that the present study did not solve all the questions regarding overlapping genes evolution and their origins. However, it did shed a light on how some of these overlaps evolved, provided a strong confirmation for lineage specific overlaps, and delivered firsthand evidence of gene overlap loss in the vertebrate lineage.

METHODS

Sequence data

Assembled sequences and annotations of seven analyzed genomes were downloaded from Ensembl [20] and stored in a local mySQL database. We used following versions of the genomes: human-24.34e (NCBI 34), chimp-24.1 (CHIMP 1), mouse-24.33 (NCBI m33), rat-24.3c (RGSC 3.1), chicken-24.1a (WASHUC 1), fugu-24.2c (FUGU 2.0), and zebrafish-24.4 (Zv 4).

Identification of the overlapping genes

For practical reasons, we applied an operational definition of a gene, as a part of the genomic region from the beginning to the end of an annotated transcript. Any two genes, defined as above, whose coordinates overlap and are transcribed from the different DNA strand, are considered as overlapping.

Identification of orthologous genes and mapping information

Orthology inference was done based on any two genomes homology information provided in Ensembl. The set of overlapping genes for a given species was always a starting point for each orthology analysis. As a result seven by seven orthology matrix was created. It is important to stress that orthology relationship provided by Ensembl is not a simple one-to-one relationship. Whenever lineage specific gene duplication is detected one-to-many orthologs are provided. In these cases, each of several orthologs was checked for the overlaps. The detailed description of the method is available at Ensembl webpage [35]. Additionally, we used conserved synteny information of the neighboring genes to enhance reliability of the orthology inference. However, not all the genes have had their orthologs listed. In these cases, we assumed that a cognate gene is missing from a given genome.

The mapping information of each orthologous gene was downloaded from the Ensemble. For each pair of overlapping genes in one genome, e.g. human, spatial relationship of their orthologs in the other six genomes was checked based on existing annotation.

Extending neighboring but not overlapping genes

We mapped TIGR gene indices [36] to all neighboring but not overlapping orthologs of human overlapping genes to check for possible extensions. In each case, we extracted genomic fragments containing a pair of neighboring genes and BLAST [37] against corresponding TGI sequences. Next we mapped transcripts to genomic fragments together with TGI sequences obtained from BLAST search. Only sequences showing similarity over 98% and fully aligning with the genomic fragment were used in order to avoid false positive hits to repetitive elements and ESTs from related genes. Results were stored in ASN.1 format and examined in Sequin [38].

Multiple alignments

MultiZ alignments of genomic sequences were obtained from

UCSC genome browser [39]. Protein multiple alignments were constructed using Clustalw [40].

DECLARATIONS

Acknowledgements

We gratefully thank Stephen Schaeffer for reading the manuscript and for his valuable comments, and Narayanan Veeraraghavan for his assistance.

REFERENCES

1. Makalowska I, Lin CF, Makalowski W: Overlapping genes in vertebrate genomes. Comput Biol Chem. 2005, 29 (1): 1-12. 10.1016/j.compbiolchem.2004.12.006.
2. Rougeulle C, Heard E: Antisense RNA in imprinting: spreading silence through Air. Trends Genet. 2002, 18 (9): 434-437. 10.1016/S0168-9525(02)02749-X.
3. Brantl S: Antisense-RNA regulation and RNA interference. Biochim Biophys Acta. 2002, 1575 (1-3): 15-25.
4. Prescott EM, Proudfoot NJ: Transcriptional collision between convergent genes in budding yeast. Proc Natl Acad Sci U S A. 2002, 99 (13): 8796-8801. 10.1073/pnas.132270899.
5. Hastings ML, Ingle HA, Lazar MA, Munroe SH: Post-transcriptional regulation of thyroid hormone receptor expression by cis-acting sequences and a naturally occurring antisense RNA. J Biol Chem. 2000, 275 (15): 11507-11513. 10.1074/jbc.275.15.11507.
6. Ogawa Y, Lee JT: Antisense regulation in X inactivation and autosomal imprinting. Cytogenet Genome Res. 2002, 99 (1-4): 59-65. 10.1159/000071575.
7. Marcelino J, Carpten JD, Suwairi WM, Gutierrez OM, Schwartz S, Robbins C, Sood R, Makalowska I, Baxevanis A, Johnstone B, Laxer RM, Zemel L, Kim CA, Herd JK, Ihle J, Williams C, Johnson M, Raman V, Alonso LG, Brunoni D, Gerstein A, Papadopoulos N, Bahabri SA, Trent JM, Warman ML: CACP, encoding a secreted proteoglycan, is mutated in camptodactyly-arthropathy-coxa vara-pericarditis syndrome. Nature genetics. 1999, 23 (3): 319-322. 10.1038/15496.
8. Lamason RL, Mohideen MA, Mest JR, Wong AC, Norton HL, Aros MC, Jurynec MJ, Mao X, Humphreville VR, Humbert JE, Sinha S, Moore JL, Jagadeeswaran P, Zhao W, Ning G, Makalowska I, McKeigue PM, O'Donnell D, Kittles R, Parra EJ, Mangini NJ, Grunwald DJ, Shriver MD, Canfield VA, Cheng KC: SLC24A5, a putative cation exchanger, affects pigmentation in zebrafish and humans. Science. 2005, 310 (5755): 1782-1786. 10.1126/science.1116238.
9. Keese PK, Gibbs A: Origins of genes: "big bang" or continuous creation?. Proc Natl Acad Sci U S A. 1992, 89 (20): 9489-9493. 10.1073/pnas.89.20.9489.
10. Shintani S, O'HUigin C, Toyosawa S, Michalova V, Klein J: Origin of gene overlap: the case of TCP1 and ACAT2. Genetics. 1999, 152 (2): 743-754.
11. Dahary D, Elroy-Stein O, Sorek R: Naturally occurring antisense: transcriptional leakage or real overlap?. Genome Res. 2005, 15 (3): 364-368. 10.1101/gr.3308405.
12. Veeramachaneni V, Makalowski W, Galdzicki M, Sood R, Makalowska I: Mammalian overlapping genes: the comparative perspective. Genome Res. 2004, 14 (2): 280-286. 10.1101/gr.1590904.
13. Dan I, Watanabe NM, Kajikawa E, Ishida T, Pandey A, Kusumi A: Overlapping of MINK and CHRNE gene loci in the course of mammalian evolution. Nucleic Acids Res. 2002, 30 (13): 2906-2910. 10.1093/nar/gkf407.
14. Kasper G, Taudien S, Staub E, Mennerich D, Rieder M, Hinzmann B, Dahl E, Schwidetzky U, Rosenthal A, Rump A: Different structural organization of the encephalopsin gene in man and mouse. Gene. 2002, 295 (1): 27-32. 10.1016/S0378-1119(02)00799-0.
15. Steigele S, Nieselt K: Open reading frames provide a rich pool of potential natural antisense transcripts in fungal genomes. Nucleic Acids Res. 2005, 33 (16): 5034-5044. 10.1093/nar/gki804.
16. Su AI, Cooke MP, Ching KA, Hakak Y, Walker JR, Wiltshire T, Orth AP, Vega RG, Sapinoso LM, Moqrich A, Patapoutian A, Hampton GM, Schultz PG, Hogenesch JB: Large-scale analysis of the human and mouse transcriptomes. Proc Natl Acad Sci U S A. 2002, 99 (7): 4465-4470. 10.1073/pnas.012025199.
17. Yelin R, Dahary D, Sorek R, Levanon EY, Goldstein O, Shoshan A, Diber A, Biton S, Tamir Y, Khosravi R, Nemzer S, Pinner E, Walach S, Bernstein J, Savitsky K, Rotman G: Widespread occurrence of antisense transcription in the human genome. Nat Biotechnol. 2003, 21 (4): 379-386. 10.1038/nbt808.
18. Lehner B, Williams G, Campbell RD, Sanderson CM: Antisense transcripts in the human genome. Trends Genet. 2002, 18 (2): 63-65. 10.1016/S0168-9525(02)02598-2.
19. Zhang Y, Liu XS, Liu QR, Wei L: Genome-wide in silico identification and analysis of cis natural antisense transcripts (cis-NATs) in ten species. Nucleic Acids Res. 2006, 34 (12): 3465-3475. 10.1093/nar/gkl473.
20. Birney E, Andrews TD, Bevan P, Caccamo M, Chen Y, Clarke L, Coates G, Cuff J, Curwen V, Cutts T, Down T, Eyras E, Fernandez-Suarez XM, Gane P, Gibbins B, Gilbert J, Hammond M, Hotz HR, Iyer V, Jekosch K, Kahari A, Kasprzyk A, Keefe D, Keenan S, Lehvaslaiho H, McVicker G, Melsopp C, Meidl P, Mongin E, Pettett R, Potter S, Proctor G, Rae M, Searle S, Slater G, Smedley D, Smith J, Spooner W, Stabenau A, Stalker J, Storey R, Ureta-Vidal A, Woodwark KC, Cameron G, Durbin R, Cox A, Hubbard T, Clamp M: An overview of Ensembl. Genome Res. 2004, 14 (5): 925-928. 10.1101/gr.1860604.
21. Gibbs RA, Weinstock GM, Metzker ML, Muzny DM, Sodergren EJ, Scherer S, Scott G, Steffen D, Worley KC, Burch PE, Okwuonu G, Hines S, Lewis L, DeRamo C, Delgado O, Dugan-Rocha S, Miner G, Morgan M, Hawes A, Gill R, Celera, Holt RA, Adams MD, Amanatides PG, Baden-Tillson H, Barnstead M, Chin S, Evans CA, Ferriera S, Fosler C, Glodek A, Gu Z, Jennings D, Kraft CL, Nguyen T, Pfannkoch CM, Sitter C, Sutton GG, Venter JC, Woodage T, Smith D, Lee HM, Gustafson E, Cahill P, Kana A, Doucette-Stamm L,

Weinstock K, Fechtel K, Weiss RB, Dunn DM, Green ED, Blakesley RW, Bouffard GG, De Jong PJ, Osoegawa K, Zhu B, Marra M, Schein J, Bosdet I, Fjell C, Jones S, Krzywinski M, Mathewson C, Siddiqui A, Wye N, McPherson J, Zhao S, Fraser CM, Shetty J, Shatsman S, Geer K, Chen Y, Abramzon S, Nierman WC, Havlak PH, Chen R, Durbin KJ, Egan A, Ren Y, Song XZ, Li B, Liu Y, Qin X, Cawley S, Worley KC, Cooney AJ, D'Souza LM, Martin K, Wu JQ, Gonzalez-Garay ML, Jackson AR, Kalafus KJ, McLeod MP, Milosavljevic A, Virk D, Volkov A, Wheeler DA, Zhang Z, Bailey JA, Eichler EE, Tuzun E, Birney E, Mongin E, Ureta-Vidal A, Woodwark C, Zdobnov E, Bork P, Suyama M, Torrents D, Alexandersson M, Trask BJ, Young JM, Huang H, Wang H, Xing H, Daniels S, Gietzen D, Schmidt J, Stevens K, Vitt U, Wingrove J, Camara F, Mar Alba M, Abril JF, Guigo R, Smit A, Dubchak I, Rubin EM, Couronne O, Poliakov A, Hubner N, Ganten D, Goesele C, Hummel O, Kreitler T, Lee YA, Monti J, Schulz H, Zimdahl H, Himmelbauer H, Lehrach H, Jacob HJ, Bromberg S, Gullings-Handley J, Jensen-Seaman MI, Kwitek AE, Lazar J, Pasko D, Tonellato PJ, Twigger S, Ponting CP, Duarte JM, Rice S, Goodstadt L, Beatson SA, Emes RD, Winter EE, Webber C, Brandt P, Nyakatura G, Adetobi M, Chiaromonte F, Elnitski L, Eswara P, Hardison RC, Hou M, Kolbe D, Makova K, Miller W, Nekrutenko A, Riemer C, Schwartz S, Taylor J, Yang S, Zhang Y, Lindpaintner K, Andrews TD, Caccamo M, Clamp M, Clarke L, Curwen V, Durbin R, Eyras E, Searle SM, Cooper GM, Batzoglou S, Brudno M, Sidow A, Stone EA, Venter JC, Payseur BA, Bourque G, Lopez-Otin C, Puente XS, Chakrabarti K, Chatterji S, Dewey C, Pachter L, Bray N, Yap VB, Caspi A, Tesler G, Pevzner PA, Haussler D, Roskin KM, Baertsch R, Clawson H, Furey TS, Hinrichs AS, Karolchik D, Kent WJ, Rosenbloom KR, Trumbower H, Weirauch M, Cooper DN, Stenson PD, Ma B, Brent M, Arumugam M, Shteynberg D, Copley RR, Taylor MS, Riethman H, Mudunuri U, Peterson J, Guyer M, Felsenfeld A, Old S, Mockrin S, Collins F: Genome sequence of the Brown Norway rat yields insights into mammalian evolution. Nature. 2004, 428 (6982): 493-521. 10.1038/nature02426.

22. Aparicio S, Chapman J, Stupka E, Putnam N, Chia JM, Dehal P, Christoffels A, Rash S, Hoon S, Smit A, Gelpke MD, Roach J, Oh T, Ho IY, Wong M, Detter C, Verhoef F, Predki P, Tay A, Lucas S, Richardson P, Smith SF, Clark MS, Edwards YJ, Doggett N, Zharkikh A, Tavtigian SV, Pruss D, Barnstead M, Evans C, Baden H, Powell J, Glusman G, Rowen L, Hood L, Tan YH, Elgar G, Hawkins T, Venkatesh B, Rokhsar D, Brenner S: Whole-genome shotgun assembly and analysis of the genome of Fugu rubripes. Science. 2002, 297 (5585): 1301-1310. 10.1126/science.1072104.
23. Sequence and comparative analysis of the chicken genome provide unique perspectives on vertebrate evolution. Nature. 2004, 432 (7018): 695-716. 10.1038/nature03154.
24. Long M, Betran E, Thornton K, Wang W: The origin of new genes: glimpses from the young and old. Nat Rev Genet. 2003, 4 (11): 865-875. 10.1038/nrg1204.
25. Ohno S: Evolution by gene duplication. 1970, Berlin; New York , Springer-Verlag
26. Brosius J: RNAs from all categories generate retrosequences that may be exapted as novel genes or regulatory elements. Gene. 1999, 238 (1): 115-134. 10.1016/S0378-1119(99)00227-9.
27. Makalowski W: Genomics. Not junk after all. Science. 2003, 300 (5623): 1246-1247. 10.1126/science.1085690.
28. van Duin M, van Den Tol J, Hoeijmakers JH, Bootsma D, Rupp IP, Reynolds P, Prakash L, Prakash S: Conserved pattern of antisense overlapping transcription in the homologous human ERCC-1 and yeast RAD10 DNA repair gene regions. Mol Cell Biol. 1989, 9 (4): 1794-1798.
29. Brosius J: On "genomencature": a comprehensive (and respectful) taxonomy for pseudogenes amd other "junk DNA". Proc Natl Acad Sci. 1992, 89 (22): 10706-10710. 10.1073/pnas.89.22.10706.
30. Thornburg BG, Gotea V, Makalowski W: Transposable elements as a significant source of transcription regulating signals. Gene. 2006, 365: 104-110. 10.1016/j.gene.2005.09.036.
31. Romanish MT, Lock WM, de Lagemaat LN, Dunn CA, Mager DL: Repeated Recruitment of LTR Retrotransposons as Promoters by the Anti-Apoptotic Locus NAIP during Mammalian Evolution. PLoS Genet. 2007, 3 (1): e10-10.1371/journal.pgen.0030010.
32. Lorenc A, Makalowski W: Transposable elements and vertebrate protein diversity. Genetica. 2003, 118 (2-3): 183-191. 10.1023/A:1024105726123.
33. Gotea V, Makalowski W: Do transposable elements really contribute to proteomes?. Trends Genet. 2006, 22 (5): 260-267. 10.1016/j.tig.2006.03.006.
34. Nei M, Rooney AP: Concerted and birth-and-death evolution of multigene families. Annual review of genetics. 2005, 39: 121-152. 10.1146/annurev.genet.39.073003.112240.
35. Ensembl: Gene Orthology/Paralogy prediction method. [http://www.ensembl.org/info/data/compara/homology_method.html]
36. Pertea G, Huang X, Liang F, Antonescu V, Sultana R, Karamycheva S, Lee Y, White J, Cheung F, Parvizi B, Tsai J, Quackenbush J: TIGR Gene Indices clustering tools (TGICL): a software system for fast clustering of large EST datasets. Bioinformatics. 2003, 19 (5): 651-652. 10.1093/bioinformatics/btg034.
37. Altschul SF, Gish W, Miller W, Myers EW, Lipman DJ: Basic local alignment search tool. J Mol Biol. 1990, 215 (3): 403-410.
38. Benson DA, Karsch-Mizrachi I, Lipman DJ, Ostell J, Wheeler DL: GenBank. Nucleic Acids Res. 2005, 33 (Database issue): D34-8. 10.1093/nar/gki063.
39. Hinrichs AS, Karolchik D, Baertsch R, Barber GP, Bejerano G, Clawson H, Diekhans M, Furey TS, Harte RA, Hsu F, Hillman-Jackson J, Kuhn RM, Pedersen JS, Pohl A, Raney BJ, Rosenbloom KR, Siepel A, Smith KE, Sugnet CW, Sultan-Qurraie A, Thomas DJ, Trumbower H, Weber RJ, Weirauch M, Zweig AS, Haussler D, Kent

WJ: The UCSC Genome Browser Database: update 2006. Nucleic Acids Res. 2006, 34 (Database issue): D590-8. 10.1093/nar/gkj144.

40. Thompson JD, Higgins DG, Gibson TJ: CLUSTAL W: improving the sensitivity of progressive multiple sequence alignment through sequence weighting, position-specific gap penalties and weight matrix choice. Nucleic Acids Res. 1994, 22 (22): 4673-4680. 10.1093/nar/22.22.4673.

CHAPTER 12 EVOLUTION OF BILATERIAN CENTRAL NERVOUS SYSTEMS

Linda Z Holland, João E Carvalho, Hector Escriva, Vincent Laudet, Michael Schubert, Sebastian M Shimeld and Jr-Kai Yu

Keywords: Central nervous system evolution Hemichordate Urbilaterian Amphioxus Tunicate Vertebrate brain Nerve cord

REVIEW

Introduction

There is general agreement that the relatively complex central nervous system (CNS) characterizing most higher metazoan animals can be traced back through evolution to a nerve net in a cnidarian-like ancestor. However, it is highly controversial whether the nervous system of the next evolutionary stage (the urbilaterian) still consisted solely of a nerve net or included a CNS. If the urbilaterian had only a nerve net, then the CNSs of protostomes and deuterostomes likely evolved independently. In contrast, the view that the urbilaterian had a CNS is consistent with the view that the CNSs of all metazoans are homologous. At present, opinion is still divided, with the majority advocating a single evolutionary origin for the CNS [1, 2, 3, 4, 5, 6, 7, 8] and the minority favoring an urbilaterian with a nerve net [9, 10].

This controversy about CNS evolution is intimately related to issues of homology, and it is useful to outline current thinking about homology at the outset. It is important not to conflate the concepts and recognition criteria for homology. There are currently three concepts underlying homology – biological [11], taxic/cladistic [12, 13] and historical [14, 15, 16]. What matters for the last, which is the most familiar and most germane for the nervous system controversy, is the historical continuity of descent from a common ancestor. The historical concept requires one to be explicit about what is being compared [17]; for example, bird wings and bat wings are homologous as vertebrate forelimbs, but not as wings. Importantly, historical homologies can become very different through divergence. The three chief criteria for recognizing homology are relative position to other body parts, special quality, and transitional stages [18]. A developmental criterion, introduced by Haeckel [19], proved difficult to apply and has been largely submerged into the previous three criteria. Importantly, there are differing views concerning the hierarchical distribution of homology across levels of biological organization. In the view of Striedter and Northcutt, homology at one level (say behavior) does not necessarily connote homology at another (say morphology) [20]. In contrast, Wagner argued that structures descended from a common ancestor are homologous even if they have diverged and have no clear morphological similarity [16]. The problems raised by the hierarchical nature of homology have been heightened by the discovery of developmental

Citation: Linda Z Holland, João E Carvalho, Hector Escriva, Vincent Laudet, Michael Schubert, Sebastian M Shimeld and Jr-Kai Yu, Evolution of bilaterian central nervous systems, DOI: 10.1186/2041-9139-4-27.

gene conservation [21] and are especially noticeable in the discussion of CNS evolution.

In the past 20 years, it has been found that developmental genes and core signaling pathways are typically conserved across phyla and that gene expression patterns during development can often be used as characters for inferring homologies. Thus, although the majority view had been that the urbilaterian had a nerve net, the balance was tipped towards an urbilaterian with a CNS by the discovery that *bone morphogenic protein* (*BMP*)/*decapentaplegic* genes were expressed dorsally in *Drosophila* and ventrally in vertebrates with the BMP antagonists *chordin/short gastrulation* expressed on the opposite side [22] (Figure 1). In this view, which is consistent with the CNSs of all higher metazoans being homologous, a dorso/ventral (D/V) inversion occurred either in basal protostomes or in the deuterostome lineage [3, 5, 23]. However, in the last 10 years, studies of gene expression and function in an enteropneust (acorn worm; phylum Hemichordata) have been interpreted as evidence that the ancestral deuterostome and, by extension, the urbilaterian had a nerve net and no CNS [9, 24, 25]. Thus, while a CNS would have arisen close to the base of the protostomes, the evolution of a CNS in deuterostomes did not occur until the base of the chordates. In the present review, we examine the detailed evidence on both sides of the controversy and evaluate its interpretations. We conclude that a stronger case can be made for the initial appearance of the CNS at the level of the urbilaterian than for independent evolution of the CNS in more than one line of metazoan descent.

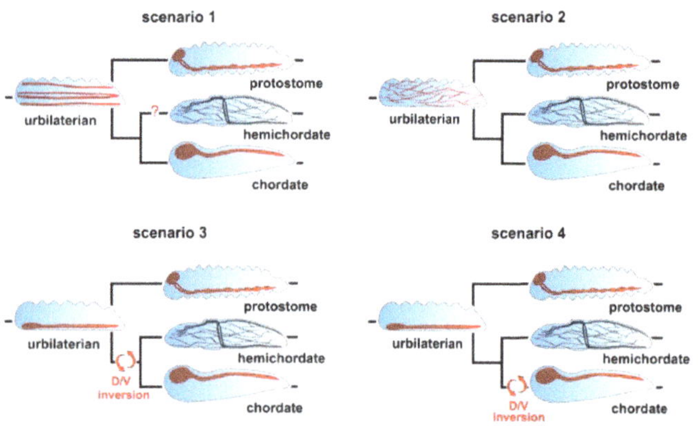

Figure 1. Four scenarios for evolution of central nervous systems in bilaterians. In scenario 1, the urbilaterian had multiple nerve cords, one of which evolved into the dorsal central nervous system (CNS) of chordates, while another nerve cord evolved into the ventral CNS of protostomes. In scenario 2, the CNSs of protostomes and deuterostomes evolved independently from an ectodermal nerve net in the bilaterian ancestor. In scenario 3, the chordate and protostome nerve cords evolved from a ventral nerve cord in the urbilaterian ancestor. A dorso/ventral (D/V) inversion occurred at the base of the deuterostomes; the dorsal nerve cord of hemichordates is thus homologous to the chordate CNS and to the protostome ventral nerve cord. In scenario 4, the protostome and chordate nerve cords evolved from the CNS of an urbilaterian ancestor, but a D/V inversion occurred at the base of the chordates. Thus, the ventral nerve cord of a hemichordate is homologous to the chordate and protostome CNSs. Scenarios after [1, 3, 7, 23, 26, 27, 28, 29].

Reconstructing the Ancestral Bilaterian

Although several features of the ancestral bilaterian in addition to the presence or absence of a CNS are widely debated, a range of molecular, developmental and comparative morphological evidence indicates that this animal was bilaterally symmetrical, with distinct anterior and posterior ends, dorsal and ventral surfaces, and left and right sides. It almost certainly had defined muscle, derived from mesoderm, allowing active locomotion and a gut with either a single opening or a separate mouth and anus [30]. Whether or not this animal had a CNS, an ectodermal nerve net or some combination of the two has been hotly debated (reviewed in [31]) (Figure 1).

One difficulty in deciding whether the ancestral bilaterian had a CNS is that the ectoderm in bilaterians is broadly neurogenic. Therefore, the distinction between the CNS and the remainder of the relatively neurogenic ectoderm is not always clear-cut. In chordates, arthropods and annelids, the distinction is most clear as there is a fully internalized concentration of neurons, axons and supporting cells along the anterior/posterior (A/P) axis (that is, a CNS) that integrates information from sensory cells both associated with the CNS (for example, eyes) and with other portions of the ectoderm and coordinates behavior. Importantly, the CNS in these organisms has an anterior concentration of discrete neural centers or "brain", which coordinates sensory inputs and responses. At the other extreme are "diffuse ectodermal nerve nets" such as in cnidarians. However, such nerve nets are not uniform; specific types of neurons may be regionally localized [32]. An additional problem in understanding the evolution of CNSs comes with the Ambulacraria (echinoderms and hemichordates), as they have both ectodermal nerve nets and nerve cords. It is controversial whether echinoderm and/or hemichordate nerve cords, neither of which has a concentration of neurons that could be termed a brain, and the CNS of chordates have a common evolutionary origin [33, 34]. Here we will use the term CNS for a nervous system that is derived from ectoderm, includes both axons and neurons and is specialized along the A/P axis with an anterior concentration of neural centers (brain), and the term "nerve cord" more broadly to include axonal tracts with few or no neurons and lacking a discrete brain. The diversity of animal nervous systems and paucity of data from some species may blur this distinction on occasion; however, we will be explicit in such instances.

What is the evidence for a CNS in the ancestral bilaterian?

It is generally agreed that bilaterians evolved from radially or bi-radially symmetrical animals, comparable in some ways to modern cnidarians. Adult cnidarians have an ectodermal nerve net with a concentration of neurons around the single gut opening (Figure 2). Therefore, if the ancestral bilaterian had already evolved a CNS, it would presumably have arisen as a concentration or amplification of neurons along one side of this nerve net, perhaps together with a reduction in numbers of neurons elsewhere in the ectoderm.

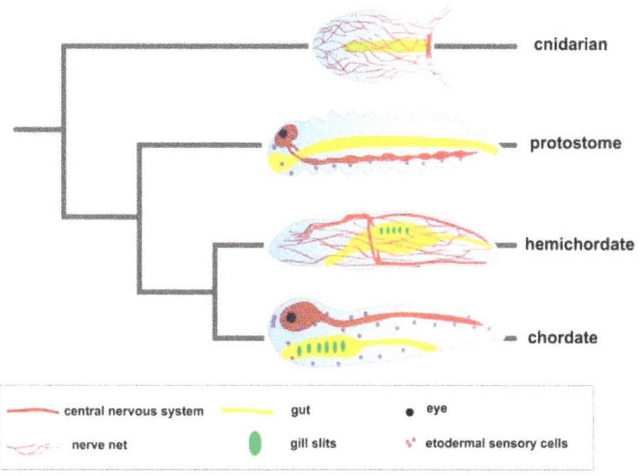

Figure 2. Comparison of metazoan body plans. A typical cnidarian polyp, a generalized protostome, hemichordate and chordate and their phylogenetic relations are shown. Special attention is given to nervous systems and neural structures of the respective animals.

Unfortunately, no extant animal is a good stand-in for this ancestral bilaterian. Extant animals that are thought to have diverged from the bilaterian lineage before it radiated into the protostomes (Ecdysozoa and Lophotrochozoa) and the deuterostomes do not show an intermediate condition between a nerve net and a CNS (Figure 2). The best candidates for such early bilaterian offshoots are the acoel and nemertodermatid flatworms, and the xenoturbellids, which in some studies have been placed basal to the deuterostomes plus protostomes but in others are placed basal in the deuterostome lineage [35, 36]. Acoels have a concentration of neurons, or a "brain", anteriorly with up to six tracts of axons extending posteriorly [37]. In contrast, xenoturbellids have an intraepithelial nerve net that lacks aggregations of neurons or axonal tracts [38]. As a result of the lack of a clear intermediate, scenarios for evolution of CNSs are necessarily based on similarities in gene expression and neuroanatomy in the two main lineages of bilaterians: protostomes (Ecdysozoa plus Lophotrochozoa) and deuterostomes (Figure 1).

Regionalization of nerve cords in Protostomes and Chordates

Because the CNSs in protostomes and deuterostomes are in different positions, develop rather differently and are morphologically somewhat diverse, possible homologies between them have been highly contentious. Complicating the picture is that some of the genetic mechanisms for specifying A/P positions in the CNS are common to the entire organism, including the general ectoderm exterior to the CNS, and are therefore not entirely useful for inferring homologies of CNSs. For example, some genetic mechanisms mediating A/P patterning in the CNS were clearly inherited from a cnidarian-like ancestor in which they patterned the entire body axis. Thus, *Six3/6* and *Irx* are expressed in the aboral region of the planula larva of the sea anemone *Nematostella vectensis*, opposite the blastopore [39] and in the anterior end of the brain of both protostomes and deuterostomes – *Six3/6* in the anterior tip of the CNS and *Irx* genes a little more posteriorly [24, 40, 41, 42, 43, 44]. In *N. vectensis* the domains of these two genes are initially congruent, while in the CNS of bilaterians the *Six3/6* domain is anterior to that of *Irx*. Therefore, although it is most parsimonious to propose that these genes were coopted into the CNS of an ancestral bilaterian, it cannot be ruled out that they were coopted independently into the CNS of protostomes and deuterostomes.

Hox genes are another example of A/P patterning genes that are not entirely useful for inferring homologies between the protostome and chordate CNS. The problem is that although they do mediate A/P patterning of the CNS in bilaterians [45, 46], they mediate A/P patterning of other tissues as well [47, 48, 49, 50, 51, 52, 53,54]. Thus, while their expression patterns have been used to infer homologies between the CNS in insects and vertebrates, it remains possible that they patterned the entire body axis of the Urbilaterian and were independently coopted into the CNSs of protostomes and deuterostomes. It is not clear when a role for nested expression of *Hox* genes in regionalization of the A/P axis evolved. They do not appear to be involved in A/P patterning in cnidarians [55, 56]. Comparisons of *Hox* genes in protostomes with up to 10 or 11 *Hox*genes and invertebrate deuterostomes with up to 15 indicate that the ancestral bilaterian had at least eight to 10 *Hox* genes [57], while cnidarians have up to six depending on the species and acoel flatworms have three, which are more or less regionally expressed in the surface ectoderm along the A/P axis [58, 59] with later expression in putative neural precursors [49]. Thus, acoels either arose before a large *Hox* cluster evolved or they lost some *Hox* genes. Even so, a role for an expanded array of *Hox* genes in specification of A/P positions in the ectoderm was evidently present in the ancestral bilaterian. Thus, although expression of the *Drosophila melanogaster Hox1* gene *labial* in a stripe at the posterior end of the tritocerebrum within the *unpg* (*Gbx*) domain has been likened to nested expression of *Hox* genes in the vertebrate hindbrain, with *Hoxb1* being expressed in a stripe in rhombomere 4, the possibility that *Hox* genes were independently coopted into the CNS in protostomes and deuterostomes cannot be ruled out.

Stronger support for a single origin of the CNS comes from similar expression in the CNSs of protostomes and chordates of genes that are not expressed in comparable patterns in other tissues. Thus, Reichert and colleagues used gene expression patterns to support the perhaps surprising idea that the three parts of the *Drosophila* brain – protocerebrum, deutocerebrum and tritocerebrum [5, 45] – are homologous to the forebrain, midbrain and hindbrain of vertebrates (Figure 2) [60]. For example, in *D. melanogaster*, the *Otx*homolog *Otd* is expressed throughout the protocerebrum and deutocerebrum, while *unpg* (homologous to *Gbx*) is expressed in the tritocerebrum, the subesophageal ganglion and the ventral nerve cord [5]. The domains of the two abut at the boundary between the deutocerebrum and tritocerebrum, similar to the abutting domains of *Otx2* and *Gbx2* at the midbrain/hindbrain boundary (MHB) in vertebrates [5, 61]. In addition, although some domains of *Pax2/5/8* genes are not similar between the CNS of flies and chordates, *Pax2/5/8* is expressed at high levels in the posterior part of the deutocerebrum (just anterior to the deutocerebrum/tritocerebrum boundary) in *D. melanogaster*, while the three vertebrate *Pax2/5/8* genes are expressed at high levels at the MHB [5, 62]. Moreover, in third instar larvae of *D. melanogaster*, the *earmuff*gene (homologous to *Fezf*) is broadly expressed in the anterior brain with a posterior boundary at the protocerebrum/deutocerebrum boundary [63]. The domain is just anterior to that of *mirror*, one of the three *Irx* homologs. Similarly, Irimia and colleagues showed that in chordates, the posterior limits of *Fez* genes (*Fez*and *Fez-like*) abut the anterior limit of *Irx1* in the forebrain [64]. In vertebrates, this is the zona limitans intrathalamica (ZLI) [65].

Compatible with a single origin of the CNS, expression of the genes mediating D/V patterning within the CNS is also conserved between protostomes and deuterostomes [66] (Figure 3). These genes are not comparably expressed in cnidarians, suggesting that they were recruited for roles in D/V patterning the CNS of an ancestral bilaterian. Notably, homologs of some key genes expressed mediolaterally in the neuroectoderm of *D. melanogaster* embryos are expressed in comparable domains in the vertebrate CNS. Thus, the *msh* gene is expressed laterally in the *D. melanogaster* neuroectoderm, with *ind* expressed in an intermediate longitudinal domain and *vnd* expressed in a medial stripe of neuroblasts (reviewed in [7, 67]). Vertebrate homologs of these three homeobox genes are comparably expressed in the developing neural tube. Two of the three *msh*orthologs (*Msx1, Msx2, Msx3*) are expressed dorsally (that is, laterally) in the roof plate of the CNS, one of the two *ind* orthologs (*Gsh1*) is expressed in the adjacent zone (alar plate), and one of the two *vnd* orthologs (*Nkx2.2*) is expressed more ventrally (that is, medially) in the basal plate.

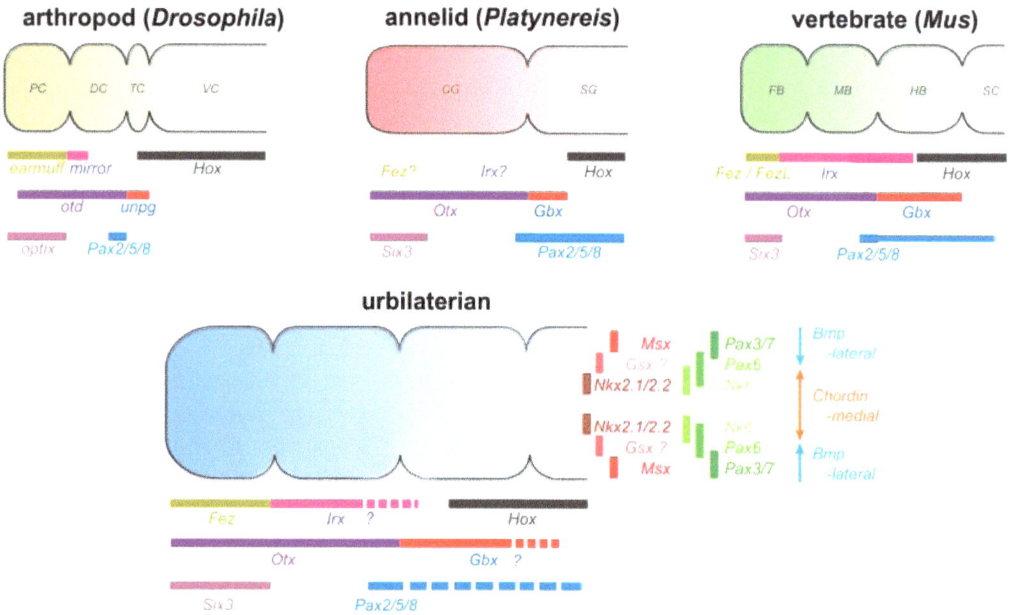

Figure 3. Anterior–posterior gene expression in central nervous systems of three extant bilaterians and the urbilaterian.Anterior–posterior regionalization of gene expression in the central nervous systems of three extant bilaterians (an arthropod, an annelid and a vertebrate) and inferred expression in the last common bilaterian ancestor, the urbilaterian. Expression of Fez and Irx in the annelid Platynereis is unknown. For the urbilaterian, both anterior–posterior and medio-lateral gene expression domains are shown. Hypothetical posterior limits of Irx and Gbx-domains in the urbilaterian brain are highlighted by a "?" and dashed lines. PC, protocerebrum; DC, deutocerebrum; TC, tritocerebrum; VC, ventral nerve cord; CG, cerebral ganglion; SG, segmental ganglia; FB, forebrain; MB, midbrain; HB, hindbrain; SC, spinal cord. Gene expression domains based on [1, 2, 9, 24, 29, 34, 42, 64],[68, 69, 70, 71, 72, 73, 74, 75, 76, 77, 78, 79, 80].

Additional evidence for homology of protostome and chordate nerve cords, and thus a bilaterian ancestor with a CNS, comes from neuroanatomy, neuronal function and gene expression. Strausfeld and Hirth found striking parallels between the central complex in the arthropod protocerebrum and the basal ganglia in the ventral forebrain of vertebrates [3]. In particular, the vertebrate striatum and pallidum have similar organization as, respectively, the insect fan-shaped body and ellipsoid body. Both the types of neurons and their connections and the functions of these regions are similar in the two organisms. Taken together, the data from comparative gene expression and anatomy provide relatively strong support for a single origin of the CNS in insects and chordates.

Parallels between the Brains of Annelids and Vertebrates

Additional evidence for a single origin of the CNS comes from comparisons between annelids and vertebrates. Not only have parallels been drawn between patterning the *Drosophila* and vertebrate brains, but Arendt and colleagues have also noted similarities between the genetic mechanisms patterning the nervous systems of the annelid *Platynereis dumerilii* and vertebrates [2, 81] (reviewed in [4]). The annelid brain varies from species to species, with the brains of some species lacking clear compartments but many others having such features as complex, neuron-rich mushroom bodies (a comparatively large part of the brain in insects and annelids that integrates olfactory information) [82]. Extensive comparisons of gene expression have been used to argue for homology between the mushroom bodies and the pallium of the vertebrate brain [2, 81]. For example, *Bf-1* (*FoxG1*) is expressed in the anterior part of the vertebrate telencephalon and the pallium as well as in the tip of the annelid brain, while *Wnt5/8* is expressed in the vertebrate pallium and in the annelid mushroom bodies, flanking more medial expression of *Hh* in both [2].

Furthermore, in *P. dumerilii Six3* and *Otx* are expressed anteriorly in the CNS (the peristomium) with the *Six3*domain extending anterior to that of *Otx*[42]. The posterior limit of the *Otx* domain abuts that of *Gbx* in the first larval segment, while the anterior boundaries of *Hox1* and *Hox4* are in the second and third larval segments [48]. *Six3/6* and *Otx* are similarly expressed in acoel flatworms [26, 83, 84], and in *D. melanogaster*all three genes are expressed in similar patterns as in *P. dumerilii*. Therefore, the annelid cerebral ganglion has been homologized with the insect protocerebrum. In addition, similar to the CNS in *Drosophila* and vertebrates, the neuroectoderm in *P. dumerilii* is divided into a series of domains with outer/dorsal expression of *Msx* and *Pax3/7* (*gooseberry*), intermediate expression of *Nk6* and *Pax6*, and medial expression of *Nkx2.1*/*Nkx2.2*[1].

Together with anatomical similarities, these data showing distinct similarities in expression of genes patterning the CNS both anteriorly/posteriorly and medio/laterally between both major lineages of protostomes (Ecdysozoa and Lophotrochozoa) and vertebrates support a single origin of the CNS in the bilaterian ancestor (Figure 2). The counterargument would be that the CNS in protostomes evolved independently coopting A/P and D/V patterning mechanisms from an ancestor that used them to pattern a body axis. However, this would mean that the extensive similarities in neuronal architecture between chordates, arthropods and annelids would have been convergently evolved, which seems most unlikely.

Was there a dorso/ventral inversion, and if so, when did it occur?

If the CNS evolved just once, then a D/V inversion must have occurred during evolution of either protostomes or deuterostomes (reviewed in [27, 85]). At present, the chief theories are as follows. The first is Anton Dohrn's idea that a D/V inversion occurred either at the base of the protostomes or within the deuterostomes [86]. The second is the idea most recently articulated by John Gerhart, Christopher Lowe and colleagues that the ancestral deuterostome was hemichordate-like with dorsal and ventral nerve cords and an ectodermal nerve net, with the chordate CNS arising directly from the nerve net [9, 24, 27] or alternatively, as proposed by van Wijhe [87] and more recently by Nomaksteinsky and colleagues [88], from the dorsal nerve cord. A third theory that the ancestral bilaterian had multiple nerve cords, with one evolving into the protostome CNS and another into the deuterostome CNS, was suggested by Gerhart [27] but has received little attention.

Major evidence supporting a D/V inversion in either basal deuterostomes, basal protostomes or basal chordates is that genes involved specifying polarity of the D/V body axis are expressed in opposite orientations in protostomes and chordates. Sasai, de Robertis and colleagues found that in both groups, BMP signaling is involved in establishing D/V polarity and in neural specification, with suppression of BMP signaling being a prerequisite for formation of a CNS [22, 23, 89]. In agreement with a D/V inversion having occurred in either the deuterostome or protostome lineages [23], *BMPs* are expressed dorsally in protostomes and hemichordates and the *BMP* antagonist *short gastrulation* (= *chordin* in deuterostomes) is expressed ventrally, while in chordates it is the opposite – *BMPs* are expressed ventrally and *chordin* dorsally. In most bilaterians, D/V orientation of the body and position of the nerve cord are coupled; however, Hejnol and Martindale have noted that expression of *BMP2/4* dorsally (opposite the future mouth) in an acoel with neurite tract(s) dorsally as well as laterally [26] supports the idea that a role for *BMP*/*chordin* in axial patterning may have preceded a role in neural patterning. Another line of evidence supporting D/V inversion comes from analysis of genes involved in left–right patterning. For example, two key regulators of this distinction, *Nodal* and *Pitx*, are expressed on the left side of chordates, but on the right in echinoderms and in some molluscs [90, 91].

In summary, conserved expression of some genes along the longitudinal axis of cnidarians and in the CNS and general ectoderm of bilaterians indicates likely cooption of roles for these genes in patterning the CNS. However, similar expression of genes involved in both D/V patterning of the CNS and in A/P regionalization of the brain in chordates and protostomes together with neuroanatomical parallels provides considerable support for the idea that the bilaterian ancestor had a CNS, which was modified or possibly lost in various protostome and deuterostome lineages.

Hemichordate Theories

Despite considerable evidence in support of a single origin of the CNS, data from hemichordates have been interpreted as indicating that the ancestral deuterostome had a nerve net, and therefore the CNSs in chordates and protostomes evolved independently. Hemichordates and echinoderms form a clade, the Ambulacraria, which branched off the deuterostome tree as a sister group to chordates. Indirect developing members of both groups have similar pelagic larvae with an apical tuft of cilia. Echinoderms, which have pentamerous symmetry, typically have an ectodermal nerve net plus radial nerves and a circumoral nerve ring. While Haag proposed that the sea urchin radial nerves are homologous to the chordate CNS [33], most authors disagree [92, 93]. Importantly, echinoderm nerve cords do not express *Hox* genes [51, 94, 95], and an extensive screen by Sly and colleagues for expression of neural patterning genes in a juvenile sea urchin failed to find evidence that the nerve ring or radial nerves are homologous to any part of the brain or nerve cord in bilaterians [93]. Moreover, *Engrailed* is very broadly expressed in the nervous systems and other tissues of the juvenile starfish and not in localized domains as in the chordate CNS [94]. Therefore, echinoderms are currently not considered relevant to the question of evolution of chordate nerve chords. In contrast, the worm-like enteropneust hemichordates, which have longitudinal nerve cords as well as a nerve net, have figured prominently in discussions of the evolution of chordates [10, 96, 97, 98].

Inferring homologies between chordate nerve cords and hemichordate nervous tissues has been complicated by large differences in morphology. Adult hemichordates have three distinct regions: proboscis, collar and trunk. There are two classes of hemichordates – enteropneusts and pterobranchs. All four families of enteropneusts (Harrimaniidae, Spengelidae, Ptychoderidae, Torquaratoridae) have an ectodermal nerve net, located in all three regions, plus dorsal and ventral nerve cords, suggesting that this organization is a basal hemichordate characteristic. In contrast, the sessile pterobranchs have anterior tentacles and a concentration of neurons at the base of the tentacles that has been termed a brain, as well as several concentrations of neurites and associated neurons extending into the tentacles, the stalk and between the gill slits [99]. Although Romer and others argued that pterobranchs were basal hemichordates [100, 101], recent molecular phylogenetic analyses do not distinguish which family is basal [102, 103], leaving open the possibility that pterobranchs are derived. Indeed, fossil tube-dwelling enteropneusts from the Cambrian were recently discovered [104].

Most of the work on neural development in hemichordates concerns indirectly developing ptychoderids and the direct developing harrimaniid *Saccoglossus kowalevskii* (reviewed by Röttinger and Lowe [105]). Miyamoto and colleagues showed that the larval nervous system in indirect ptychoderids does not carry over into the adult; in late larvae, the larval nervous system is gradually replaced by the adult one [106]. Therefore, it is the development of the adult nervous system that is pertinent for understanding evolution of the CNS.

Hemichordates and the Argument of an Ectodermal nerve net versus a CNS: theory one

There are two competing theories concerning the evolutionary relationship between the nerve net and nerve cords of hemichordates and the chordate CNS (Figure 1). One theory, most recently articulated by Kaul and Stach [107], proposes that one of the hemichordate nerve cords, typically the dorsal one, is homologous to the chordate CNS. This theory implies that the ancestral deuterostome and perhaps also the ancestral bilaterian had a CNS. The chief basis for this idea is that the collar nerve cord neurulates, suggestive of neurulation in vertebrates [88, 108]. Anterior and posterior to the collar, the nerve cord is continued by basiepithelial tracts of neurites, which are concentrated dorsally [99]. However, there is nothing that resembles a brain. In the direct developing S. *kowalevskii*, neurulation in the collar nerve cord progresses from posterior to anterior, and there are posterior and anterior neuropores [107]. The nerve cord continues posteriorly as a superficial tract of nerve cell bodies overlying nerve cell fibers and rostral of the anterior neuropore as a wide, superficial tract of both neurons and nerve fibers [88]. In addition to the longitudinal nerve cords, there is a peribranchial nerve ring, which develops from ventral to dorsal, as well as a collar nerve ring at the collar–trunk boundary. Although initially neither nerve cord was thought to contain nerve cell bodies, studies with electron microscopy and with specific nerve cell markers have demonstrated nerve cell bodies and glia in the dorsal nerve cord and at least some neurons associated with the ventral one [107, 109, 110]. Ventrally in the dorsal cord, there is a neuropil. Bullock [111] and Brown and colleagues [110] have suggested that the large neurons may be homologous to Mauthner cells of the lamprey and Rhode cells of amphioxus. The developing collar and ventral nerve cords as well as the peripharyngeal cord of both *Ptychodera flava* and *S. kowalevskii* express nerve cell-specific genes including *Elav*, *synaptogamin* and also genes for peptides and proteins specific for subsets of nerve cells including *VAChT*, *serotonin*, *Hb9*, *Drg11* and *GABA*. Serotonergic neurons are restricted to the peripheral nervous system, while those labeling with *Drg11*, *Hb9* and cholinergic neurons are preferentially in the collar nerve cord [88, 112].

Although most of the ptychoderid ectoderm is non-neural [88], basiepithelial nerve cells are moderately numerous in the proboscis [106]. Nomaksteinsky and colleagues suggested that the more even distribution of neurons in the basiepithelial nerve net of *S. kowalevskii* might represent a transient larval nervous system and that part of the diffuse nervous system of developing *S. kowalevskii* larvae, especially in the proboscis, will become the peripheral nervous system [88]. Based on the morphological and gene expression data, they concluded that it was 'implausible that the enteropneust skin is homologous to the chordate CNS'. Instead, they argued that the relatively few neurons in the adult "non-neural" ectoderm constituted a peripheral nervous system, and that either the dorsal or ventral nerve cord (they could not decide which one) was homologous to that of chordates [88]. A comprehensive study of developmental gene expression in the developing nerve cords of hemichordates is sorely needed; to date only a few pictures showing expression of genes including *Dlx*, several *Hox* genes, *Tbx2/3*, *PoxN*, *Pitx* and *Olig* in the dorsal and/or ventral midline of embryos of *S. kowalevskii* have been published [9, 24, 25, 113], but whether the tissue expressing these genes is the developing nerve cord or overlying ectoderm is not clear. There are no studies of developmental gene expression in indirectly developing species such as *P. flava* due to a long pelagic larval period [114, 115].

A problem with homologizing the dorsal nerve cord of hemichordates with the chordate CNS is the finding by Lowe and colleagues [113] that, as in protostomes, *BMP2-4* and *BMP5-8* are expressed dorsally in *S. kowalevskii*, while *chordin* is expressed ventrally. Consistent with a role in D/V patterning, excess BMP4 protein radializes the embryos and eliminates *chordin* expression, indicating an evolutionarily conserved role of BMP in D/V patterning. This suggests that if either nerve cord in hemichordates is homologous to the chordate CNS, it is the ventral nerve cord, which does not neurulate, and a D/V inversion occurred at the base of the chordates. This is consistent with the gill slits being dorsal and the stomochord, a dorsal/anterior extension of the gut, having been shown to be unrelated to the notochord [116]. Confusing the issue further, in amphioxus and vertebrates, *Nodal* expression dorsally acts in opposition to *BMP* expression ventrally [117], while in sea urchin embryos, BMPs and Nodal oppose each other in patterning the oral/aboral axis (Nodal ventralizes; BMP dorsalizes), suggesting that a role for Nodal in opposing BMPs was present at the base of the deuterostomes and that a D/V inversion occurred in chordates. To some extent this is similar in *S. kowalevskii*, in that perturbation of Nodal signaling results in D/V patterning defects [118]. However, treatment with the Nodal inhibitor SB431542 eliminates expression of both *BMP2/4* and *chordin* and anteriorizes embryos, indicating that Nodal posteriorizes embryos, the opposite of the situation in chordates [118]. These results suggest that the role of Nodal [119] may have been altered in *S. kowalevskii*. Whether a role for BMP/Nodal opposition in D/V patterning was present in the ancestral bilaterian is uncertain. Nodal is involved in left/right patterning in a mollusk [120], but possible roles in D/V or A/P patterning in protostomes have apparently not been investigated. In summary, since it neurulates, the dorsal nerve cord of hemichordates has been proposed as homologous to the chordate CNS. However, the rather scanty data on gene expression are more compatible with homology of the ventral nerve cord and the chordate CNS. More data are clearly needed on both anatomy and gene expression in the hemichordate nerve cords.

Hemichordates and the Argument of an Ectodermal nerve net versus a CNS: theory two

In spite of the evidence supporting the idea that the ancestral bilaterian had a CNS, there is an alternative theory – namely that the ancestral bilaterian and the ancestral deuterostome had ectodermal nerve nets, from which the chordate CNS evolved. The dorsal and ventral nerve cords of hemichordates are therefore not only a hemichordate invention, but are unrelated to the chordate CNS (reviewed in [34]). This theory, most recently articulated by Lowe and colleagues [9, 24, 113], is based on developmental gene expression and gene interactions in the direct-developing hemichordate, *S. kowalevskii*. It proposes that the chordate CNS evolved from the ectodermal nerve net of a hemichordate-like ancestral deuterostome and maintains that the hemichordate nerve net contains signaling centers evolutionarily related to the anterior neural ridge (ANR), the ZLI and the isthmic organizer (ISO) in vertebrates. As a corollary, part or all of these signaling centers have been lost in the invertebrate chordates (amphioxus and tunicates) [9]. This idea deserves careful consideration because it not only argues that the considerable similarities of gene expression in protostome and chordate nerve cords represent convergent evolution, but it assigns a key position to hemichordates in evolution of the vertebrate CNS.

de Beer [121] was one of the first to recognize the hierarchical nature of homology when he noted similar morphological features in two different animals could develop under the control of different genes; a phenomenon now known as genetic piracy [122]. The converse is also known – where parts of homologous gene networks are involved in the development of apparently nonhomologous structures [123]. When such disconnects are discovered, some would pay more attention to structure [124], and others would pay more attention to the genes (as a deep homology) [125]. In discussing neural evolution in higher deuterostomes, Lowe and colleagues strongly favor genes over morphological features as arbiters of homology. Thus, they maintain that structures with very different morphology (for example, the proboscis of a hemichordate and the forebrain of a vertebrate) are not morphologically homologous [25, 126] even though conserved gene expression patterns may indicate a common evolutionary origin. Thus, while those authors find that homologous genetic programs operate at the anterior end of the hemichordate proboscis and vertebrate ANR, at the boundary between the hemichordate proboscis and collar and at the vertebrate ZLI, and at the boundary between the hemichordate collar and trunk and at the vertebrate ISO located at the MHB, they do not refer to these three regions of the hemichordate ectoderm and vertebrate brain as homologs [9]. Even so, as discussed by Wagner [16], these regions could be homologous even though morphologically divergent.

Lowe and colleagues, in their series of papers concerning genes and structures involved in neural evolution of deuterostomes, have been somewhat inconsistent in their treatment of the subject of homology. In their initial work, which concerned expression of 22 genes with restricted ectodermal domains along the A/P axis (Table 1), they concluded that the surface ectoderm of *S. kowalevskii* and the chordate CNS have a common ancestry [24]. These authors noted that *S. kowalevskii* 'shows pervasive neurogenesis with no large, contiguous non-neurogenic subregion, as occurs in chordates' and concluded that the deuterostome ancestor had a diffuse ectodermal nerve net that evolved into the vertebrate CNS [24]. One difficulty with this argument is that, as Aronowicz and Lowe [25] later noted, the surface ectoderm of invertebrate chordates outside the neural tube contains widespread ectodermal sensory neurons (Figure 1) [127, 128], while in vertebrates the large pan-placodal region outside the neural plate is highly neurogenic [68]. In addition, although the authors maintained that most of 22 genes they studied are not expressed in the "epidermal ectoderm" in chordates [24], in the 10 years since this paper was published, it has become clear that all except possibly *Vax*, *Rx* and *Nkx2-1* are expressed in ectodermal sensory cells in chordates (Table 1) [68, 129, 130, 131]; expression of *Vax*in amphioxus and of *Nkx2-1* in the tunicate *Ciona intestinalis* have not been determined. For example, *Sox1/2/3* (*SoxB*) and *Hu/Elav* are expressed in ectodermal sensory cells in chordates [129, 130], *Hox1*, *Hox3*and *Hox4* are expressed in nested patterns in the amphioxus ectoderm and at especially high levels in ectodermal sensory cells [132], ascidian *Hox1* is expressed in development of ectodermal sensory cells [133], and some *Hox* genes are expressed in placode derivatives in vertebrates [134]. In addition, *Gbx* is expressed in the "non-neural" ectoderm in amphioxus [135] and in developing placodes in vertebrates [136], while *Gbx* and *Otx2* mutually repress one another in development of the otic placode as they do in A/P patterning of the CNS [137]. Therefore, as noted above, the ectodermal expression patterns of genes along the A/P axis do not clearly distinguish CNS from ectoderm outside the CNS. Indeed, Aronowicz and Lowe noted that genes such as *Otx* and *Hox* appear to be involved in patterning neurogenic tissues generally [25]. Thus, expression of these A/P patterning genes alone does not distinguish between theories that the ancestral bilaterian and ancestral deuterostome had an ectodermal nerve net or whether they both had a CNS. As noted above, genes expressed exclusively in the CNS of bilaterians (for example, *earmuff/Fezf/Fezl* and D/V patterning genes) are generally more informative for inferring the course of evolution of the CNS [64]. Indeed, domains of homologs of three genes mediating lateral to medial (D/V) patterning in the chordate and protostome CNSs (*Hh*, *Nkx2.2*, *Msx*) [138, 139, 140] are not expressed in comparable patterns in the hemichordate ectoderm [113]. Thus, if the chordate CNS evolved from a nerve net in the ancestral deuterostome, expression of these lateral to medial genes would represent convergent evolution in the CNSs of protostomes and chordates.

Table 1. Gene expression in nervous tissues of - *Saccoglossus kowalevskii* , *Branchiostoma floridae*, *Ciona intestinalis* and vertebrates

Gene	*S. kowalevskii*		*B. floridae*		*C. intestinalis*		Vertebrate		References
	s.e.	CNC[a]	s.e.	CNS	Ectoderm	CNS	Placode	CNS	
SoxB	+	+	+	+	+	+	+	+	[131]
Hu/Elav	+	+	+	+	?	?	+	+	
Nrp/Musashi	+	+	+	+	?	+	+	+	[141]
Vax	+	–	?	?	gene	lost	–	retina, forebrain	[142]
Rx	+	–	–	+	–	+	–	+	[69, 76]
Six3	+	+/–	+	+	–	+	–	+	[43, 131, 143]
Nkx2.1	+	–	–	+	?	?	–	+	[140, 144]
Bf-1	+	+/–	+	+	+	?	+	+	[131, 145, 146]
Dlx	+	+	+	+	+	+	+	+	[77]
Pax6	+	+	+	+	–	+	+	+	[75]
Tll/Tlx	+	+/–	+	+	?	?	+	+	[147, 148]
BarH	+	+	?	?	?	?	+	+	[149]
Emx	+	+	?	?	?	?	+	+	[150]
Otp	+	?	?	?	+	+	?	+	[131, 151]
Dbx	+	+	?	?	?	?	–	+	[152]
Lim1/5	+	+	+	+	?	?	+	+	[153]
Irx	+	+	+	+	?	?	+	+	[154]
Otx	+	+	–	+	+	+	+	+	[78, 137]
En	+	+	+	+	–	+	+ lamprey	+	[78, 155]
Gbx	+	+	+	+	gene	lost	+	+	[142]
Hox1	+	+	+	+	+	+	+	+	[156, 157, 158]
Hox3	+	–	+	+	+	+	+	+	[134, 156, 158]
Hox4	+	?	+	+	–	–	–	+	[156]
Hox7/8	+	?	?	?	gene	Lost	–	+	[156]
Hox11/13	+	–	?	?	Hox12+	Hox12+	Hox11+	+	[156, 159]

Gene expression in ectoderm and nervous tissue of the hemichordate *Saccoglossus kowalevskii*, the cephalochordate *Branchiostoma floridae*, the tunicate *Ciona intestinalis* and vertebrates. Note: En expression in placodes only documented for lamprey among the vertebrates. CNC, collar nerve cord; CNS, central nervous system; s.e., surface ectoderm. [a]In the absence of sections, which would distinguish between expression in the surface ectoderm and in the collar nerve cord, all genes with ectodermal expression in the region of the dorsal portion of the collar are listed.

The similarities in patterning the ANR and the anterior part of the hemichordate proboscis, the ZLI and the boundary between the proboscis and collar, and the ISO and the boundary between the collar and trunk lead to the conclusion that in the ancestral deuterostome the role for these signaling centers, which ultimately gave rise to the vertebrate ANR, ZLI and ISO, was to regionalize the general body plan [9]. However, Pani and colleagues also proposed that amphioxus partially lost the ANR, and completely lost both the ZLI and ISO, while the tunicate *C. intestinalis* lost the ZLI and partially lost both the ANR and ISO [9]. They conclude that in 'certain cases hemichordates will be a more informative group than basal chordates for reconstructing stem chordate characters and understanding the origins of vertebrate developmental genetic processes' [9]. As this is quite an extreme view and is at odds with conclusions based on morphology, gene expression and gene function in amphioxus, tunicates and vertebrates, the evidence merits close examination.

How much of the ANR gene network is present in S. kowalevskii? Is the evidence sufficient that the vertebrate ANR evolved from the anteriormost ectoderm of the ancestral deuterostome?

The vertebrate ANR is characterized by expression of *Fgf8*, *Six3*, *Pax6*, *Otx2*, *Sox2* with *Dlx5* expressed in adjacent non-neural ectoderm and *Bf-1* (*FoxG-1*) expressed in the rostral forebrain (Figure 4) [160], while *Sfrp1a* is expressed in the anterior/ventral part of the developing neural tube [161]. In addition, transplantation of the ANR laterally expanded the telencephalon and promoted expression of *Bf-1* (*FoxG1*), demonstrating that the ANR is an organizer [162]. In *S. kowalevskii*, homologs of these genes are expressed in the proboscis ectoderm, although their relative domains are rather divergent from those of their homologs in the vertebrate forebrain. For example, *Six3*, *Sfrp1/5* and *Fgf8/17/18* are strongly expressed in the anterior proboscis ectoderm, and *Sox1/2/3* is broadly expressed in the proboscis and anterior part of the collar [9]. However, *Pax6* is not expressed at all in the anteriormost proboscis ectoderm but more posteriorly throughout much of the proboscis and collar, while *Otx* is chiefly expressed in the collar ectoderm with only very weak expression at the anterior end of the proboscis. In addition, *Dlx* is very patchily expressed in the proboscis ectoderm, but appears to be highly expressed where the dorsal nerve cord will form. *Hh* is expressed in ectoderm at the tip of the proboscis in *S. kowalevskii*[9], while in vertebrates it is expressed in the basal plate of the forebrain and midbrain and in the floor plate of the anterior hindbrain as well as in the prechordal plate [163].

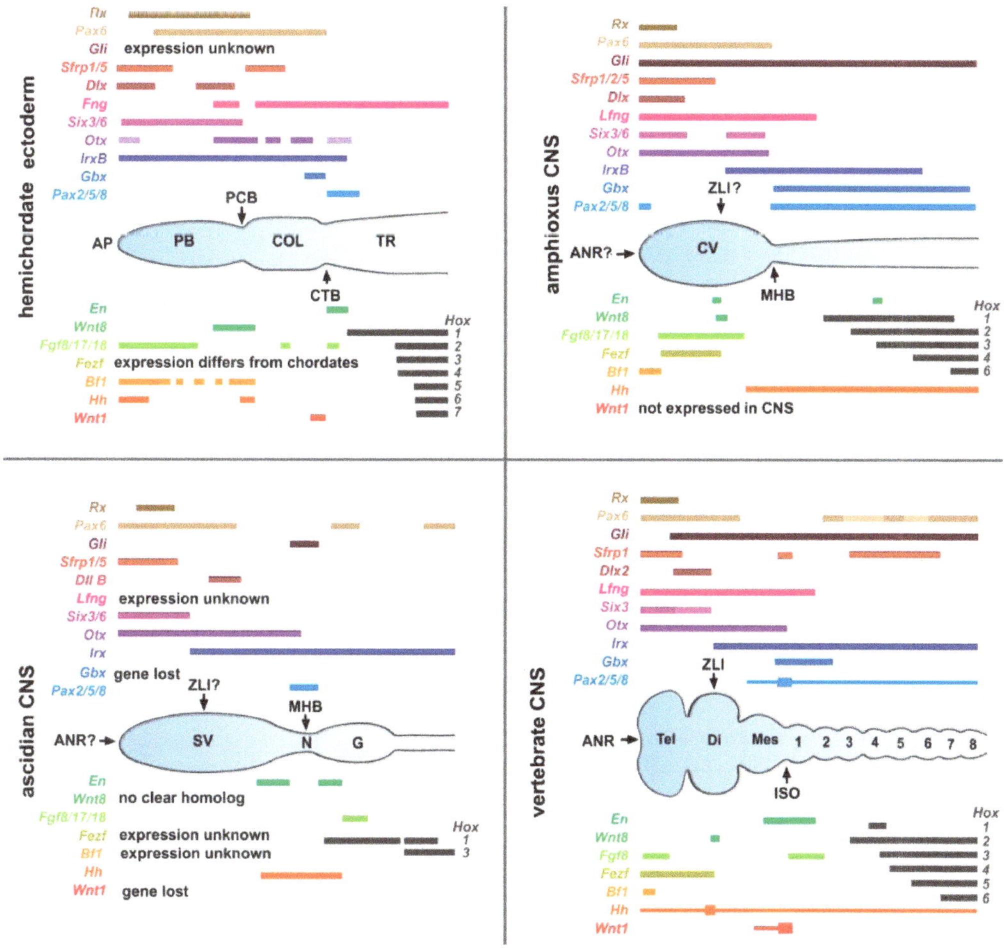

Figure 4. Anterior–posterior gene expression in hemichordate ectoderm and central nervous systems of three chordate subphyla. Anterior–posterior regionalization of gene expression domains in the ectoderm of the hemichordate *Saccoglossus kowalevskii* as well as in the central nervous system (CNS) of representatives of the three chordate subphyla (that is, amphioxus, ascidian tunicates and vertebrates). Question marks on the diagrams of the amphioxus and ascidian CNS indicate that organizer properties of the regions marked 'ZLI?' and 'ANR?' have not been tested. AP, anterior proboscis; PB, proboscis; PCB, proboscis/collar boundary; COL, collar; CTB, collar/trunk boundary; TR, trunk; ANR, anterior neural ridge; ZLI, zona limitans intrathalamica; MHB, midbrain/hindbrain boundary; SV, sensory vesicle; N, neck; G, ganglion; ISO, isthmic organizer; Tel, telencephalon; Di, diencephalon; Mes, mesencephalon. Gene expression domains based on [9, 24, 34, 64, 68, 69, 70, 71, 72, 74, 75, 76, 77, 78, 79, 80].

Experimental evidence for homology of gene networks patterning the ANR and anterior proboscis ectoderm in *S. kowalevskii* came from manipulation of Fgf and Hh signaling. Although inhibition of *Fgf* or Fgf signaling scarcely affected expression of the proboscis domain of the anterior marker *Rx*, it did eliminate *FoxG1* (*BF-1*) expression in the proboscis. In addition, knockdown of *Hh* eliminated expression of the anterior marker *Fgf-Sk1*. Thus, inhibition of Fgf or Hh signaling affects development of the proboscis and expression of some anterior markers. However, Green and colleagues found that the major role of *Fgf8/17/18* in development appears to be in mesoderm induction [164]. Therefore, the effects of inhibition of Fgf signaling on anterior development could be secondary to those on mesoderm.

Whether this evidence is sufficient to indicate that the vertebrate ANR evolved from the anterior part of a diffuse ectodermal nerve net in the deuterostome ancestor is open to question. A major difficulty in identifying homologous gene networks is deciding how much of two gene networks must be conserved for them to be considered homologous [165]. This is a particular problem when the morphology is not conserved. Gene networks can include several thousand genes, and it is well known that core parts of signaling pathways are often coopted for patterning nonhomologous structures [166]. Thus, there are several alternative explanations for similarities in gene expression between the hemichordate proboscis ectoderm and the vertebrate ANR. One is that similar expression of a suite of genes including *Fgfs* in these two regions may simply be indicative of ancient roles in specification of the anterior end of embryos in general. For example, Sinigaglia and colleagues reported that *Six3/6*, *FoxQ2*, *Irx*, *SoxB1* and *Fgf* are expressed in the aboral region in the cnidarian *N. vectensis*, while *Wnts* are expressed at the opposite end of the embryo [39]. *Six3/6* is initially involved in specification of the anterior end of the embryo and later in neurogenesis. Fgf signaling is required for development of the apical tuft of cilia. Therefore, roles for these genes in patterning the hemichordate proboscis may simply reflect an inheritance from prebilaterians. Another possibility is that if, as comparisons of protostomes and chordates indicate, these genes were coopted for patterning the anterior CNS of the ancestral bilaterian, then the ancestral deuterostome might have had a more extensive CNS than modern hemichordates. If so, the forebrain may have been lost as the proboscis evolved, and the genetic pathways for anterior neural development coopted into the proboscis ectoderm. Expression of *Rx*, *Fgf8/17/18* and other genes in the hemichordate proboscis is compatible with either theory. In chordates, *Rx* expression is apparently restricted to the anterior end of the forebrain (it is not expressed in sensory ectoderm) [69], while *Fgf8/17/18* is expressed in the ANR and telencephalon of vertebrates as well as throughout the forebrain of amphioxus. In contrast, *Hh* is expressed in the anterior tip of the notochord in amphioxus and in the floorplate but not in the anterior CNS. Given that *Fgf8* and *Hh* have been coopted in vertebrate limbs, which evolved in gnathostomes, for patterning the apical ectodermal ridge (AER) and zone of polarizing activity (ZPA) [167], it is certainly possible that these genes plus *Rx* and other genes involved in neuronal specification may have been coopted for patterning the proboscis ectoderm of hemichordates.

Another possibility, proposed by Nomaksteinsky and colleagues [88], is that the ectoderm of developing *S. kowalevskii* may represent a transient larval nervous system unrelated to that of the adult. Evidence is that *SoxB* genes, which mark the neural plate in chordates, are broadly expressed in the ectoderm of indirectly developing embryos of the hemichordate *Ptychodera flava*, with expression becoming localized during development to the stomodaeum, ciliary bands and apical ciliary tuft [114]. *Dlx* is expressed in the aboral ectoderm, with a concentration towards the apical tuft, and *Fz5/8* and *FoxQ2* are expressed in the apical tuft [168]. Similarly in *S. kowalevskii*, *Six3*, *Fz5/8* and *Dlx* are broadly expressed in the proboscis ectoderm and *Bf-1*(*FoxG1*) in the apical region [9, 24]. The tornaria larva of indirect developing hemichordates has much in common with the dipleurula type of echinoderm larvae, which also have an apical tuft and a band of cilia around the mouth. Moreover, gene expression in the apical tuft and surrounding ectoderm is highly conserved in the sea urchin and hemichordate larvae, and there is considerable similarity with expression in the proboscis ectoderm of *S. kowalevskii* embryos [169]. In sea urchins, *FoxQ2*, *Fzl5/8*, *Sfrp1/5*, *Dkk3* and *Six3*are expressed at the apical end of the larva, with Wnt signaling acting to restrict expression of these genes to the anterior end of the larva [169, 170]. Similarly, in *S. kowalevskii* knocking down the Wnt receptor *Fz5/8*results in a posterior expansion of the domains of anterior markers [9], indicating that Wnt signaling is also regulating expression of genes in the proboscis ectoderm as it is in the apical ectoderm of sea urchin larvae.

How much of the ZLI gene network is present in S. kowalevskii Does the evidence support evolution of the vertebrate ZLI from a boundary partitioning the anterior/posterior axis of the ancestral deuterostome?

The vertebrate ZLI is positioned and regulated, at least in part, by anterior expression of *Otx* and *Fezf* abutting a posterior domain of *Irx*[171, 172]. Organizer properties are conferred by *Hh*, which is expressed at the ZLI and which is regulated by *Wnt8b* (Figure 4) [173]. In fish, knockdown of *Fezf2*, which is expressed anterior to *Irx*, eliminates the prethalamus and causes mis-specification of the ZLI. Correspondingly, knockdown of *Otx*inhibits expression of *Shh* at the ZLI and reduces expression of *Ptc1* and *Wnt8b*[172]. Evidence for organizer properties of the vertebrate ZLI comes from grafting experiments and implants of Fgf8-coated beads, which induce the rostral neuroepithelium to develop an ectopic and polarized mesencephalon/metencephalon (reviewed in [174]).

The region of the hemichordate ectoderm expressing homologs of ZLI genes lies at the boundary between the proboscis and the collar [9] and is marked by the posterior limit of *Rx* expression and bands of *Fgf8/17/18*, *FoxG*, *Otx*, *Wnt8*, *Hh*, *Ptch*, *FoxA* and *Dlx*. Congruent expression of *Wnt8* and *Hh* in *S. kowalevskii* is consistent with a possible role of *Wnt8* regulating *Hh* as it does at the ZLI in vertebrates. However, domains of *Fezf* and *Irx*, which abut at the ZLI in vertebrates, do not abut at the proboscis/collar boundary in *S. kowalevskii.*Expression of *Fezf* was not shown, but the major domain of *Irx* appears to be congruent with that of *engrailed*in the posterior part of the collar [24].

For *S. kowalevskii*, the experimental evidence presented by Pani and colleagues for the proboscis/collar boundary region acting as a signaling center was that knockdown of *Otx* expression reduced the intensity of the stripe of *Hh* expression just posterior to the proboscis/collar boundary, while inhibition of Hh signaling downregulated *Dlx* both at the proboscis/collar boundary and more anteriorly and also reduced the size of the proboscis [9]. These results show that *Otx* may act upstream of *Hh* at the proboscis/collar boundary, while *Hh*is vital for normal development. However, *Otx* and *Shh* also interact in patterning the vertebrate midbrain, not just the ZLI, and knockdown of *Otx* results in dorsal expansion of the *Shh* domain and a dorsal and anterior rotation of the MHB [175]. Thus, the relationship between *Otx* and *Hh* and perhaps some other parts of the gene network may be evolutionarily conserved between *S. kowalevskii* and vertebrates, but whether tissue at the proboscis/collar boundary in an ancestral deuterostome evolved into the ZLI is open to question given the markedly different expression of key genes such as *Fezf* and *Irx* as well as the lack of anatomical similarity.

How much of the ISO gene network is present in S. kowalevskii Does the evidence support evolution of the vertebrate ISO from a boundary partitioning the anterior/posterior axis of the ancestral deuterostome?

In the vertebrate CNS, the MHB or isthmus functions as an organizer, and is therefore termed the isthmic organizer or ISO. The MHB is positioned by opposition of anterior *Otx* and posterior *Gbx*. Organizer properties are conferred by the action of a suite of genes including *En*, *Fgf8/17/18*, *Wnt1* and *Pax2/5/8* (Figure 4) [176]. *Otx* and *Fgf8* are expressed in the midbrain and *Gbx* in the anterior hindbrain, with the *Wnt1*, *Engrailed* and *Pax2/5/8* domains spanning the boundary; mutual repression of *Otx* and *Gbx* positions the ISO, and Wnt signaling is required for expression of *Engrailed*. Pani and colleagues also presented evidence for a homologous gene network in *S. kowalevskii* operating at the boundary between the collar and trunk [9]. *Engrailed*, *Fgf8/17/18*, *Wnt1* and *Gbx* are expressed near this boundary. In addition, clonal suppression of -*catenin* effectively inhibited *Engrailed* expression, suggesting that Wnt signaling regulates *Engrailed*, as it does at the ISO in vertebrates, while suppression of *Fgf8/17/18* reduced *Engrailed* expression at the collar/trunk boundary. However, despite these similarities, there are several problems in interpreting the gene network at the collar trunk boundary in *S. kowalevskii* as homologous to that patterning the vertebrate ISO. Importantly, as the authors note, the 'spatial arrangements of *Otx* and *Wnt1*, and *Gbx* and *Fgf8/17/18* are reversed in *S. kowalevskii* compared to the ISO in vertebrates' [9]. Moreover, suppression of Fgf signaling had no effect on expression of *Pax2/5/8*. A problem in interpreting these experiments is that in vertebrates, *Fgf8* has an early role in induction of neural tissue, while patterning the ISO is a relatively late role. In *S. kowalevskii, Fgf8/17/18*is expressed from the blastula stage, and by the neurula stage expression is restricted to anterior ectoderm. Knockdown experiments show that *Fgf8/17/18* is required for mesoderm induction [164]. Therefore, effects of *Fgf8/17/18* inhibition in *S. kowalevskii* on *Engrailed* expression may be secondary to the loss of mesoderm and not directly related to effects on A/P patterning. Additional evidence for the common ancestry of the collar/trunk boundary and the vertebrate ISO was that *Hox* genes are expressed in nested patterns in the trunk ectoderm. However, it appears that at least *Hox1* and perhaps *Hox4* are also expressed in the dorsal part of the collar (Figure five in [24]), indicating that Hox expression in the ectoderm is probably not congruent with that in the collar, and raising the question of the genetic mechanisms patterning the collar nerve cord and how they might compare with those in chordates.

As for the ANR and ZLI, drawbacks to interpreting the boundary of the collar and trunk in hemichordates with the vertebrate ISO as having common ancestry not only include differences in expression of key genes expressed in these regions and also the general phenomenon of cooption of parts of gene networks for new functions. For example, in the AER of the vertebrate limb bud, *Wnt3a* induces expression of *Fgfs*, while ectopic *Engrailed* (*En-1*) induces ectopic *Fgf8* expression [177, 178]. Moreover, in amphioxus, *engrailed* is co-expressed with *Wnt8*, but not at the MHB, suggesting that coexpression of *Wnt* and *engrailed* is not wedded to the MHB. The questions therefore become: how conserved must gene networks be in order for one to be reasonably certain that two morphologically rather different structures have a common ancestry; and if they do share an ancestry, can one distinguish whether portions of the gene networks operating in the CNS of a deuterostome ancestor were transferred to the hemichordate ectoderm as the hemichordate CNS became reduced or whether the ancestral deuterostome lacked a CNS and used these gene networks to partition the A/P axis?

How much of the ANR, ZLI and ISO do invertebrate chordates have?

In addition to asserting that the *S. kowalevskii* ectoderm is not only homologous to the vertebrate CNS but also has homologs of the ANR, ZLI and ISO, Pani and colleagues maintain that the invertebrate chordates, amphioxus and tunicates, have lost all or part of these three regions [9]. Here we draw attention to data that are not consistent with such a view (Figure 4).

Although transplantation experiments in vertebrates demonstrated that the ANR functions as an organizer [163], such transplantation experiments are not feasible for hemichordate, amphioxus or *C. intestinalis*embryos due to their small size. However, it is clear that much of the gene network for specification of the ANR and conferring organizer properties upon it is present in amphioxus. *Fgf8/17/18* is expressed in the entire forebrain of amphioxus, and *Bf-1* (*FoxG1*) is expressed at the tip of the forebrain [145] as are *Pax2/5/8* and *Six3/6*[43]. *Otx* and *Pax6* are expressed in comparable patterns with strong expression in the anterior forebrain [135, 179], and the Wnt antagonist *Sfrp1/2/5* is expressed in the anteriormost dorsal ectoderm, including the most anterior neuroectoderm [180]. Moreover, as in vertebrates, *Dlx* is expressed in ectoderm outside the neural tube as well as in the edges of the anterior neural plate [70], while Hh is expressed in the underlying tip of the notochord [181, 182]. Taken together, these expression patterns indicate that amphioxus has most of the components for an ANR comparable to that in vertebrates.

Tunicates have undergone some radical changes to their genome and their anatomy in evolution. For example, the *Gbx* gene has been lost, and there are relatively few neurons – none in the tail nerve cord of ascidians. *Otx* is expressed anteriorly in the CNS, but the evidence for an ANR in *C. intestinalis* is relatively weak. In addition to *Otx*, *Pax6* is expressed in the anterior CNS, but not *Fgf8/17/18* and *Pax2/5/8*, *Hh*, or *Gli*[183]. Expression of *Bf-1* is not known. Therefore, it may be that as tunicates adopted early decision of cell fates and decreased the number of cells in the nerve cord, the need for an anterior brain organizer diminished.

For the ZLI, patterns of gene expression indicate that amphioxus probably has much of the genetic mechanism in place. In amphioxus, Irimia and colleagues showed that anterior expression of *Fezf* abuts posterior expression of *IrxB* about the midpoint of the forebrain [64] – approximately where *Wnt8* is expressed [184]. Similarly, *Wnt8b* expression at the ZLI in vertebrates is flanked by *Lfng*, while in amphioxus the *Fng* domain appears to be posterior to and possibly abutting that of the *Wnt8* domain at the late neurula stage [184, 185]. *Dlx*, *Nkx2-2* and *Gli*, which mediates Hh signaling, are also expressed in this region, while *Fgf8/17/18* is expressed throughout the forebrain [70, 71, 186]. *Hh* is expressed in the floor plate as it is in vertebrates, but it is unclear at the early neurula stage whether it is expressed congruently with *Engrailed* and *Wnt8* or a little more posteriorly. Later expression is limited to a zone posterior to the forebrain [181]. Inhibition of Fgf signaling at the late blastula does not inhibit expression of neural plate markers, but eliminates that of *Otx* in the cerebral vesicle (forebrain) and reduces its size [187], indicating that Fgf signaling is essential for development of the forebrain. Similarly, upregulation of Wnt/β-catenin signaling reduces expression of *Otx* in the forebrain and eliminates expression of the anterior marker *FoxQ2*[188], showing that suppression of Wnt/β-catenin signaling by inhibitors such as Sfrp1/2/5 [180] is essential for forebrain development.

For ascidians, expression of *Irx* and *Fezf* is not known. *C. intestinalis* has two *Wnt* genes that are possibly related to *Wnt8* but which have long branches in phylogenetic analyses and for which assignation remain unclear [189]. Expression of these genes has not been characterized. However, *Gli*, *Hh* and *Fgf8/17/18* are not expressed in the anterior CNS. Even so, inhibition of Fgf signaling blocks expression of the anterior marker *Six3/6*, indicating that Fgf signaling is required for development of the anterior brain [190]. Thus, while *C. intestinalis* likely lacks most of the gene network specifying the vertebrate ZLI and conferring organizer properties, until the expression patterns of *Irx* and *Fezf* are known the presence of part of the network for specification of the ZLI cannot be ruled out.

Amphioxus also appears to have in place the genetic mechanism for positioning the ISO, but this region does not express homologs of the genes that confer organizer properties on the vertebrate ISO. In the amphioxus brain, as in that of vertebrates, the domain of *Otx* expression in the forebrain/midbrain abuts that of *Gbx* in the hindbrain [135]. In vertebrates, organizer properties are conferred on the ISO by expression of *Fgf8*, *Wnt1*, *Engrailed* and *Pax2/5/8* genes at this boundary. In amphioxus, *Wnt1* and *Engrailed* are not expressed between the *Otx* and *Gbx* domains, although *Fgf8/17/18* is expressed anterior to and abutting this boundary, while *Pax2/5/8* is expressed posterior to and abutting it [187, 191, 192]. *Engrailed* is expressed together with *Wnt8*approximately at the boundary between the *Fezf* and *IrxB* domains. Interestingly, there is also a stripe of *Engrailed* in the general ectoderm that is in register with that in the CNS [193].

The tunicate *C. intestinalis* appears to have more of the ISO network in place than amphioxus does, but it has lost a key gene – *Gbx*. *Pax2/5/8* and *Fgf8/17/18* are expressed in the neck region in a few cells just posterior to the posterior limit of *Otx*. *Engrailed* is expressed in two domains, one near the *Pax2/5/8* domain and the other overlapping with that of *Otx* [194].

The emerging picture is that amphioxus has most, if not all of the genetic mechanism for the ANR in place and part of that for the ZLI and ISO while the ancestral tunicate probably added to these gene networks but subsequently lost parts of them as the CNS became simplified. The most parsimonious explanation is that the ancestral bilaterian had a CNS in which portions of these gene networks were used to establish divisions in the CNS and perhaps to some extent in the ectoderm generally, and these gene networks became modified in the protostome and deuterostome lineages. Under this scenario, the basic gene networks for specifying the ANR, ZLI and MHB began to be established in the ancestral chordate or even earlier in the ancestral deuterostome or ancestral bilaterian, and additional genes were recruited to these networks before tunicates branched off the lineage leading to vertebrates. In the tunicate lineage, in conjunction with shrinking genomes, a switch from regulative to determinate development and a reduction of cell numbers in the CNS, some parts of the ancestral tunicate networks for these brain regions were lost. Modifications occurring in the hemichordate lineage may have involved changes in the CNS and/or cooption of additional genes and portions of gene networks to the surface ectoderm.

CONCLUSIONS

The bilaterian CNS typically develops from a longitudinal strip of ectoderm that is then internalized. Although the remainder of the ectoderm is usually termed "non-neural", it also contains numerous neurons – often sensory neurons – which send signals to the CNS. Therefore, the ectoderm really should be thought of as "more neural" and "less neural" rather than "neural" and "non-neural". This is highly relevant for discussion of how a CNS evolved from an ectodermal nerve net, since part of the genetic mechanism(s) for patterning the bilaterian CNS are shared with the ectoderm as a whole.

The basic A/P patterning mechanism in bilaterians with *Six3/6* expressed anteriorly and *Wnt* genes posteriorly was inherited from an ancestral cnidarian-like animal. Additional genes (for example, *Hox* genes) were recruited to pattern the A/P axis of the ancestral bilaterian. Consequently, even though the ectoderm in annelids and arthropods is segmented, and the ectoderm of hemichordates and chordates is not, several genes that mediate A/P patterning of the CNS are expressed in similar patterns along the A/P axis of the non-neural ectoderm in early embryos/larvae of both protostomes, chordates and hemichordates. For example, the patterns of *Otx*, *Gbx* and *Hox* genes in the early ectoderm of the annelid *P. dumerilii* are conserved in the neuroectoderm of amphioxus embryos with an anterior *Otx* domain abutting that of *Gbx*, anterior to nested domains of *Hox* expression [48]. Expression of *Hox* genes and *Gbx* in the "non-neural ectoderm" in amphioxus is similar to that in the CNS [131, 134, 192]. In contrast, in the hemichordate ectoderm, while *Otx* and *Gbx*domains lie anterior to nested *Hox* domains, that of *Gbx* lies in between two domains of *Otx*[9].

More critical for distinguishing a CNS from the remainder of the ectoderm is the expression of the lateral/medial patterning genes (*msh/Msx*, *ind/Gsh1*, *vnd/Nkx2.2*), because expression of these genes is similar in the CNS of protostomes and chordates, but they are not expressed in comparable patterns in the "non-neural" ectoderm of either group.

The idea that the hemichordate ectoderm has organizing centers patterned by gene networks equivalent to those operating at the ANR, ZLI and ISO in the vertebrate CNS is based on similarities in gene expression as well as in some conserved gene interactions. However, some key genes involved in establishing these organizing centers in the vertebrate CNS are not similarly expressed in the hemichordate ectoderm. For example, in the *S. kowalevskii* ectoderm, the *Fezf* and *Irx* domains, which mark the vertebrate ZLI, do not abut at the collar/proboscis interface, while patterns of *Otx* and *Wnt1* as well as those of *Gbx* and *Fgf8/17/18*, which mark the ISO, are reversed. Moreover, since parts of gene networks are often coopted for roles in different tissues, the fact that two genes interact does not suffice to demonstrate that a given region acts as an organizer. Thus, without transplantation experiments or at least the demonstration that a given morphogen such as *Fgf8/17/18* can induce a change in cell fate when ectopically expressed, conclusions that homologs of these organizing centers are present in hemichordates are premature.

In addition, since several genes involved in specification of the ISO and ZLI are expressed in comparable positions in the amphioxus and vertebrate CNS as well as in the protostome CNS, claims that amphioxus has completely lost these regions are unwarranted. A more parsimonious explanation is that amphioxus has part of the machinery in place, upon which tunicates and vertebrates elaborated. Tunicates do appear to have more of the genetic machinery for specification of the ISO than amphioxus does, but have lost some key components, such as the *Gbx* gene.

The Ambulacraria (hemichordates and echinoderms), which are the sister group to chordates in the deuterostomes, are interesting in regard to what evolution can do. Both hemichordates and echinoderms have ectodermal nerve nets plus nerve cords. Gene expression indicates that the nerve cords of echinoderms, which have evolved pentamerous symmetry, are probably not homologous to chordate or protostome nerve cords. *Hox* genes, for example, are expressed in the developing coeloms of the adult, but not in the nerve cords or ectoderm [48, 51]. It has been suggested that one or the other nerve cord of hemichordates is evolutionarily related to the chordate CNS, with historical opinions generally favoring the dorsal cord. However, dorsal expression of *BMPs* suggests that this homolog would be the ventral nerve cord even though it is the dorsal nerve cord that neurulates. If this is true, then a D/V inversion would have occurred at the base of the chordate lineage.

In sum, similar expression patterns of developmental genes involved in both A/P and D/V patterning in the protostome and chordate nerve cords as well as anatomical and functional similarities support the view that the ancestral bilaterian had a CNS. In the Ambulacraria, echinoderms may have lost this CNS, while it remains to be seen whether either of the hemichordate nerve cords is homologous to that of protostomes and chordates. Similarities of gene expression in the hemichordate ectoderm and chordate CNS may reflect conservation of A/P patterning in the ectoderm as a whole and in the CNS of bilaterians and/or a merging of the anterior portion of the brain and the proboscis ectoderm during evolution of hemichordates. A comprehensive study of expression of developmental genes in the hemichordate nerve cords would help to resolve these questions. In any event, similarities in gene expression between the CNS of amphioxus and vertebrates indicate that amphioxus has at least part of the genetic mechanisms for the ANR, ZLI and MHB in place and suggest that vertebrates subsequently elaborated upon these gene networks.

DECLARATIONS

Acknowledgements

The authors thank Peter WH Holland, ND Holland and Nacho Maeso for critically reading the manuscript and for useful suggestions. JKY was supported by the National Science Council, Taiwan (NSC99-2627-B-001-003, NSC101-2923-B-001-004-MY2), and by the Career Development Award from Academia Sinica, Taiwan (AS-98-CDA-L06). MS was supported by funds from the ANR (ANR-09-BLAN-0262-02 and ANR-11-JSV2-002-01), CNRS and UPMC.

REFERENCES

1. Arendt D, Denes AS, Jékely G, Tessmar-Raible K: The evolution of nervous system centralization. Philos Trans R Soc Lond B Biol Sci. 2008, 363: 1523-1528. 10.1098/rstb.2007.2242.
2. Tomer R, Denes AS, Tessmar-Raible K, Arendt D: Profiling by image registration reveals common origin of annelid mushroom bodies and vertebrate pallium. Cell. 2010, 142: 800-809. 10.1016/j.cell.2010.07.043.
3. Strausfeld NJ, Hirth F: Deep homology of arthropod central complex and vertebrate basal ganglia. Science. 2013, 340: 157-161. 10.1126/science.1231828.
4. Strausfeld NJ: Brain homology: Dohrn of a new era?. Brain Behav Evol. 2010, 76: 165-167. 10.1159/000322150.
5. Hirth F, Kammermeier L, Frei E, Walldorf U, Noll M, Reichert H: An urbilaterian origin of the tripartite brain: developmental genetic insights from *Drosophila*. Development. 2003, 130: 2365-2373. 10.1242/dev.00438.
6. Sen S, Reichert H, VijayRaghavan K: Conserved roles of *ems/Emx* and *otd/Otx* genes in olfactory and visual system development in Drosophila and mouse. Open Biol. 2013, 3: 120177-10.1098/rsob.120177.
7. Bailly X, Reichert H, Hartenstein V: The urbilaterian brain revisited: novel insights into old questions from new flatworm clades. Dev Genes Evol. 2013, 223: 149-157. 10.1007/s00427-012-0423-7.
8. De Robertis EM: The molecular ancestry of segmentation mechanisms. Proc Natl Acad Sci U S A. 2008, 105: 16411-16412. 10.1073/pnas.0808774105.
9. Pani AM, Mullarkey EE, Aronowicz J, Assimacopoulos S, Grove EA, Lowe CJ: Ancient deuterostome origins of vertebrate brain signalling centres. Nature. 2012, 483: 289-294. 10.1038/nature10838.
10. Gerhart J, Lowe C, Kirschner M: Hemichordates and the origin of chordates. Curr Opin Genet Dev. 2005, 15: 461-467. 10.1016/j.gde.2005.06.004.
11. Wagner GP: The biological homology concept. Annu Rev Ecol Sys. 1989, 20: 51-69. 10.1146/annurev.

es.20.110189.000411.

12. Cracraft J: Phylogeny and evo-devo: characters, homology, and the historical analysis of the evolution of development. Zoology. 2005, 108: 345-356. 10.1016/j.zool.2005.09.003.

13. Richter S: Homologies in phylogenetic analyses – concept and tests. Theory Biosci. 2005, 124: 105-120.

14. Jenner RA: Unburdening evo-devo: ancestral attractions, model organisms, and basal baloney. Dev Genes Evol. 2006, 216: 385-394. 10.1007/s00427-006-0084-5.

15. Ghiselin MT: The nomenclature of correspondence: a new look at 'homology' and 'analogy'. Evolution, Brain and Behavior: Persistent Problems. Edited by: Masterton RB, Hodos W, Jerison H. 1976, New York: Wiley, 2

16. Wagner GP: The developmental genetics of homology. Nat Rev Genet. 2007, 8: 473-479.

17. Hodos W: The concept of homology and the evolution of behavior. In Evolution, Brain and Behavior: Persistent Problems. Volume 2. Edited by Masterton RB, Hodos W, Jerison H. 1976, New York: Wiley

18. Remane A: Methodological problems of hominid phylogeny. 3. Phylogeny of lifestyles and the origin of upright gait. Z Morphol Anthropol. 1956, 48: 28-54.

19. Haeckel E: The gastraea-theory, the phylogenetic classification of the animal kindgom and the homology of the germ-lamellae. Q J Microsc Sci. 1874, 14: 142-165. 223–247

20. Striedter GF, Northcutt RG: Biological hierarchies and the concept of homology. Brain Behav Evol. 1991, 38: 177-189. 10.1159/000114387.

21. Bolker JA, Raff RA: Developmental genetics and traditional homology. Bioessays. 1996, 18: 489-494. 10.1002/bies.950180611.

22. Sasai Y, Lu B, Steinbeisser H, De Robertis EM: Regulation of neural induction by the Chd and Bmp-4 antagonistic patterning signals *Xenopus* in. Nature. 1995, 376: 333-336. 10.1038/376333a0.

23. De Robertis EM: Evo-devo: variations on ancestral themes. Cell. 2008, 132: 185-195. 10.1016/j.cell.2008.01.003.

24. Lowe CJ, Wu M, Salic A, Evans L, Lander E, Stange-Thomann N, Gruber CE, Gerhart J, Kirschner M: Anteroposterior patterning in hemichordates and the origins of the chordate nervous system. Cell. 2003, 113: 853-865. 10.1016/S0092-8674(03)00469-0.

25. Aronowicz J, Lowe CJ: *Hox* gene expression in the hemichordate *Saccoglossus kowalevskii* and the evolution of deuterostome nervous systems. Integr Comp Biol. 2006, 46: 890-901. 10.1093/icb/icl045.

26. Hejnol A, Martindale MQ: Acoel development indicates the independent evolution of the bilaterian mouth and anus. Nature. 2008, 456: 382-386. 10.1038/nature07309.

27. Gerhart J: Inversion of the chordate body axis: are there alternatives?. Proc Natl Acad Sci U S A. 2000, 97: 4445-4448. 10.1073/pnas.97.9.4445.

28. Gerhart J: The deuterostome ancestor. J Cell Physiology. 2006, 209: 677-685. 10.1002/jcp.20803.

29. Sprecher SG, Reichert H: The urbilaterian brain: developmental insights into the evolutionary origin of the brain in insects and vertebrates. Arthropod Struct Dev. 2003, 32: 141-156. 10.1016/S1467-8039(03)00007-0.

30. Martindale Mark Q: Evolution of development: the details are in the entrails. Curr Biol. 2013, 23: R25-R28. 10.1016/j.cub.2012.11.023.

31. Northcutt RG: Evolution of centralized nervous systems: two schools of evolutionary thought. Proc Natl Acad Sci USA. 2012, 109 (Suppl 1): 10626-10633.

32. Koizumi O, Sato N, Goto C: Chemical anatomy of hydra nervous system using antibodies against hydra neuropeptides: a review. Hydrobiologia. 2004, 530–531: 41-47.

33. Haag ES: Echinoderm rudiments, rudimentary bilaterians, and the origin of the chordate CNS. Evol Dev. 2005, 7: 280-281. 10.1111/j.1525-142X.2005.05032.x.

34. Holland ND: Early central nervous system evolution: an era of skin brains?. Nat Rev Neurosci. 2003, 4: 617-627. 10.1038/nrn1175.

35. Philippe H, Brinkmann H, Copley RR, Moroz LL, Nakano H, Poustka AJ, Wallberg A, Peterson KJ, Telford MJ: Acoelomorph flatworms are deuterostomes related to *Xenoturbella*. Nature. 2011, 470: 255-258. 10.1038/nature09676.

36. Hejnol A, Obst M, Stamatakis A, Ott M, Rouse GW, Edgecombe GD, Martinez P, Baguñà J, Bailly X, Jondelius U, Wiens M, Müller WE, Seaver E, Wheeler WC, Martindale MQ, Giribet G, Dunn CW: Assessing the root of bilaterian animals with scalable phylogenomic methods. Proc Roy Soc B: Biol Sci. 2009, 276: 4261-4270. 10.1098/rspb.2009.0896.

37. Bery A, Cardona A, Martinez P, Hartenstein V: Structure of the central nervous system of a juvenile acoel, Symsagittifera roscoffensis. Dev Genes Evol. 2010, 220: 61-76. 10.1007/s00427-010-0328-2.

38. Raikova OI, Reuter M, Jondelius U, Gustafsson MKS: An immunocytochemical and ultrastructural study of the nervous and muscular systems of *Xenoturbella westbladi* (Bilateria inc. sed.). Zoomorphology. 2000, 120: 107-118. 10.1007/s004350000028.

39. Sinigaglia C, Busengdal H, Leclère L, Technau U, Rentzsch F: The bilaterian head patterning gene *six3/6* controls aboral domain development in a cnidarian. PLoS Biol. 2013, 11: e1001488-10.1371/journal.pbio.1001488.

40. Yaguchi S, Yaguchi J, Angerer RC, Angerer LM: A Wnt–FoxQ2–Nodal pathway links primary and secondary axis specification in sea urchin embryos. Dev Cell. 2008, 14: 97-107. 10.1016/j.devcel.2007.10.012.

41. Yu J-K, Holland N, Holland L: *AmphiFoxQ2*, a novel winged helix/forkhead gene, exclusively marks the anterior end of the amphioxus embryo. Dev Genes Evol. 2003, 213: 102-105.

42. Steinmetz PR, Urbach R, Posnien N, Eriksson J, Kostyuchenko RP, Brena C, Guy K, Akam M, Bucher G, Arendt D: *Six3*demarcates the anterior-most developing brain region in bilaterian animals. EvoDevo. 2010, 1: 14-

10.1186/2041-9139-1-14.

43. Kozmik Z, Holland ND, Kreslova J, Oliveri D, Schubert M, Jonasova K, Holland LZ, Pestarino M, Benes V, Candiani S: *Pax–Six–Eya–Dach* network during amphioxus development: conservation *in vitro* but context specificity *in vivo*. Dev Biol. 2007, 306: 143-159. 10.1016/j.ydbio.2007.03.009.

44. Santagata S, Resh C, Hejnol A, Martindale MQ, Passamaneck YJ: Development of the larval anterior neurogenic domains of *Terebratalia transversa* (Brachiopoda) provides insights into the diversification of larval apical organs and the spiralian nervous system. EvoDevo. 2012, 3: 3-10.1186/2041-9139-3-3.

45. Hirth F, Reichert H: Conserved genetic programs in insect and mammalian brain development. BioEssays. 1999, 21: 677-684. 10.1002/(SICI)1521-1878(199908)21:8<677::AID-BIES7>3.0.CO;2-8.

46. Lichtneckert R, Reichert H: Anteroposterior regionalization of the brain: genetic and comparative aspects. Adv Exp Med Biol. 2008, 628: 32-41. 10.1007/978-0-387-78261-4_2.

47. Sasakura Y, Mita K, Ogura Y, Horie T: Ascidians as excellent chordate models for studying the development of the nervous system during embryogenesis and metamorphosis. Dev Growth Differ. 2012, 54: 420-437. 10.1111/j.1440-169X.2012.01343.x.

48. Steinmetz PRH, Kostyuchenko RP, Fischer A, Arendt D: The segmental pattern of *otx*, *gbx*, and *Hox* genes in the annelid *Platynereis dumerilii*. Evol Dev. 2011, 13: 72-79. 10.1111/j.1525-142X.2010.00457.x.

49. Lee PN, Callaerts P, de Couet HG, Martindale MQ: Cephalopod *Hox* genes and the origin of morphological novelties. Nature. 2003, 424: 1061-1065. 10.1038/nature01872.

50. Tihanyi BVT, Regos A, Ari E, Müller F, Takács-Vellai K: The *C. elegans Hox* gene *ceh-13* regulates cell migration and fusion in a non-colinear way. Implications for the early evolution of *Hox* clusters. BMC Dev Biol. 2010, 10: 78-10.1186/1471-213X-10-78.

51. Arenas-Mena C, Cameron AR, Davidson EH: Spatial expression of Hox cluster genes in the ontogeny of a sea urchin. Development. 2000, 127: 4631-4643.

52. Barak H, Preger-Ben Noon E, Reshef R: Comparative spatiotemporal analysis of *Hox* gene expression in early stages of intermediate mesoderm formation. Dev Dynam. 2012, 241: 1637-1649. 10.1002/dvdy.23853.

53. Hara Y, Yamaguchi M, Akasaka K, Nakano H, Nonaka M, Amemiya S: Expression patterns of *Hox* genes in larvae of the sea lily *Metacrinus rotundus*. Dev Genes Evol. 2006, 216: 797-809. 10.1007/s00427-006-0108-1.

54. Seo H-C, Edvardsen RB, Maeland AD, Bjordal M, Jensen MF, Hansen A, Flaat M, Weissenbach J, Lehrach H, Wincker P, Reinhardt R, Chourrout D: Hox cluster disintegration with persistent anteroposterior order of expression in *Oikopleura dioica*. Nature. 2004, 431: 67-71. 10.1038/nature02709.

55. Galliot B, Quiquand M, Ghila L, de Rosa R, Miljkovic-Licina M, Chera S: Origins of neurogenesis, a cnidarian view. Dev Biol. 2009, 332: 2-24. 10.1016/j.ydbio.2009.05.563.

56. Ryan JF, Mazza ME, Pang K, Matus DQ, Baxevanis AD, Martindale MQ, Finnerty JR: Pre-bilaterian origins of the Hox cluster and the Hox code: evidence from the sea anemone. *Nematostella vectensis*. PLoS ONE. 2007, 2: e153-10.1371/journal.pone.0000153.

57. de Rosa R, Grenier JK, Andreeva T, Cook CE, Adoutte A, Akam M, Carroll SB, Balavoine G: Hox genes in brachiopods and priapulids and protostome evolution. Nature. 1999, 399: 772-776. 10.1038/21631.

58. Hejnol A, Martindale MQ: Coordinated spatial and temporal expression of Hox genes during embryogenesis in the acoel *Convolutriloba longifissura*. BMC Biol. 2009, 7: 65-10.1186/1741-7007-7-65.

59. Moreno E, Nadal M, Baguna J, Martinez P: Tracking the origins of the bilaterian Hox patterning system: insights from the acoel flatworm *Symsagittifera roscoffensis*. Evol Dev. 2009, 11: 574-581. 10.1111/j.1525-142X.2009.00363.x.

60. Reichert H: A tripartite organization of the urbilaterian brain: developmental genetic evidence from *drosophila*. Brain Res Bull. 2005, 66: 491-494. 10.1016/j.brainresbull.2004.11.028.

61. Inoue F, Kurokawa D, Takahashi M, Aizawa S: Gbx2 directly restricts Otx2 expression to forebrain and midbrain, competing with Class III POU factors. Mol Cell Biol. 2012, 32: 2618-2627. 10.1128/MCB.00083-12.

62. Püschel AW, Westerfield M, Dressler GR: Comparative analysis of Pax-2 protein distributions during neurulation in mice and zebrafish. Mech Dev. 1992, 38: 197-208. 10.1016/0925-4773(92)90053-M.

63. Pfeiffer BD, Jenett A, Hammonds AS, Ngo T-TB, Misra S, Murphy C, Scully A, Carlson JW, Wan KH, Laverty TR, Mungall C, Svirskas R, Kadonaga JT, Doe CQ, Eisen MB, Celniker SE, Rubin GM: Tools for neuroanatomy and neurogenetics in *Drosophila*. Proc Natl Acad Sci U S A. 2008, 105: 9715-9720. 10.1073/pnas.0803697105.

64. Irimia M, Piñeiro C, Maeso I, Gómez-Skarmeta JL, Casares F, Garcia-Fernàndez J: Conserved developmental expression of *Fezf* in chordates and *Drosophila* and the origin of the Zona Limitans Intrathalamica (ZLI) brain organizer. EvoDevo. 2010, 1: 7-10.1186/2041-9139-1-7.

65. Hirata T, Nakazawa M, Muraoka O, Nakayama R, Suda Y, Hibi M: Zinc-finger genes *Fez* and *Fez-like* function in the establishment of diencephalon subdivisions. Development. 2006, 133: 3993-4004. 10.1242/dev.02585.

66. Mizutani CM, Bier E: EvoD/Vo: the origins of BMP signalling in the neuroectoderm. Nat Rev Genet. 2008, 9: 663-677.

67. Urbach R, Technau GM: Dorsoventral patterning of the brain: a comparative approach. Volume 628. Edited by: Technau GM. 2008, New York: Springer, 42-56.

68. Holland LZ: Chordate roots of the vertebrate nervous system: expanding the molecular toolkit. Nat Rev

Neurosci. 2009, 10: 736-746.

69. Vopalensky P, Pergner J, Liegertova M, Benito-Gutierrez E, Arendt D, Kozmik Z: Molecular analysis of the amphioxus frontal eye unravels the evolutionary origin of the retina and pigment cells of the vertebrate eye. Proc Natl Acad Sci U S A. 2012, 109: 15383-15388. 10.1073/pnas.1207580109.

70. Holland ND, Panganiban G, Henyey EL, Holland LZ: Sequence and developmental expression of *AmphiDll*, an amphioxus Distal-less gene transcribed in the ectoderm, epidermis and nervous system: insights into evolution of craniate forebrain and neural crest. Development. 1996, 122: 2911-2920.

71. Shimeld SM, van den Heuvel M, Dawber R, Briscoe J: An amphioxus *Gli* gene reveals conservation of midline patterning and the evolution of Hedgehog signalling diversity in chordates. PLoS ONE. 2007, 2: e864-10.1371/journal.pone.0000864.

72. Robertshaw E, Kiecker C: Phylogenetic origins of brain organisers. Scientifica. 2012, 2012: 1-14.

73. Seimiya M, Gehring WJ: The *Drosophila* homeobox gene *optix* is capable of inducing ectopic eyes by an eyeless-independent mechanism. Development. 2000, 127: 1879-1886.

74. Beccari L, Marco-Ferreres R, Bovolenta P: The logic of gene regulatory networks in early vertebrate forebrain patterning. Mech Dev. 2013, 130: 95-111. 10.1016/j.mod.2012.10.004.

75. Mazet F, Hutt JA, Millard J, Shimeld SM: Pax gene expression in the developing central nervous system of *Ciona intestinalis*. Gene Exp Patt. 2003, 3: 743-745. 10.1016/S1567-133X(03)00137-6.

76. D'Aniello S, D'Aniello E, Locascio A, Memoli A, Corrado M, Russo MT, Aniello F, Fucci L, Brown ER, Branno M: The ascidian homolog of the vertebrate homeobox gene *Rx* is essential for ocellus development and function. Differentiation. 2006, 74: 222-234. 10.1111/j.1432-0436.2006.00071.x.

77. Caracciolo A, Di Gregorio A, Aniello F, Di Lauro R, Branno M: Identification and developmental expression of three *Distal-less* homeobox containing genes in the ascidian *Ciona intestinalis*. Mech Dev. 2000, 99: 173-176. 10.1016/S0925-4773(00)00474-3.

78. Imai KS, Satoh N, Satou Y: Region specific gene expressions in the central nervous system of the ascidian embryo. Mech Dev. 2002, 119 (Suppl): S275-S277.

79. Islam AFMT, Moly PK, Miyamoto Y, Kusakabe TG: Distinctive expression patterns of Hedgehog pathway genes in the *Ciona intestinalis* larva: implications for a role of Hedgehog signaling in postembryonic development and chordate evolution. Zoolog Sci. 2010, 27: 84-90. 10.2108/zsj.27.84.

80. Retaux S, Kano S: Midline signaling and evolution of the forebrain in chordates: a focus on the lamprey hedgehog case. Integr Comp Biol. 2010, 50: 98-109. 10.1093/icb/icq032.

81. Denes AS, Jékely G, Steinmetz PRH, Raible F, Snyman H, Prud'homme B, Ferrier DEK, Balavoine G, Arendt D: Molecular architecture of annelid nerve cord supports common origin of nervous system centralization in Bilateria. Cell. 2007, 129: 277-288. 10.1016/j.cell.2007.02.040.

82. Heuer CM, Mueller CHG, Todt C, Loesel R: Comparative neuroanatomy suggests repeated reduction of neuroarchitectural complexity in Annelida. Front Zool. 2010, 7: 13-10.1186/1742-9994-7-13.

83. Sikes JM, Bely AE: Making heads from tails: development of a reversed anterior–posterior axis during budding in an acoel. Dev Biol. 2010, 338: 86-97. 10.1016/j.ydbio.2009.10.033.

84. Achatz JG, Martinez P: The nervous system of *Isodiametra pulchra* (Acoela) with a discussion on the neuroanatomy of the Xenacoelomorpha and its evolutionary implications. Front Zool. 2012, 9: 27-10.1186/1742-9994-9-27.

85. Lacalli TC: Protochordate body plan and the evolutionary role of larvae: old controversies resolved?. Can J Zool. 2005, 83: 216-224. 10.1139/z04-162.

86. Dohrn A: Der Ursprung der Wirbelthiere und das Princip des Functionswechsels. 1875, Leipzig: Wilhelm Englmann

87. Van-Wijhe JW: Die Terminalbeuge des Vorderhirns bei den Chordaten und die prootischen Nerven der Kranioten bei Amphioxus. Erste Mitteilung plus Zweite Mitteilung: Schluss des Riechorgans. Proc Koninklijke Akademie Van Wetenschappen Te Amsterdam. 1931, 34: 131-141. (Erste Mitteilung); & 636–653 (Zweite Mitteilung)

88. Nomaksteinsky M, Röttinger E, Dufour HD, Chettouh Z, Lowe CJ, Martindale MQ, Brunet J-F: Centralization of the deuterostome nervous system predates chordates. Curr Biol. 2009, 19: 1264-1269. 10.1016/j.cub.2009.05.063.

89. De Robertis EM, Sasai Y: A common plan for dorsoventral patterning in Bilateria. Nature. 1996, 380: 37-40. 10.1038/380037a0.

90. Duboc V, Röttinger E, Lapraz F, Besnardeau L, Lepage T: Left-right asymmetry in the sea urchin embryo is regulated by Nodal signaling on the right side. Dev Cell. 2005, 9: 147-158. 10.1016/j.devcel.2005.05.008.

91. Grande C, Patel NH: Nodal signalling is involved in left-right asymmetry in snails. Nature. 2009, 457: 1007-1011. 10.1038/nature07603.

92. Nielsen C: Homology of echinoderm radial nerve cords and the chordate neural tube?. Evol Dev. 2006, 8: 1-2. 10.1111/j.1525-142X.2006.05068.x.

93. Sly BJ, Hazel JC, Popodi EM, Raff RA: Patterns of gene expression in the developing adult sea urchin central nervous system reveal multiple domains and deep-seated neural pentamery. Evol Dev. 2002, 4: 189-204. 10.1046/j.1525-142X.2002.02002.x.

94. Cisternas P, Byrne M: Expression of Hox4 during development of the pentamerous juvenile sea star, *Parvulastra exigua*. Dev Genes Evol. 2009, 219: 613-618. 10.1007/s00427-010-0318-4.

95. Morris VB, Byrne M: Involvement of two Hox genes and Otx in echinoderm body-plan morphogenesis in the sea urchin *Holopneustes purpurescens*. J Exp Zool

Part B: Mol Dev Evol. 2005, 304B: 456-467. 10.1002/jez.b.21065.

96. Lacalli TC: Head organization and the head/trunk relationship in protochordates: problems and prospects. Integr Comp Biol. 2008, 48: 620-629. 10.1093/icb/icn012.

97. Nieuwenhuys R: Deuterostome brains: synopsis and commentary. Brain Res Bull. 2002, 57: 257-270. 10.1016/S0361-9230(01)00668-2.

98. Tagawa K, Humphreys T, Satoh N: *T-brain* expression in the apical organ of hemichordate tornaria larvae suggests its evolutionary link to the vertebrate forebrain. J Exp Zool. 2000, 288: 23-31. 10.1002/(SICI)1097-010X(20000415)288:1<23::AID-JEZ3>3.0.CO;2-II.

99. Stach T, Gruhl A, Kaul-Strehlow S: The central and peripheral nervous system of *Cephalodiscus gracilis* (Pterobranchia, Deuterostomia). Zoomorphology. 2012, 131: 11-24. 10.1007/s00435-011-0144-x.

100. Romer AS: Major steps in vertebrate evolution. Science. 1967, 158: 1629-1637. 10.1126/science.158.3809.1629.

101. Williams JB: Sessile lifestyle and origin of chordates. New Zealand J Zool. 1996, 23: 111-133. 10.1080/03014223.1996.9518072.

102. Osborn KJ, Kuhnz LA, Priede IG, Urata M, Gebruk AV, Holland ND: Diversification of acorn worms (Hemichordata, Enteropneusta) revealed in the deep sea. Proc Roy Soc B: Biol Sci. 2012, 279: 1646-1654. 10.1098/rspb.2011.1916.

103. Cannon JT, Rychel AL, Eccleston H, Halanych KM, Swalla BJ: Molecular phylogeny of hemichordata, with updated status of deep-sea enteropneusts. Mol Phylog Evol. 2009, 52: 17-24. 10.1016/j.ympev.2009.03.027.

104. Caron J-B, Morris SC, Cameron CB: Tubicolous enteropneusts from the Cambrian period. Nature. 2013, 495: 503-506. 10.1038/nature12017.

105. Röttinger E, Lowe CJ: Evolutionary crossroads in developmental biology: hemichordates. Development. 2012, 139: 2463-2475. 10.1242/dev.066712.

106. Miyamoto N, Nakajima Y, Wada H, Saito Y: Development of the nervous system in the acorn worm *Balanoglossus simodensis*: insights into nervous system evolution. Evol Dev. 2010, 12: 416-424. 10.1111/j.1525-142X.2010.00428.x.

107. Kaul S, Stach T: Ontogeny of the collar cord: neurulation in the hemichordate *Saccoglossus kowalevskii*. J Morph. 2010, 271: 1240-1259. 10.1002/jmor.10868.

108. Bateson W: The later stages in the development of *Balanoglossus kowalevskii*, with a suggestion as to the affinities of the Enteropneusta. Q J Microsc Sci. 1886, 126: 511-533.

109. Rehkämper G, Welsch U, Dilly PN: Fine structure of the ganglion of *Cephalodiscus gracilis* (Pterobranchia, Hemichordata). J Comp Neurol. 1987, 259: 308-315. 10.1002/cne.902590210.

110. Brown FD, Prendergast A, Swalla BJ: Man is but a worm: chordate origins. Genesis. 2008, 46: 605-613. 10.1002/dvg.20471.

111. Bullock TH: The giant nerve fiber system in balanoglossids. J Comp Neurol. 1944, 80: 355-367. 10.1002/cne.900800305.

112. Nielsen C, Hay-Schmidt A: Development of the enteropneust *Ptychodera flava*: ciliary bands and nervous system. J Morph. 2007, 268: 551-570. 10.1002/jmor.10533.

113. Lowe CJ, Terasaki M, Wu M, Freeman RM, Runft L, Kwan K, Haigo S, Aronowicz J, Lander E, Gruber C, Smith M, Kirschner M, Gerhart J: Dorsoventral patterning in hemichordates: Insights into early chordate evolution. PLoS Biol. 2006, 4: e291-10.1371/journal.pbio.0040291.

114. Taguchi S, Tagawa K, Humphreys T, Satoh N: Group B *Sox* genes that contribute to specification of the vertebrate brain are expressed in the apical organ and ciliary bands of hemichordate larvae. Zool Sci. 2002, 19: 57-66. 10.2108/zsj.19.57.

115. Tagawa K, Satoh N, Humphreys T: Molecular studies of hemichordate development: a key to understanding the evolution of bilateral animals and chordates. Evol Dev. 2001, 3: 443-454. 10.1046/j.1525-142X.2001.01050.x.

116. Peterson KJ, Cameron RA, Tagawa K, Satoh N, Davidson EH: A comparative molecular approach to mesodermal patterning in basal deuterostomes: the expression pattern of *Brachyury* in the enteropneust hemichordate *Ptychodera flava*. Development. 1999, 126: 85-95.

117. Onai T, Yu J-K, Blitz IL, Cho KWY, Holland LZ: Opposing Nodal/Vg1 and BMP signals mediate axial patterning in embryos of the basal chordate amphioxus. Dev Biol. 2010, 344: 377-389. 10.1016/j.ydbio.2010.05.016.

118. Wlizla M: Evolution of Nodal signaling in deuterostomes: insights from Saccoglossus kowalevskii. PhD dissertation. University of Chicago. 2011, 195-

119. Saudemont A, Haillot E, Mekpoh F, Bessodes N, Quirin M, Lapraz F, Duboc V, Röttinger E, Range R, Oisel A, Besnardeau L, Wincker P, Lepage T: Ancestral regulatory circuits governing ectoderm patterning downstream of Nodal and BMP2/4 revealed by gene regulatory network analysis in an echinoderm. PLoS Genet. 2010, 6: e1001259-10.1371/journal.pgen.1001259.

120. Grande C, Patel NH: Lophotrochozoa get into the game: the nodal pathway and left/right asymmetry in Bilateria. Cold Spring Harb Symp Quant Biol. 2009, 74: 281-287. 10.1101/sqb.2009.74.044.

121. de Beer G: Homology, An Unsolved Problem, Volume 11. 1971, Oxford: Oxford University Press

122. Roth VL: The biological basis of homology. Ontogeny and Systematics. Edited by: Humphries CJ. 1988, New York: Columbia University Press

123. Panganiban G, Irvine SM, Lowe CJ, Roehl HH, Corley LS, Sherbon B, Grenier JK, Fallon JF, Kimble J, Walker M, Wray GA, Swalla BJ, Martindale MQ, Carroll SB: The origin and evolution of animal appendages. Proc Natl Acad Sci USA. 1997, 94: 5162-5166. 10.1073/pnas.94.10.5162.

124. Northcutt RG: Changing views of brain evolution.

Brain Res Bull. 2001, 55: 663-674. 10.1016/S0361-9230(01)00560-3.

125. Shubin N, Tabin C, Carroll SB: Deep homology and the origins of evolutionary novelty. Nature. 2009, 457: 818-823. 10.1038/nature07891.

126. Lowe CJ: Molecular genetic insights into deuterostome evolution from the direct-developing hemichordate *Saccoglossus kowalevskii*. Philos Trans R Soc Lond B Biol Sci. 2008, 363: 1569-1578. 10.1098/rstb.2007.2247.

127. Stokes MD, Holland ND: Embryos and larvae of a lancelet, *Branchiostoma floridae*, from hatching through metamorphosis: growth in the laboratory and external morphology. Acta Zool. 1995, 76: 105-120. 10.1111/j.1463-6395.1995.tb00986.x.

128. Bone Q, Ryan KP: Cupular sense organs in *Ciona* (Tunicata: Ascidiacea). J Zool, London. 1978, 186: 417-429.

129. Wood HB, Episkopou V: Comparative expression of the mouse *Sox1*, *Sox2* and *Sox3* genes from pre-gastrulation to early somite stages. Mech Dev. 1999, 86: 197-201. 10.1016/S0925-4773(99)00116-1.

130. Satoh G, Wang Y, Zhang P, Satoh N: Early development of amphioxus nervous system with special reference to segmental cell organization and putative sensory cell precursors: a study based on the expression of pan-neuronal marker gene *Hu/elav*. J Exp Zool. 2001, 291: 354-364. 10.1002/jez.1134.

131. Ghost Database. [http://ghost.zool.kyoto-u.ac.jp/SearchGenomekh.html]

132. Schubert M, Holland ND, Escriva H, Holland LZ, Laudet V: Retinoic acid influences anteroposterior positioning of epidermal sensory neurons and their gene expression in a developing chordate (amphioxus). Proc Natl Acad Sci U S A. 2004, 101: 10320-10325. 10.1073/pnas.0403216101.

133. Sasakura Y, Kanda M, Ikeda T, Horie T, Kawai N, Ogura Y, Yoshida R, Hozumi A, Satoh N, Fujiwara S: Retinoic acid-driven *Hox1* is required in the epidermis for forming the otic/atrial placodes during ascidian metamorphosis. Development. 2012, 139: 2156-2160. 10.1242/dev.080234.

134. Watari-Goshima N, Chisaka O: Chicken HOXA3 gene: its expression pattern and role in branchial nerve precursor cell migration. Int J Biol Sci. 2011, 7: 87-101.

135. Castro LFC, Rasmussen SLK, Holland PWH, Holland ND, Holland LZ: A *Gbx* homeobox gene in amphioxus: insights into ancestry of the ANTP class and evolution of the midbrain/hindbrain boundary. Dev Biol. 2006, 295: 40-51. 10.1016/j.ydbio.2006.03.003.

136. Sánchez-Calderón H, Martín-Partido G, Hidalgo-Sánchez M: Differential expression of *Otx2, Gbx2, Pax2,* and *Fgf8* in the developing vestibular and auditory sensory organs. Brain Res Bull. 2002, 57: 321-323. 10.1016/S0361-9230(01)00725-0.

137. Steventon B, Mayor R, Streit A: Mutual repression between Gbx2 and Otx2 in sensory placodes reveals a general mechanism for ectodermal patterning. Dev Biol. 2012, 367: 55-65. 10.1016/j.ydbio.2012.04.025.

138. Sharman AC, Shimeld SM, Holland PWH: An amphioxus *Msx* gene expressed predominantly in the dorsal neural tube. Dev Genes Evol. 1999, 209: 260-263. 10.1007/s004270050251.

139. Osborne PW, Benoit G, Laudet V, Schubert M, Ferrier DE: Differential regulation of ParaHox genes by retinoic acid in the invertebrate chordate amphioxus (*Branchiostoma floridae*). Dev Biol. 2009, 327: 252-262. 10.1016/j.ydbio.2008.11.027.

140. Venkatesh TV, Holland ND, Holland LZ, Su M-T, Bodmer R: Sequence and developmental expression of amphioxus *AmphiNk2–1*: insights into the evolutionary origin of the vertebrate thyroid gland and forebrain. Dev Genes Evol. 1999, 209: 254-259. 10.1007/s004270050250.

141. Kawashima T, Murakami AR, Ogasawara M, Tanaka KJ, Isoda R, Sasakura Y, Nishikata T, Okano H, Makabe KW: Expression patterns of *musashi* homologs of the ascidians, *Halocynthia roretzi* and *Ciona intestinalis*. Dev Genes Evol. 2000, 210: 162-165. 10.1007/s004270050024.

142. Wada S, Tokuoka M, Shoguchi E, Kobayashi K, Di Gregorio A, Spagnuolo A, Branno M, Kohara Y, Rokhsar D, Levine M, Saiga H, Satoh N, Satou Y: A genomewide survey of developmentally relevant genes in *Ciona intestinalis*. II. Genes for homeobox transcription factors. Dev Genes Evol. 2003, 213: 222-234. 10.1007/s00427-003-0321-0.

143. Ghanbari H, Seo H-C, Fjose A, Brändli AW: Molecular cloning and embryonic expression of *Xenopus Six* homeobox genes. Mech Dev. 2001, 101: 271-277. 10.1016/S0925-4773(00)00572-4.

144. Osorio J, Mazan S, Rétaux S: Organisation of the lamprey (*Lampetra fluviatilis*) embryonic brain: insights from LIM-homeodomain, *Pax* and *hedgehog* genes. Dev Biol. 2005, 288: 100-112. 10.1016/j.ydbio.2005.08.042.

145. Toresson H, Martinez-Barbera JP, Bardsley A, Caubit X, Krauss S: Conservation of *BF-1* expression in amphioxus and zebrafish suggests evolutionary ancestry of anterior cell types that contribute to the vertebrate telencephalon. Dev Genes Evol. 1998, 208: 431-439. 10.1007/s004270050200.

146. Hatini V, Ye X, Balas G, Lai E: Dynamics of placodal lineage development revealed by targeted transgene expression. Dev Dynam. 1999, 215: 332-343. 10.1002/(SICI)1097-0177(199908)215:4<332::AID-AJA5>3.0.CO;2-R.

147. Kaltenbach SL, Yu J-K, Holland ND: The origin and migration of the earliest-developing sensory neurons in the peripheral nervous system of amphioxus. Evol Dev. 2009, 11: 142-151. 10.1111/j.1525-142X.2009.00315.x.

148. Logan C, Wingate RJ, McKay IJ, Lumsden A: *Tlx-1* and *Tlx-3* homeobox gene expression in cranial sensory ganglia and hindbrain of the chick embryo: markers of patterned connectivity. J Neurosci. 1998, 18: 5389-5402.

149. Chellappa R, Li S, Pauley S, Jahan I, Jin K, Xiang M: *Barhl1* regulatory sequences required for cell-specific gene expression and autoregulation in the inner

ear and central nervous system. Mol Cell Biol. 2008, 28: 1905-1914. 10.1128/MCB.01454-07.

150. Kawahara A, Dawid IB: Developmental expression of zebrafish *emx1* during early embryogenesis. Gene Exp Patt. 2002, 2: 201-206. 10.1016/S1567-133X(02)00062-5.

151. Bardet SM, Martinez-de-la-Torre M, Northcutt RG, Rubenstein JLR, Puelles L: Conserved pattern of OTP-positive cells in the paraventricular nucleus and other hypothalamic sites of tetrapods. Brain Res Bull. 2008, 75: 231-235. 10.1016/j.brainresbull.2007.10.037.

152. Ma P, Zhao S, Zeng W, Yang Q, Li C, Lv X, Zhou Q, Mao B: *Xenopus Dbx2* is involved in primary neurogenesis and early neural plate patterning. Biochem Biophy Res Com. 2011, 412: 170-174.

153. Langeland JA, Holland LZ, Chastain RA, Holland ND: An amphioxus LIM-homeobox gene, *AmphiLim1/5*, expressed early in the invaginating organizer region and later in differentiating cells of the kidney and central nervous system. Int J Biol Sci. 2006, 2: 110-116.

154. Feijóo CG, Saldias MP, De la Paz JF, Gómez-Skarmeta JL, Allende ML: Formation of posterior cranial placode derivatives requires the *Iroquois* transcription factor irx4a. Mol Cell Neurosci. 2009, 40: 328-337. 10.1016/j.mcn.2008.11.003.

155. Matsuura M, Nishihara H, Onimaru K, Kokubo N, Kuraku S, Kusakabe R, Okada N, Kuratani S, Tanaka M: Identification of four *Engrailed* genes in the Japanese lamprey, Lethenteron japonicum. Dev Dyn. 2008, 237: 1581-1589. 10.1002/dvdy.21552.

156. Ikuta T, Yoshida N, Satoh N, Saiga H: *Ciona intestinalis Hox* gene cluster: its dispersed structure and residual colinear expression in development. Proc Natl Acad Sci U S A. 2004, 101: 15118-15123. 10.1073/pnas.0401389101.

157. Makki N, Capecchi MR: Hoxa1 lineage tracing indicates a direct role for *Hoxa1* in the development of the inner ear, the heart, and the third rhombomere. Dev Biol. 2010, 341: 499-509. 10.1016/j.ydbio.2010.02.014.

158. Takio Y, Kuraku S, Murakami Y, Pasqualetti M, Rijli FM, Narita Y, Kuratani S, Kusakabe R: *Hox* gene expression patterns in *Lethenteron japonicum* embryos – insights into the evolution of the vertebrate Hox code. Dev Biol. 2007, 308: 606-620. 10.1016/j.ydbio.2007.05.009.

159. Uchiyama K, Otsuka R, Hanaoka K: *cHox11L2*, a *Hox11* related gene, is expressed in the peripheral nervous system and subpopulation of the spinal cord during chick development. Neurosci Lett. 1999, 273: 97-100. 10.1016/S0304-3940(99)00637-0.

160. Sanchez-Arrones L, Stern CD, Bovolenta P, Puelles L: Sharpening of the anterior neural border in the chick by rostral endoderm signalling. Development. 2012, 139: 1034-1044. 10.1242/dev.067934.

161. Tendeng C, Houart C: Cloning and embryonic expression of five distinct sfrp genes in the zebrafish *Danio rerio*. Gene Expr Patterns. 2006, 6: 761-771. 10.1016/j.modgep.2006.01.006.

162. Eagleson GW, Dempewolf RD: The role of the anterior neural ridge and Fgf-8 in early forebrain patterning and regionalization in *Xenopus laevis*. Comp Biochem Physiol B Biochem Mol Biol. 2002, 132: 179-189. 10.1016/S1096-4959(01)00521-8.

163. Vieira C, Martinez S: Sonic hedgehog from the basal plate and the zona limitans intrathalamica exhibits differential activity on diencephalic molecular regionalization and nuclear structure. Neuroscience. 2006, 143: 129-140. 10.1016/j.neuroscience.2006.08.032.

164. Green SA, Norris RP, Terasaki M, Lowe CJ: FGF signaling induces mesoderm in the hemichordate *Saccoglossus kowalevskii*. Development. 2013, 140: 1024-1033. 10.1242/dev.083790.

165. Holland ND, Holland LZ. Amphioxus and the utility of molecular genetic data for hypothesizing body part homologies between distantly related animals. Am Zool. 1999, 39: 630-640.

166. Holland LZ: Evolution of new characters after whole genome duplications: insights from amphioxus. Sem Cell Dev Biol. 2013, 24: 101-109. 10.1016/j.semcdb.2012.12.007.

167. Capdevila J, Belmonte JCI: Patterning mechanisms controlling vertebrte limb development. Ann Rev Cell Dev Biol. 2001, 17: 87-132. 10.1146/annurev.cellbio.17.1.87.

168. Röttinger E, Martindale MQ: Ventralization of an indirect developing hemichordate by NiCl2 suggests a conserved mechanism of dorso-ventral (D/V) patterning in Ambulacraria (hemichordates and echinoderms). Dev Biol. 2011, 354: 173-190. 10.1016/j.ydbio.2011.03.030.

169. Range R, Angerer R, Angerer L: Integration of canonical and noncanonical Wnt signaling pathways patterns the neuroectoderm along the anterior-posterior axis of sea urchin embryos. PLoS Biol. 2013, 11: e1001467-10.1371/journal.pbio.1001467.

170. Emily-Fenouil F, Ghiglione C, Lhomond G, Lepage T, Gache C: *GSK3beta/shaggy* mediates patterning along the animal–vegetal axis of the sea urchin embryo. Development. 1998, 125: 2489-2498.

171. Jeong J-Y, Einhorn Z, Mathur P, Chen L, Lee S, Kawakami K, Guo S: Patterning the zebrafish diencephalon by the conserved zinc-finger protein Fezl. Development. 2007, 134: 127-136. 10.1242/dev.02705.

172. Scholpp S, Foucher I, Staudt N, Peukert D, Lumsden A, Houart C: *Otx1l*, *Otx2* and *Irx1b* establish and position the ZLI in the diencephalon. Development. 2007, 134: 3167-3176. 10.1242/dev.001461.

173. Martinez-Ferre A, Navarro-Garberi M, Bueno C, Martinez S: Wnt signal specifies the intrathalamic limit and its organizer properties by regulating Shh induction in the alar plate. J Neurosci. 2013, 33: 3967-3980. 10.1523/JNEUROSCI.0726-12.2013.

174. Martinez-Ferre A, Martinez S: Molecular regionalization of the diencephalon. Front Neurosci. 2012, 6: 73-

175. Puelles E, Annino A, Tuorto F, Usiello A, Acampora D, Czerny T, Brodski C, Ang S-L, Wurst W, Simeone A: *Otx2* regulates the extent, identity and fate of neuronal progenitor domains in the ventral midbrain.

Development. 2004, 131: 2037-2048. 10.1242/dev.01107.

176. Partanen J: FGF signalling pathways in development of the midbrain and anterior hindbrain. J Neurochem. 2007, 101: 1185-1193. 10.1111/j.1471-4159.2007.04463.x.

177. Kengaku M, Capdevila J, Rodriguez-Esteban C, De La Peña J, Johnson RL, Belmonte JCI, Tabin CJ: Distinct WNT pathways regulating AER formation and dorsoventral polarity in the chick limb bud. Science. 1998, 280: 1274-1277. 10.1126/science.280.5367.1274.

178. Tanaka M, Shigetani Y, Sugiyama S, Tamura K, Nakamura H, Ide H: Apical ectodermal ridge induction by the transplantation of *En-1*-overexpressing ectoderm in chick limb bud. Dev Growth Differ. 1998, 40: 423-429. 10.1046/j.1440-169X.1998.t01-2-00007.x.

179. Glardon S, Holland LZ, Gehring WJ, Holland ND: Isolation and developmental expression of the amphioxus Pax-6 gene (*AmphiPax-6*): insights into eye and photoreceptor evolution. Development. 1998, 125: 2701-2710.

180. Yu J-K, Satou Y, Holland ND, Shin-I T, Kohara Y, Satoh N, Bronner-Fraser M, Holland LZ: Axial patterning in cephalochordates and the evolution of the organizer. Nature. 2007, 445: 613-617. 10.1038/nature05472.

181. Shimeld SM: The evolution of the hedgehog gene family in chordates: insights from amphioxus *hedgehog*. Dev Genes Evol. 1999, 209: 40-47. 10.1007/s004270050225.

182. Shimamura K, Rubenstein JL: Inductive interactions direct early regionalization of the mouse forebrain. Development. 1997, 124: 2709-2718.

183. Imai KS, Stolfi A, Levine M, Satou Y: Gene regulatory networks underlying the compartmentalization of the *Ciona* central nervous system. Development. 2009, 136: 285-293. 10.1242/dev.026419.

184. Schubert M, Holland LZ, Panopoulou GD, Lehrach H, Holland ND: Characterization of amphioxus *AmphiWnt8*: insights into the evolution of patterning of the embryonic dorsoventral axis. Evol Dev. 2000, 2: 85-92. 10.1046/j.1525-142x.2000.00047.x.

185. Mazet F, Shimeld S: Characterisation of an amphioxus *Fringe* gene and the evolution of the vertebrate segmentation clock. Dev Genes Evol. 2003, 213: 505-509. 10.1007/s00427-003-0351-7.

186. Holland LZ, Venkatesh TV, Gorlin A, Bodmer R, Holland ND: Characterization and developmental expression of *AmphiNk2-2*, an NK2 class homeobox gene from amphioxus (Phylum Chordata; Subphylum Cephalochordata). Dev Genes Evol. 1998, 208: 100-105. 10.1007/s004270050159.

187. Bertrand S, Camasses A, Somorjai I, Belgacem MR, Chabrol O, Escande M-L, Pontarotti P, Escriva H: Amphioxus FGF signaling predicts the acquisition of vertebrate morphological traits. Proc Natl Acad Sci U S A. 2011, 108: 9160-9165. 10.1073/pnas.1014235108.

188. Onai T, Lin H-C, Schubert M, Koop D, Osborne PW, Alvarez S, Alvarez R, Holland ND, Holland LZ: Retinoic acid and Wnt/β-catenin have complementary roles in anterior/posterior patterning embryos of the basal chordate amphioxus. Dev Biol. 2009, 332: 223-233. 10.1016/j.ydbio.2009.05.571.

189. Hotta K, Takahashi H, Ueno N, Gojobori T: A genome-wide survey of the genes for planar polarity signaling or convergent extension-related genes in *Ciona intestinali* s and phylogenetic comparisons of evolutionary conserved signaling components. Gene. 2003, 317: 165-185.

190. Wagner E, Levine M: FGF signaling establishes the anterior border of the *Ciona* neural tube. Development. 2012, 139: 2351-2359. 10.1242/dev.078485.

191. Kozmik Z, Holland ND, Kalousova A, Paces J, Schubert M, Holland LZ: Characterization of an amphioxus paired box gene, *AmphiPax2/5/8*: developmental expression patterns in optic support cells, nephridium, thyroid-like structures and pharyngeal gill slits, but not in the midbrain-hindbrain boundary region. Development. 1999, 126: 1295-1304.

192. Meulemans D, Bronner-Fraser M: The amphioxus SoxB family: implications for the evolution of vertebrate placodes. Int J Biol Sci. 2007, 3: 356-364.

193. Holland LZ, Kene M, Williams NA, Holland ND: Sequence and embryonic expression of the amphioxus engrailed gene (*AmphiEn*): the metameric pattern of transcription resembles that of its segment-polarity homolog in Drosophila. Development. 1997, 124: 1723-1732.

194. Ikuta T, Saiga H: Dynamic change in the expression of developmental genes in the ascidian central nervous system: revisit to the tripartite model and the origin of the midbrain–hindbrain boundary region. Dev Biol. 2007, 312: 631-643. 10.1016/j.ydbio.2007.10.005.

CHAPTER 13 WHAT CAN VERTEBRATES TELL US ABOUT SEGMENTATION?

Anthony Graham, Thomas Butts, Andrew Lumsden and Clemens Kiecker

Keywords: Segmentation Metamerism Somites Rhombomeres Pharyngeal arches Vertebrates Evolution

INTRODUCTION

Within the bilateria there are a number of clades, such as the arthropods, annelids and the chordates, which display the serial repetition of parts along the long body axis. These animals have been classified as being segmented [1]. There has, however, long been contention as to whether or not these instances of segmentation are homologous. A major problem that pervades this issue is how segmentation can be defined. A recent article by Hannibal and Patel, in this journal, makes the point that if we cannot understand exactly what is meant by segmentation and which structures are genuinely segmented then we have no hope of understanding how segmentation evolved and of clarifying the relationships between different types of segmentation [2]. These authors further suggest that any discussions of segmentation be clarified to explain exactly what is being studied and it is through this route that one might arrive at a better understanding of what segmentation is.To begin to understand segmentation in its different manifestations and to consider whether or not these are homologous processes across animal taxa we need to comprehend the segmentation process; that is, how segments arise and are allocated during development. With regard to this point, a study of diverse segmented structures in vertebrates is particularly useful. There are three clear distinct instances of segmentation being used as a developmental strategy in vertebrates and this affords one the opportunity to effectively and readily compare and contrast these processes (Figure 1). Moreover, we have a wealth of information about how these segmental systems are established. The first, and arguably the most well-known, is the formation of the somites from the paraxial mesoderm. The second is the subdivision of the hindbrain to generate rhombomeres. The last, and least discussed, is the formation of the pharyngeal arches, which are evident as serial bulges on the lateral surface of the embryonic head.

Citation: Anthony Graham, Thomas Butts, Andrew Lumsden and Clemens Kiecker, What can vertebrates tell us about segmentation?DOI: 10.1186/2041-9139-5-24.

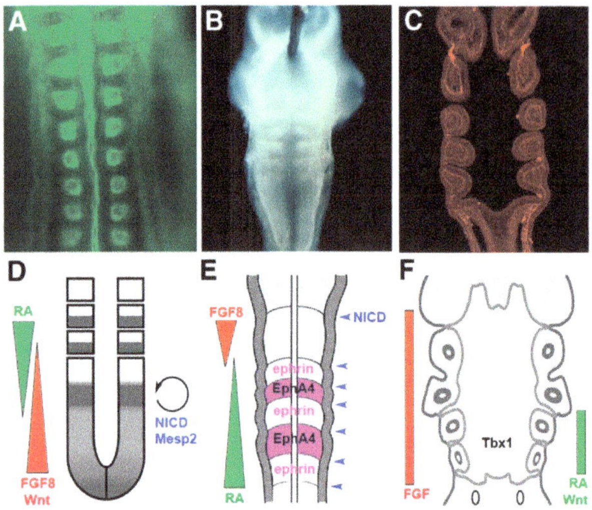

Figure 1. Segmentation in vertebrates is governed by divergent molecular mechanisms: somites, rhombomeres and pharyngeal arches. (A) A whole mount phalloidin staining of a stage 11 chick embryo viewed dorsally. The somites are readily visible as iterated blocks on either side of the neural tube. (B) An image of a dissected hindbrain region of chick embryo at 3 days of development. The roofplate has been removed to show the rhombomeres. (C) A longitudinal section through the pharyngeal region of a dogfish embryo. The embryo has been stained with an anti-laminin antibody. The pharyngeal segments are clearly shown. (D) Schematic diagram of somitogenesis. As the embryonic axis elongates, the presomitic mesoderm is patterned by opposing gradients of FGF8/Wnt activity (high posterior-low anterior) and retinoic acid (RA; low posterior-high anterior). Cycling expression of Notch pathway components (NICD=Notch Intracellular Domain) and of the transcription factor *Mesp2* acts as a pacemaker in this process. (E) Diagram of rhombomere (r) formation. R identity is regulated by a posteriorly increasing gradient of RA and by FGF8 signalling from the midbrain-hindbrain boundary. R boundary formation involves Notch activation (NICD) and they are maintained through cell sorting driven by alternating expression of EphA4 (in r3/5) and ephrin ligands (in r2, 4, 6). (F) The formation of all pharyngeal arches depends on FGF signalling, that of the posterior arches also on Wnt and RA signalling and on the transcription factor *Tbx1*. In all panels anterior is to the top.

REVIEW

In this review, we consider how these segmentation processes are controlled and whether or not they have any relationship with each other. We also probe the evolutionary history of each of these segmental systems to gain insights into how they emerged during evolution and again we have a relatively solid knowledge of their origins. Additionally, we make no terminological distinction between 'segmented' and 'metameric' structures. All of the structures we discuss in vertebrates can be readily described as both segmented and metameric, and we make the argument that the only useful definition of such terms is morphological. Finally, we consider how the lessons we have learned from our consideration of segmentation in vertebrates feed into more general discussions of segmentation within the bilateria.

Somites: Segmentation of the Paraxial Mesoderm

The formation and regionalisation of the somites underpin the segmental organisation of the trunk of vertebrates [3, 4]. Somites are blocks of mesoderm that bud off from the anterior end of the presomitic mesoderm as it retreats posteriorly and which will give rise to the vertebrae, skeletal muscle, connective tissue and regions of the dermis. The key events in the formation of the somites occur within the presomitic mesoderm and involve an interaction between a segmentation clock and opposing morphogen gradients. A component of the clock is the Notch pathway which acts to produce cycling gene expression such that a wave of expression, with a periodicity that matches somite formation, sweeps across the presomitic mesoderm [4]. The gradients take the form of a posterior to anterior gradient of Fgf8 [5, 6] and nuclear b-catenin [7] and an anterior to posterior gradient of retinoic acid [8, 9]. The intersection of these gradients defines the determination front, the point at which a cohort of cells at the same phase of oscillation is allocated to form a somite. This final step also involves the induction of *mesp2* expression, which suppresses Notch activity and thus facilitates the transition from cycling status to somite [10]. Thus, a significant facet of somitogenesis is that it is a drawn out process, largely concomitant with axis elongation, with many somites being formed over a protracted period. Central to this is the segmentation clock, which regulates the allocation of cells within the presomitic mesoderm to somites, and the continued posterior growth of the embryo.

However, although Notch signalling has been shown to oscillate in the presomitic mesoderm of all vertebrate models, its role in actually driving segmentation has been questioned. A recent study has even demonstrated that somites can form in the complete absence of the molecular clock indicating that a likely role of Notch may be in synchronising the segmentation clock rather than governing segmentation *per se*[11, 12, 13].

Another important feature of somites is that their segmental organisation is translated into the muscle segments of the body and the iterated vertebrae of the backbone. However, the positional relationship between the somites and the vertebrae is not a simple one to one correspondence but involves a resegmentation process [14, 15]. In amniotes, the somites are subdivided into rostral and caudal halves with the rostral half sclerotome of each somite combining with the caudal half sclerotome of the immediately anterior somite to form each vertebra [16]. In zebrafish, studies have shown that while there is some evidence for resegmentation it is less strictly organised, with cells from the rostral or caudal half somites contributing to two consecutive vertebrae rather than exclusively to one [17].

This rostrocaudal organisation further impinges upon the neural crest cells that will form the peripheral ganglia and motor axons of the spinal nerves [18]. In amniotes, the caudal half sclerotome of each somite expresses a number of inhibitory molecules, including ephrins and semaphorins, while the neural crest and motor axon growth cones express the cognate receptors, and thus these populations are repelled by the caudal halves and only traverse the anterior half sclerotome [19, 20, 21, 22, 23]. This results in the secondarily derived segmentation of the peripheral nervous system and fixes the formation of the spinal nerves and ganglia in register with the vertebral column.

Finally, somites are also regionally distinct and those at different axial levels will generate vertebrae with different characteristics, such as cervical *versus* thoracic *versus* lumbar. Their anteroposterior identity is assigned within the presomitic mesoderm, prior to somite formation, and this involves differential *Hox* gene expression [24]. The number of somites formed varies greatly between different vertebrates; mice generate 65 somites, chicks 50, pythons greater than 500 and zebrafish about 31.

Rhombomeres: Segmentation of the Hindbrain

Segmentation of the hindbrain is morphologically evident as a series of bulges and constrictions that subdivide this region of the brain and the number of segments, or rhombomeres (r), formed is largely invariant across the vertebrates [25, 26, 27]. The most anterior segment is r1, which lies just posterior of the mesencephalon, and the most posterior is continuous with the spinal cord. These structures define the ground plan upon which the development of the hindbrain is organised. Rhombomeres are, on some levels, serially homologous to each other but they also exhibit individual identities. The segments will generate similar cohorts of interneurons, but trigeminal motor neurons, for example, will only form in r2 and r3 and facial motor neurons only in r4 and r5 [25, 28]. This two-segment periodicity also affects neurogenesis and axonal projection, which are initiated, and more advanced, in the even numbered rhombomeres (r2, r4 and r6) than in the odd numbered rhombomeres (r3 and r5) [29]. The even-numbered rhombomeres and the odd numbered also have distinct cell surface properties [30].

Rhombomeres differ from the somites in that they are not discrete separate structures formed by budding. Rather, they constitute a structural continuum that forms by the internal subdivision of a prespecified territory, the hindbrain primordium. This is underpinned by graded retinoid and wnt signalling, and the action of a range of different transcription factors that are needed to establish specific rhombomeres [31], including Krox-20 (r3/r5), Hnf1b/vHNF1 (r5/r6) and Hoxa1 (r5). The formation of the rhombomeres results in the emergence of lineage restricted compartments that are formed by cell segregation [32]. Prior to boundary formation cells can cross between proto-rhombomeric domains, but they cannot cross once a boundary has formed. Central to this phenomenon and indeed to the formation of the rhombomeres are cell-sorting events mediated by Eph/ephrin signalling [33]. The Eph receptors, such as *EphA4*, are expressed in r3 and r5 and the restriction of *Krox20* expression to these segments is likely to be involved in driving the expression of these receptors, while the ephrin ligands are expressed in r2, r4 and r6. This establishes bidirectional signalling interfaces between neighbouring rhombomeres which result in cell sorting such that like surrounds like. Thus cells within each rhombomere differentiate themselves from cells in their neighbouring segments, and cell-tight boundaries form. Finally, rhombomeres also differ from somites in that they do not simply form in anteroposterior sequence. The first rhombomeres to form are those in the central region of the hindbrain, r3 to r5, and this is then followed by the more anterior and posterior segments [34].

Once the segments have formed the individual identities of the rhombomeres become apparent and these are under the control of differential *Hox* gene expression. For example, r2 expresses only *Hoxa2*, while r4 expresses *Hoxa2*, *Hoxb2* and *Hoxb1* and r6 *Hoxa2*, *Hoxb2*, *Hoxa3* and *Hoxb3*. The differential Hox gene expression domains are in part directed by their differential transcriptional response to graded retinoic acid activity [35].

The segmentation of the hindbrain also has consequences for more peripheral tissues. A conserved feature of all vertebrates is the presence of three streams of neural crest emanating from the hindbrain. The most rostral is the trigeminal, which comprises neural crest cells from the midbrain and r1 and r2, the middle is the hyoid, which is primarily populated by neural crest cells from r4 with a small contribution from the flanking rhombomeres, and the most posterior is the post-otic which is composed of neural crest cells migrating from the caudal hindbrain [36, 37]. These different streams come to populate distinct pharyngeal arches and they act to organise the sensorimotor innervation [36, 38].

Pharyngeal arches: Segmentation of the Pharynx

The pharyngeal arches are a series of bulges on the lateral surface of the head of vertebrate embryos. These structures are a key characteristic of the phylotypic stage and indeed lend that stage its name, the pharyngula. The development of these structures is complex as they comprise a number of disparate cell populations from all three germ layers [39]. Within each arch there is a mesodermal core, that will form muscle and endothelium and this is surrounded by neural crest cells, which will generate the skeletal and connective tissue components. These two mesenchymal populations are enveloped externally by ectoderm, which will generate epidermis and neurogenic placodes, and internally by the pharyngeal endoderm, which will produce the lining of the pharynx, taste buds and specialised structures such as the thyroid, thymus and parathyroids. Thus, the pharyngeal arches constitute an iterated series with each generating the same basic set of components.

An early key event in the development of the arches is the formation of outpocketings within the pharyngeal endoderm, the pharyngeal pouches [40, 41]. Mutants that fail to segment the endoderm fail to form pharyngeal arches [42]. The pharyngeal pouches form at distinct positions along the anteroposterior axis and they will contact the overlying ectoderm, which invaginates to meet them, and form the pharyngeal clefts. The points of contact between the pouches and clefts define the anterior and posterior limits of the arches. Thus the neural crest and mesoderm migrate into pre-existing epithelial segments formed by segmentation of the endoderm.

The formation of the pharyngeal pouches requires signals from the surrounding mesodermal and neural tissues and it has been shown that Fgf function is necessary for the formation of all of the pouches [43]. There are, however, also significant differences between the development of the anterior and posterior pouches. First, the two most anterior pouches form at the same time, while the more posterior are generated sequentially [40, 41]. The development of the posterior pouches is also under the control of distinct signalling pathways. It has been shown that the formation of the posterior but not the anterior pouches requires retinoic acid and wnt signalling from the adjacent lateral mesoderm [44, 45, 46, 47]. *Tbx1* has also been identified as a key transcription factor for the generation of the posterior but not the anterior pharyngeal pouches [48], that drives proliferation within the endoderm. In *Tbx1* mutants the posterior pharyngeal pouches fail to form. Finally, it is also important to note that the number of pharyngeal pouches formed, and thus the corresponding number of arches, varies among the vertebrates, with a general trend towards reduction. Thus while lampreys form eight pouches and nine arches, many chondrichthyans and all actinopterygians form six pouches and seven arches, while amphibia form five pouches and six arches and amniotes four pouches and five arches.

Although segmentation of the endoderm provides the framework for pharyngeal development, interactions between the populations that contribute to the arches is also important for their full realisation. In particular, neural crest cells play a key role as they are the source of the skeletal elements that will underpin the later identity of the arches. In gnathostomes, the most anterior arch will form the jaws, the second the hyoid and the more posterior arches are either gill bearing in fish or subsumed into the larynx in amniotes. Again, the identity of the skeletal elements is dependent upon *Hox* genes.

Segmentation in Vertebrates is achieved through Diverse, Distinct Unrelated Mechanisms

With the formation of the somites, the subdivision of the hindbrain and the generation of the pharyngeal arches one can observe the iteration of parts along the long axis of the body. However, as is apparent from our consideration of how these processes are directed, they differ fundamentally from each other. There are few, if any, common mechanisms. Somites are generated sequentially over a long period using a clock and wave front mechanism acting within the presomitic mesoderm, which is to some extent replenished during axis extension. The rhombomeres form via internal subdivision of a specified region and over a very short period. The rhombomeres also do not emerge in an anteroposterior sequence with the first segments that are formed being r3, r4 and r5 [34]. Finally, the pharyngeal segments form through the outpocketing of the pharyngeal endoderm to form the pharyngeal pouches which contact the overlying ectoderm and thus delineate the anterior and posterior limits of the arches. There are, however, some superficial similarities such as the utilisation of some of same molecules, but these are often not employed in the same way. Thus while a posterior to anterior gradient of FGF is important for somitogenesis [5], this is not true of pharyngeal or hindbrain segmentation. In the pharynx, FGF activity is required for the formation the pouches themselves [43] and in the hindbrain FGF signalling has localised roles in the development of particular regions; early FGF is needed for r4 formation and FGF signalling from the isthmus impinges upon r1/r2 identity [49, 50]. Another similarity lies in the identity of the segments being dependent on *Hox* genes, but this reflects the more general role of these genes in anteroposterior patterning of the body. *Hox* genes assign identity to segmented and unsegmented regions, such as the lateral plate mesoderm, alike.

Finally, it is also important to appreciate that these different segmented systems have different anatomical/functional outcomes and thus serve distinct functions. The segmentation of the paraxial mesoderm to generate individually packaged somites underpins the locomotory strategies of the vertebrates, resulting in the formation of separate bilateral bocks of muscle lying either side of an articulated backbone. This arrangement is essential for lateral undulatory locomotion of fish and many tetrapods. In contrast, the segmentation of the hindbrain generates subdivisions within a contiguous region which allows for seamless connections between the different hindbrain nuclei and ongoing connections, and for through traffic that connects higher brain centres with the spinal cord. This organisation of the hindbrain is vital to its function in co-ordinating respiratory activity, motor output and processing sensory information including that from the auditory/vestibular and lateral line systems. Lastly, the segmentation of the pharynx relates to its activities in feeding and respiration. The two most anterior pharyngeal segments of the gnathostomes will contribute to the jaw apparatus and the more posterior segments will form gills with an abundant vasculature and thus perform respiratory functions.

The evolutionary Origins of Vertebrate Segmentation

To further clarify the relationships between the vertebrate segmentation processes; somitogenesis, rhombomeric subdivisions of the hindbrain and pharyngeal arches, it is important to ask about their evolutionary origins.

Vertebrates are chordates and one defining feature of this phylum is the presence of segmented muscle blocks. We might therefore expect somitogenesis to be a shared characteristic of the chordates. Yet somites are lacking in urochordates and while they do form in cephalochordates this process seems to be somewhat distinct from that described in vertebrates [51]. The more anterior somites in amphioxus form as bilateral pairs by enterocoelus evagination of the wall of the archenteron while the posterior somites form by schizocoely, alternating between the left and right sides. It has also been shown that, while the very anterior somites are dependent upon FGF signalling, most of the other somites forming by enterocoely and those formed by schizocoely are FGF insensitive. However, the lineages leading to the extant representatives of the chordate subphyla diverged a very long time ago and thus ancestral characteristics may have been lost or obscured. Moving outside the chordates, and considering other deuterostomes one cannot find evidence for somites. Thus, it is reasonable to assume that somitogenesis evolved with the chordates but has undergone major modification in the different chordate lineages.

Gene expression studies in other chordates and in hemichordates have established that a region of the nervous system expressing anterior *Hox* genes, and thus homologous to the vertebrate hindbrain domain, exists in these groups [52]. There are, however, no indications that rhombomeres exist outside vertebrates. There is neither morphological nor molecular evidence to support segmentation of the nervous system in an analogous region. For example, amphioxus has a single *Krox20* gene but it is not expressed segmentally in the developing nervous system [53].

Pharyngeal segmentation is, however, relatively ancient. As with somites, pharyngeal gill slits are characteristic of the chordates and it is clear that the simple perforations of the pharynx seen in other chordates such as amphioxus are homologous to the endodermal segmentation of the vertebrate pharynx. The pharyngeal pouches of vertebrates express a *Pax-Six-Eya* regulatory network as does the pharyngeal endoderm in amphioxus [54, 55]. Similarly, the amphioxus *Tbx1/10* gene is also expressed in the pharyngeal segments, mirroring the expression of *Tbx1* in vertebrates [56]. Recent results from the hemichordate Saccoglossus kowalevskii have shown that pharyngeal segmentation is likely to be a general feature of deuterostomes [57]. In this species, the formation of the gill pores by the endoderm is also associated with the expression of *Pax1/9*, *Eya* and *Six*. Furthermore, although echinoderms lack gill slits, this is likely to result from secondary loss as fossil evidence has shown that the earliest echinoderms were bilateral and did possess gill slits [58], which further indicates that pharyngeal segmentation is a characteristic of the deuterostomes.

An important point that emerges from this is that there was no ancestral process of segmentation that was co-opted by each of these processes. The three segmental systems of the vertebrates each arose *de novo* at different points during evolution (Figure 2). The most ancient is pharyngeal segmentation, and that is a feature of the deuterostomes, with somitogenesis following with the emergence of the chordates and finally rhombomere formation and the evolution of the vertebrates.

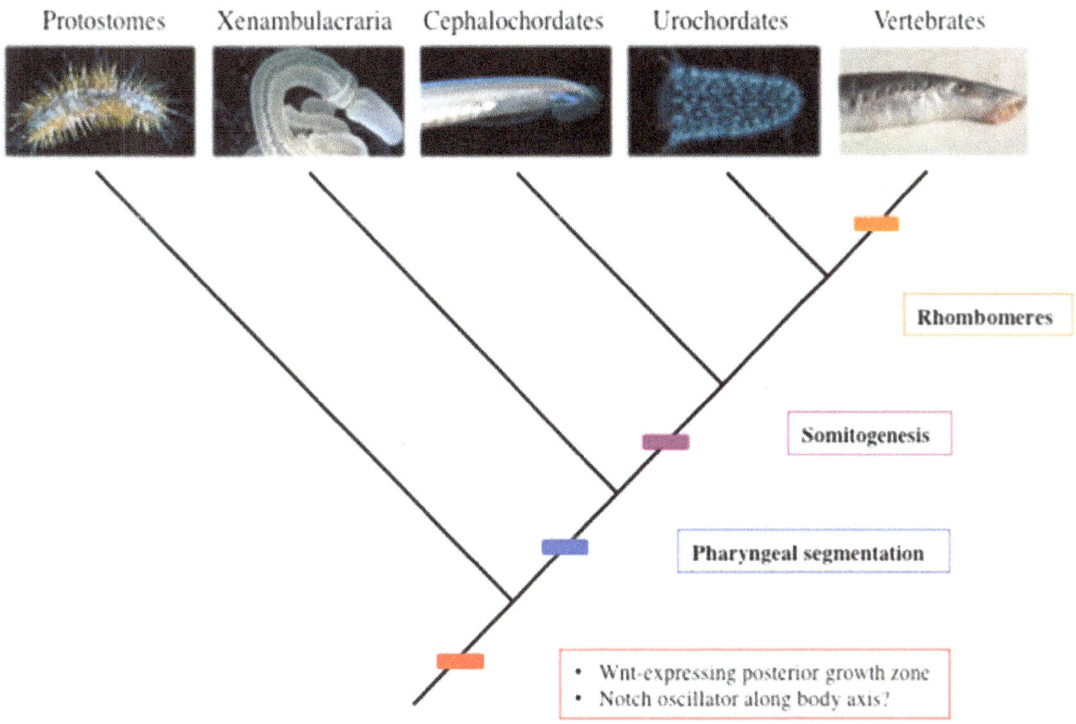

Figure 2. The evolutionary history of segmentation in the vertebrate lineage. Three instances of segmentation are found in extant vertebrates that are conserved with different invertebrate groups. While the ancestor of all Bilateria probably developed with a posterior growth zone expressing Wnts, and perhaps oscillating Notch signalling along the A/P axis, overt morphological segmentation appears later in the phylogeny. Pharyngeal segmentation can be dated to the deuterostome ancestor, while somitogenesis dates to the chordate ancestor and rhombomeric organisation of the hindbrain to the vertebrate stem.

Single or manifold evolutionary origins of segmentation?

There have been many discussions as to the evolutionary origin(s) of segmentation and there are two key issues here that must be confronted. The first is whether or not there is any evidence to support homology between the manifestations of segmentation seen in vertebrates with those displayed by other bilaterian clades. Attempts to homologize between segmentation in vertebrates and that seen in arthropods and annelids, have been strongly affected by their time. In the late 1980s and early 1990s comparisons were invariably drawn between the mechanisms underpinning the segmentation of the hindbrain and those directing segmentation in Drosophila. Both involved specification via transcription factor hierarchies and both resulted in the formation of lineage restricted compartments. However, as our molecular understanding of somitogenesis advanced it became more common to draw comparisons between that process and other modes of arthropod segmentation. For example, it was noted that segmentation in spiders involves Notch and Delta signalling [59]. Yet, for both of these comparisons, our extensive knowledge of the developmental processes underpinning rhombomere formation and somitogenesis would indicate that the highlighted commonalities are but simply superficial similarities.The mechanisms underpinning the segmentation of the Drosophila embryo are quite different from those in vertebrates. Drosophila segments are formed within a syncytium by a transcription factor cascade. While rhombomeres are formed via cell sorting, using Eph-ephrin signalling, lying downstream of a very different system of signalling molecules and transcriptional effectors. Furthermore, rhombomeres evolved with the vertebrates (Figure 2); they are clearly lacking in other deuterostomes, and thus it is unlikely that they could have a common origin with the segmental patterning systems of arthropods. The formation of somites and the segments of many arthropods do share the characteristic that they are generated sequentially and that this is tied to axis elongation. But, as such, some of the shared features associated with somite formation and arthropod segmentation may indicate more general conserved bilaterian features, such as the involvement of a posterior wnt-secreting growth zone [60, 61]. It is also very possible that the association between notch signalling and segments is the result of this very ancient signalling pathway becoming subservient to the segmentation processes in animals that organise their body plan in such a manner, and indeed as we have pointed out Notch signalling is not required for somite formation in vertebrates. Finally, it should be stressed again that somites are a chordate feature (Figure 2) and are not found in other deuterostomes and thus it seems again unlikely that somitogenesis could have a common origin with any mode of arthropod segmentation.

The second issue is the relative paucity of segmentation within the bilateria and its implications. As Hannibal and Patel point out, if segmentation is difficult to evolve, this would suggest that it had a single origin but that it was subsequently lost by the great majority of animal phyla and only retained in a few [2]. One conclusion following from this hypothesis would be that segmentation is readily dispensable in the generation of a functional body plan. By contrast, if segmentation is relatively easy to evolve then one would expect to observe unrelated, non-homologous, instances of segmentation in different phyla. Our discussion of the segmented systems of vertebrates would point us towards the latter option. We find that there is no single process of segmentation and, that in the lineage leading to the vertebrates, segmented structures evolved at least three times independently, in different germ layers and using different mechanics, at least three times. Thus it would seem that it is relatively easy to evolve segmentation.

CONCLUSIONS

Hannibal and Patel make the excellent point that there is no merit in talking about segmentation without being explicit about what is being discussed. Thus with regards to segmentation in vertebrates, it is unhelpful to talk generally of segmentation and to lump together the processes of somitogenesis, rhombomere formation and pharyngeal arch development; these are chalk and cheese comparisons. It is more correct and useful to discuss how somites form, how rhombomeres emerge and how pharyngeal arches are generated. It is likewise uninformative and potentially confounding to talk of individual taxa as 'segmented' or 'unsegmented', or even 'pseudosegmented'; it is much more sensible to talk about segmented structures or body regions individually.

Furthermore, as Hannibal and Patel note, it is incredibly difficult to arrive at a precise definition of segmentation and we would argue that this is because there is no single process of segmentation. Consequently, all definitions of segmentation are superficial; that is, repetition of structures along the main body axis - there is nothing deeper to be indicated. An analogous situation is that of wings - what is a wing? It is a structure that allows an animal to fly. Wings are a feature of flies, birds and bats but the definition of a wing has to be superficial because it describes non-homologous structures. Thus, many of the problems that arise with the concept of segmentation, and that we have discussed here, ultimately reflect a problem of terminology. The names that we apply to biological processes do not necessarily indicate anything beyond being useful appellations. Of course this is the problem of homoplasy and the only route to resolving this is to map any given biological process to the phylogeny.

DECLARATIONS

Acknowledgements

We would like to thank Peter Hardy and Victoria Shone for contributing images to Figure 1, and Malcolm Logan for his comments on the manuscript.

REFERENCES

1. Tautz D: Segmentation. Dev Cell. 2004, 7: 301-312. 10.1016/j.devcel.2004.08.008.
2. Hannibal RL, Patel NH: What is a segment?. EvoDevo. 2013, 4: 35-10.1186/2041-9139-4-35.
3. Pourquie O: Vertebrate segmentation: from cyclic gene networks to scoliosis. Cell. 2011, 145: 650-663. 10.1016/j.cell.2011.05.011.
4. Maroto M, Bone RA, Dale JK: Somitogenesis. Development. 2012, 139: 2453-2456. 10.1242/dev.069310.
5. Dubrulle J, McGrew MJ, Pourquie O: FGF signaling controls somite boundary position and regulates segmentation clock control of spatiotemporal Hox gene activation. Cell. 2001, 106: 219-232. 10.1016/S0092-8674(01)00437-8.
6. Dubrulle J, Pourquie O: fgf8 mRNA decay establishes a gradient that couples axial elongation to patterning in the vertebrate embryo. Nature. 2004, 427: 419-422. 10.1038/nature02216.
7. Aulehla A, Wiegraebe W, Baubet V, Wahl MB, Deng C, Taketo M, Lewandoski M, Pourquie O: A beta-catenin gradient links the clock and wavefront systems in mouse embryo segmentation. Nat Cell Biol. 2008, 10: 186-193. 10.1038/ncb1679.
8. del Corral DR, Olivera-Martinez I, Goriely A, Gale E, Maden M, Storey K: Opposing FGF and retinoid pathways control ventral neural pattern, neuronal differentiation, and segmentation during body axis extension. Neuron. 2003, 40: 65-79. 10.1016/S0896-6273(03)00565-8.
9. Goldbeter A, Gonze D, Pourquie O: Sharp developmental thresholds defined through bistability by antagonistic gradients of retinoic acid and FGF signaling. Dev Dynam. 2007, 236: 1495-1508. 10.1002/dvdy.21193.
10. Takahashi Y, Koizumi K, Takagi A, Kitajima S, Inoue T, Koseki H, Saga Y: Mesp2 initiates somite segmentation through the Notch signalling pathway. Nat Genet. 2000, 25: 390-396. 10.1038/78062.
11. Dias AS, de Almeida I, Belmonte JM, Glazier JA, Stern CD: Somites without a clock. Science. 2014, 343: 791-795. 10.1126/science.1247575.

12. Ozbudak EM, Lewis J: Notch signalling synchronizes the zebrafish segmentation clock but is not needed to create somite boundaries. PLoS Genet. 2008, 4: e15-10.1371/journal.pgen.0040015.

13. Soza-Ried C, Ozturk E, Ish-Horowicz D, Lewis J: Pulses of Notch activation synchronise oscillating somite cells and entrain the zebrafish segmentation clock. Development. 2014, 141: 1780-1788. 10.1242/dev.102111.

14. Verbout AJ: A critical review of the 'neugliederung' concept in relation to the development of the vertebral column. Acta Biotheor. 1976, 25: 219-258. 10.1007/BF00046818.

15. Bagnall KM, Sanders EJ, Higgins SJ, Leam H: The effects of somite removal on vertebral formation in the chick. Anat Embryol (Berl). 1988, 178: 183-190. 10.1007/BF02463652.

16. Aoyama H, Asamoto K: The developmental fate of the rostral/caudal half of a somite for vertebra and rib formation: experimental confirmation of the resegmentation theory using chick-quail chimeras. Mech Dev. 2000, 99: 71-82. 10.1016/S0925-4773(00)00481-0.

17. Morin-Kensicki EM, Melancon E, Eisen JS: Segmental relationship between somites and vertebral column in zebrafish. Development. 2002, 129: 3851-3860.

18. Keynes RJ, Stern CD: Segmentation in the vertebrate nervous system. Nature. 1984, 310: 786-789. 10.1038/310786a0.

19. Krull CE, Lansford R, Gale NW, Collazo A, Marcelle C, Yancopoulos GD, Fraser SE, Bronner-Fraser M: Interactions of Eph-related receptors and ligands confer rostrocaudal pattern to trunk neural crest migration. Curr Biol. 1997, 7: 571-580. 10.1016/S0960-9822(06)00256-9.

20. Wang HU, Anderson DJ: Eph family transmembrane ligands can mediate repulsive guidance of trunk neural crest migration and motor axon outgrowth. Neuron. 1997, 18: 383-396. 10.1016/S0896-6273(00)81240-4.

21. Bernhardt RR, Goerlinger S, Roos M, Schachner M: Anterior-posterior subdivision of the somite in embryonic zebrafish: implications for motor axon guidance. Dev Dyn. 1998, 213: 334-347. 10.1002/(SICI)1097-0177(199811)213:3<334::AID-AJA9>3.0.CO;2-4.

22. Eickholt BJ, Mackenzie SL, Graham A, Walsh FS, Doherty P: Evidence for collapsin-1 functioning in the control of neural crest migration in both trunk and hindbrain regions. Development. 1999, 126: 2181-2189.

23. Gammill LS, Gonzalez C, Gu C, Bronner-Fraser M: Guidance of trunk neural crest migration requires neuropilin 2/semaphorin 3 F signaling. Development. 2006, 133: 99-106. 10.1242/dev.02187.

24. Iimura T, Denans N, Pourquie O: Establishment of Hox vertebral identities in the embryonic spine precursors. Curr Top Dev Biol. 2009, 88: 201-234.

25. Lumsden A, Keynes R: Segmental patterns of neuronal development in the chick hindbrain. Nature. 1989, 337: 424-428. 10.1038/337424a0.

26. Kuratani S, Horigome N, Ueki T, Aizawa S, Hirano S: Stereotyped axonal bundle formation and neuromeric patterns in embryos of a cyclostome, Lampetra japonica. J Comp Neurol. 1998, 391: 99-114. 10.1002/(SICI)1096-9861(19980202)391:1<99::AID-CNE9>3.0.CO;2-M.

27. Kuratani S, Horigome N: Developmental morphology of branchiomeric nerves in a cat shark, Scyliorhinus torazame, with special reference to rhombomeres, cephalic mesoderm, and distribution patterns of cephalic crest cells. Zool Sci. 2000, 17: 893-909. 10.2108/zsj.17.893.

28. Clarke JDW, Lumsden A: Segmental repetition of neuronal phenotype sets in the chick-embryo hindbrain. Development. 1993, 118: 151-162.

29. Eickholt BJ, Graham A, Lumsden A, Wizenmann A: Rhombomere interactions control the segmental differentiation of hindbrain neurons. Mol Cell Neurosci. 2001, 18: 141-148. 10.1006/mcne.2001.1014.

30. Wizenmann A, Lumsden A: Segregation of rhombomeres by differential chemoaffinity. Mol Cell Neurosci. 1997, 9: 448-459. 10.1006/mcne.1997.0642.

31. Xu Q, Wilkinson DG: Boundary formation in the development of the vertebrate hindbrain. Wiley Interdiscip Rev Dev Biol. 2013, 2: 735-745. 10.1002/wdev.106.

32. Fraser S, Keynes R, Lumsden A: Segmentation in the chick embryo hindbrain is defined by cell lineage restrictions. Nature. 1990, 344: 431-435. 10.1038/344431a0.

33. Xu Q, Alldus G, Holder N, Wilkinson DG: Expression of truncated Sek-1 receptor tyrosine kinase disrupts the segmental restriction of gene expression in the Xenopus and zebrafish hindbrain. Development. 1995, 121: 4005-4016.

34. Lumsden A: The cellular basis of segmentation in the developing hindbrain. Trends Neurosci. 1990, 13: 329-335. 10.1016/0166-2236(90)90144-Y.

35. Maden M: Retinoic acid in the development, regeneration and maintenance of the nervous system. Nat Rev Neurosci. 2007, 8: 755-765. 10.1038/nrn2212.

36. Lumsden A, Sprawson N, Graham A: Segmental origin and migration of neural crest cells in the hindbrain region of the chick embryo. Development. 1991, 113: 1281-1291.

37. Schilling TF, Kimmel CB: Segment and cell type lineage restrictions during pharyngeal arch development in the zebrafish embryo. Development. 1994, 120: 483-494.

38. Begbie J, Graham A: Integration between the epibranchial placodes and the hindbrain. Science. 2001, 294: 595-598. 10.1126/science.1062028.

39. Graham A, Richardson J: Developmental and evolutionary origins of the pharyngeal apparatus. EvoDevo. 2012, 3: 24-10.1186/2041-9139-3-24.

40. Veitch E, Begbie J, Schilling TF, Smith MM, Graham A: Pharyngeal arch patterning in the absence of neural crest. Curr Biol. 1999, 9: 1481-1484. 10.1016/S0960-9822(00)80118-9.

41. Crump JG, Swartz ME, Kimmel CB: An integrin-dependent role of pouch endoderm in hyoid cartilage development. PLoS Biol. 2004, 2: E244-10.1371/

journal.pbio.0020244.

42. Piotrowski T, Nusslein-Volhard C: The endoderm plays an important role in patterning the segmented pharyngeal region in zebrafish (Danio rerio). Dev Biol. 2000, 225: 339-356. 10.1006/dbio.2000.9842.
43. Crump JG, Maves L, Lawson ND, Weinstein BM, Kimmel CB: An essential role for Fgfs in endodermal pouch formation influences later craniofacial skeletal patterning. Development. 2004, 131: 5703-5716. 10.1242/dev.01444.
44. Niederreither K, Subbarayan V, Dolle P, Chambon P: Embryonic retinoic acid synthesis is essential for early mouse post-implantation development. Nat Genet. 1999, 21: 444-448. 10.1038/7788.
45. Quinlan R, Gale E, Maden M, Graham A: Deficits in the posterior pharyngeal endoderm in the absence of retinoids. Dev Dyn. 2002, 225: 54-60. 10.1002/dvdy.10137.
46. Begemann G, Schilling TF, Rauch GJ, Geisler R, Ingham PW: The zebrafish neckless mutation reveals a requirement for raldh2 in mesodermal signals that pattern the hindbrain. Development. 2001, 128: 3081-3094.
47. Choe CP, Collazo A, le Trinh A, Pan L, Moens CB, Crump JG: Wnt-dependent epithelial transitions drive pharyngeal pouch formation. Dev Cell. 2013, 24: 296-309. 10.1016/j.devcel.2012.12.003.
48. Xu H, Cerrato F, Baldini A: Timed mutation and cell-fate mapping reveal reiterated roles of Tbx1 during embryogenesis, and a crucial function during segmentation of the pharyngeal system via regulation of endoderm expansion. Development. 2005, 132: 4387-4395. 10.1242/dev.02018.
49. Irving C, Mason I: Signalling by FGF8 from the isthmus patterns anterior hindbrain and establishes the anterior limit of Hox gene expression. Development. 2000, 127: 177-186.
50. Maves L, Jackman W, Kimmel CB: FGF3 and FGF8 mediate a rhombomere 4 signaling activity in the zebrafish hindbrain. Development. 2002, 129: 3825-3837.
51. Bertrand S, Camasses A, Somorjai I, Belgacem MR, Chabrol O, Escande ML, Pontarotti P, Escriva H: Amphioxus FGF signaling predicts the acquisition of vertebrate morphological traits. Proc Natl Acad Sci U S A. 2011, 108: 9160-9165. 10.1073/pnas.1014235108.
52. Wada H, Garcia-Fernandez J, Holland PW: Colinear and segmental expression of amphioxus Hox genes. Dev Biol. 1999, 213: 131-141. 10.1006/dbio.1999.9369.
53. Knight RD, Panopoulou GD, Holland PW, Shimeld SM: An amphioxus Krox gene: insights into vertebrate hindbrain evolution. Dev Genes Evol. 2000, 210: 518-521. 10.1007/s004270000092.
54. Holland LZ, Holland ND: Expression of AmphiHox-1 and AmphiPax-1 in amphioxus embryos treated with retinoic acid: insights into evolution and patterning of the chordate nerve cord and pharynx. Development. 1996, 122: 1829-1838.
55. Kozmik Z, Holland ND, Kreslova J, Oliveri D, Schubert M, Jonasova K, Holland LZ, Pestarino M, Benes V, Candiani S: Pax-Six-Eya-Dach network during amphioxus development: conservation in vitro but context specificity in vivo. Dev Biol. 2007, 306: 143-159. 10.1016/j.ydbio.2007.03.009.
56. Mahadevan NR, Horton AC, Gibson-Brown JJ: Developmental expression of the amphioxus Tbx1/ 10 gene illuminates the evolution of vertebrate branchial arches and sclerotome. Dev Genes Evol. 2004, 214: 559-566. 10.1007/s00427-004-0433-1.
57. Gillis JA, Fritzenwanker JH, Lowe CJ: A stem-deuterostome origin of the vertebrate pharyngeal transcriptional network. Proc Biol Sci. 2012, 279: 237-246. 10.1098/rspb.2011.0599.
58. Smith AB: The pre-radial history of the echinoderms. Geol J. 2005, 40: 255-280. 10.1002/gj.1018.
59. Stollewerk A, Schoppmeier M, Damen WG: Involvement of Notch and Delta genes in spider segmentation. Nature. 2003, 423: 863-865. 10.1038/nature01682.
60. Petersen CP, Reddien PW: Smed-betacatenin-1 is required for anteroposterior blastema polarity in planarian regeneration. Science. 2008, 319: 327-330. 10.1126/science.1149943.
61. Janssen R, Le Gouar M, Pechmann M, Poulin F, Bolognesi R, Schwager EE, Hopfen C, Colbourne JK, Budd GE, Brown SJ, Prpic N-M, Kosiol C, Vervoort M, Damen WGM, Balavoine G, McGregor AP: Conservation, loss, and redeployment of Wnt ligands in protostomes: implications for understanding the evolution of segment formation. BMC Evol Biol. 2010, 10: 374-10.1186/1471-2148-10-374

CHAPTER 14

GLOBAL CHARACTERIZATION OF GENES IN VERTEBRATES

Bei Huang, Zhi T Qi, Zhen Xu, Pin Nie

BACKGROUND

Interferon regulatory factors (IRFs) were identified originally as transcription factors in the regulation of interferon expression [1]. Over the last a few decades, these factors have been the focus of many immunological and medical studies [2, 3], and it has been shown that they have diversified functions in immune responses, haematopoietic differentiation and immune modulation [4, 5, 6]. These transcription factors possess a unique 'tryptophan cluster' DNA-binding domain (DBD) [7], which is responsible for binding to the IFN-regulatory factor element present in the IFN-β promoter [8]. Since the identification of the first IRF, IRF-1, as a protein binding to the *cis*-acting DNA elements of the IFN-β gene, a total of ten members has been identified in vertebrates with functions from activators (e.g., IRF-1, IRF-3, IRF-5, IRF-9 and IRF-10), to repressors (e.g., IRF-8); and some of them (e.g., IRF-2, IRF-4, and IRF-7) also exert the two functions [3, 4, 9, 10]. Previous research has largely focused on the function of individual IRFs in mammals [8, 11, 12]. However, the composition and the function of these IRF family members are less investigated in other classes of vertebrates [13]. In fish, a few members, such as IRF-1, IRF-2 and IRF-7 have been cloned in some aquacultured species [14, 15, 16, 17, 18], and recently a total of eleven IRF family members has been identified in zebrafish *Dario rerio* [19].

In an attempt to clearly delineate the evolution of IRF family lineage, we analyzed the draft genomes of human (*Homo sapiens*), mouse (*Mus musculus*), dog (*Canis familiaris*), chicken (*Gallus gallus*), anole lizard (*Anolis carolinesis*), frog (*Xenopus tropicalis*), zebrafish (*Dario rerio*), fugu (*Takifugu rubripes*), stickleback (*Gasterosteus aculeatus*), medaka (*Oryzias latipes*) to systematically identify all IRF members in each species. The searched results were further compared with literature and the location of each member in the IRF family was mapped physically on chromosomes in human, mouse, dog, chicken, anole lizard, frog and zebrafish representing classes of mammal, avian, reptile, amphibian and piscine. In addition, any possible IRF-like genes were identified in some non-vertebrate deuterostomes, including the hemichordate, acorn worm *Saccoglossus kowalevskii*, the echinoderm, sea urchin *Strongylocentrotus purpuratus*, the cephalochordate, lancelet (amphioxus) *Branchiostoma floridae* and the urochordate, sea squirt *Ciona intestinalis*. The conservation in the genome organization of individual IRF family members

Citation: Bei Huang, Zhi T Qi, Zhen Xu and Pin Nie. Global characterization of genes in vertebrate, 2010, 11:22. DOI: 10.1186/1471-2172-11-22

and the synteny in their gene loci were analyzed. The possible evolutionary mechanisms in the origination of IRF family members were then discussed.

RESULTS

A total of ten members in the IRF family, i.e., IRF- 1 to 10 have been identified in vertebrates, with another member named as IRF-11 in teleost fish, as also in another study [19]. All these members, despite their diversified functions, share a conserved motif of a five tryptophan pentad repeat, with three of them, W-x-[DNH]-x(5)-[LIVF]-x-[IV]-P-W-x-H-x(9,10)-[DE]-x(2)-[LIVF]-F-[KRQ]-x-[WR]-A, contacting with DNA sequence and recognizing the AANNGAAA sequence [7]. In mammals, IRF-10 was not found in human and mouse. In zebrafish, stickleback, frog, anole lizard, chicken and dog, all ten members were detected. However in chicken, IRF-3 and IRF-9 were not found. In addition, any possible IRF-like genes were identified in non-vertebrate deuterostomes, with the finding of three IRF genes in hemichordate, the acorn worm *Saccoglossus kowalevskii*, two in echinoderm, the sea urchin *Strongylocentrotus purpuratus*, ten in cephalochordate, the lancelet *Branchiostoma floridae* and nine in urochordate, the sea squirt *Ciona intestinalis*.

IRF-1 to IRF-11 in vertebrates

IRF-1

Genomic organizations of IRF- 1 to 11 in vertebrates were illustrated in Figures 1, 2, 3, 4, 5, 6, 7, 8, 9, 10, 11. IRF-1 was identified in all vertebrates examined in the present study. The size of genomic structure of IRF-1 varied from 2163 to 9028 bp, with the longest observed in anole lizard, and shortest in stickleback (Figure 1). The IRF-1 all had a nine exon and eight intron structure in genome, except that in zebrafish eight exons were present. It is interesting to note that the size of the first two exons is the same in all vertebrates, and the first three is the same from frog to mammal. The first four exons were also the same in size in chicken and mammals, with the equal size in the seventh and eighth exons in mammals (Figure 1).

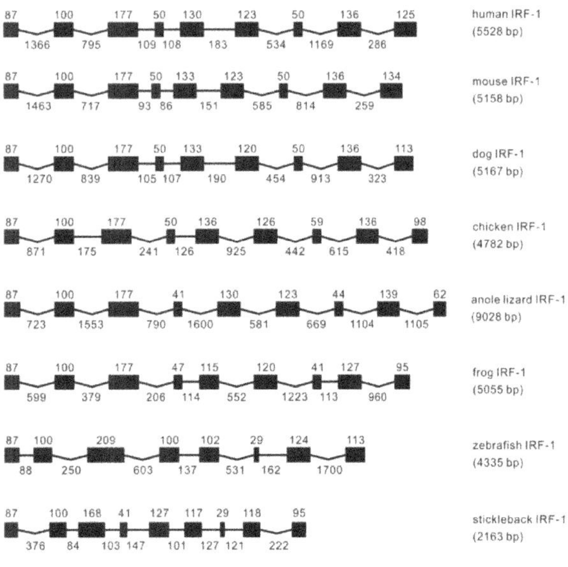

Figure 1. Genomic structure of IRF-1 in different classes of vertebrates. The gene names listed on the right are as the same as those in Table 1 and in Additional file 1. Exons are expressed as black boxes, and introns as lines. The size of exons is indicated above the boxes in bp, and the size of introns below the lines also in bp. The length of exons indicated as black boxes and of introns indicated as straight lines is proportional to their bp sizes, but the concave-lines are non-proportional. Human (*Homo sapiens*), mouse (*Mus musculus*) and dog (*Canis familiaris*), chicken (*Gallus gallus*), anole lizard (*Anolis carolinensis*), frog (*Xenopus tropicalis*), and zebrafish (*Dario rerio*) and stickleback (*Gasterosteus aculeatu* s) represent vertebrates in classes of mammal, avian, reptile, amphibian and piscine, as also indicated in Figures 2 to 11.

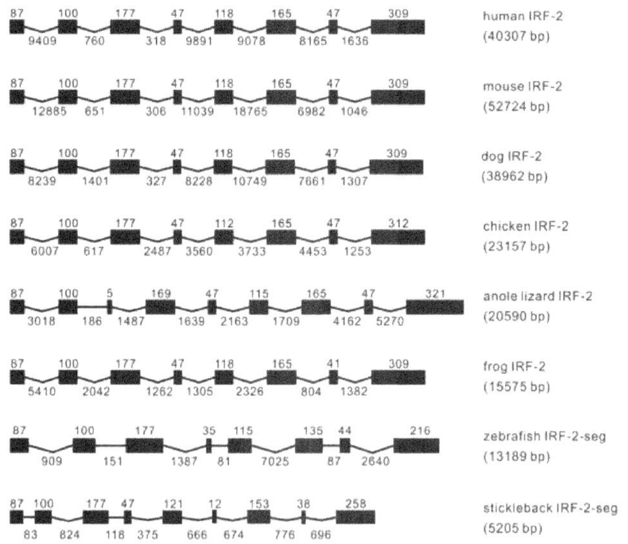

Figure 2. Genomic structure of IRF-2 in different classes of vertebrates. The gene names, and exon-intron organizations, as well as the vertebrates are expressed in the same way as appeared in Figure 1, and so for Figures 3 to 10. The gene name followed with -seg indicates that the gene only has a sequence segment in database, as also indicated in Table 1 and in Additional file 1, and also in other figures.

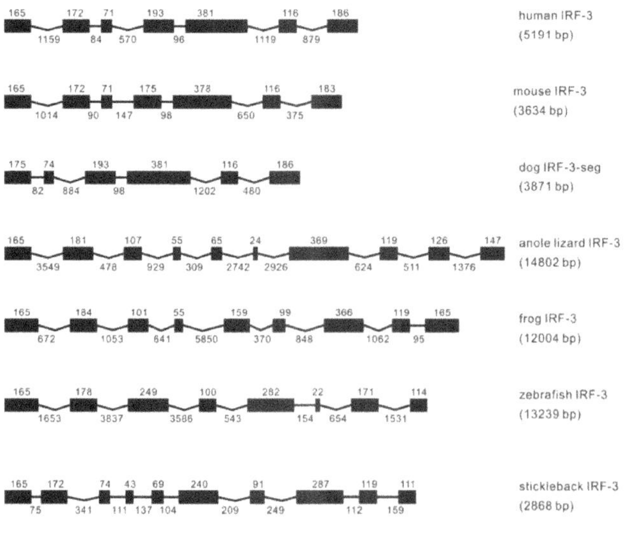

Figure 3. Genomic structure of IRF-3 in different classes of vertebrates, except that in chicken the IRF-3 was not found.

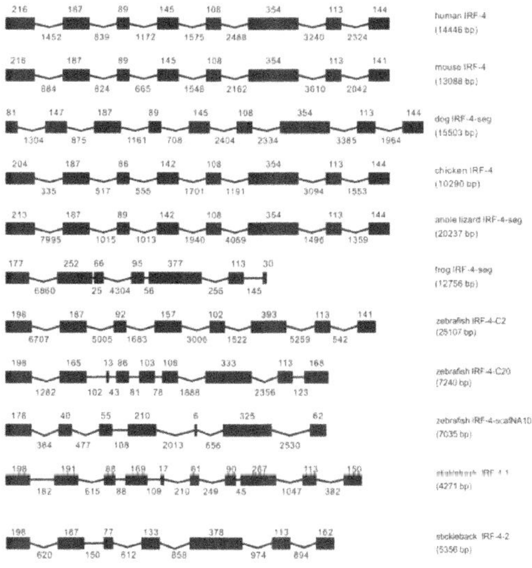

Figure 4. Genomic structure of IRF-4 in different classes of vertebrates. Multi-copy genes of IRF-4 were found in teleost fish.

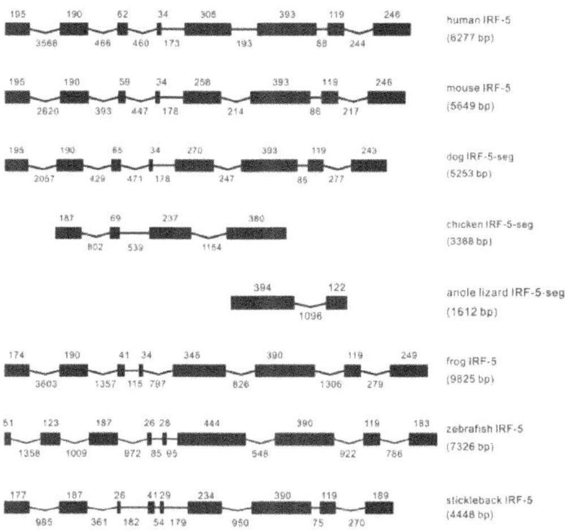

Figure 5. Genomic structure of IRF-5 in different classes of vertebrates.

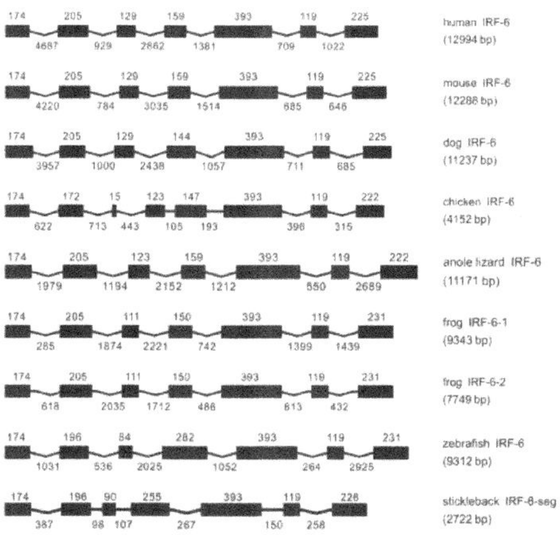

Figure 6. Genomic structure of IRF-6 in different classes of vertebrates. Two identical IRF-6 genes were found in frog.

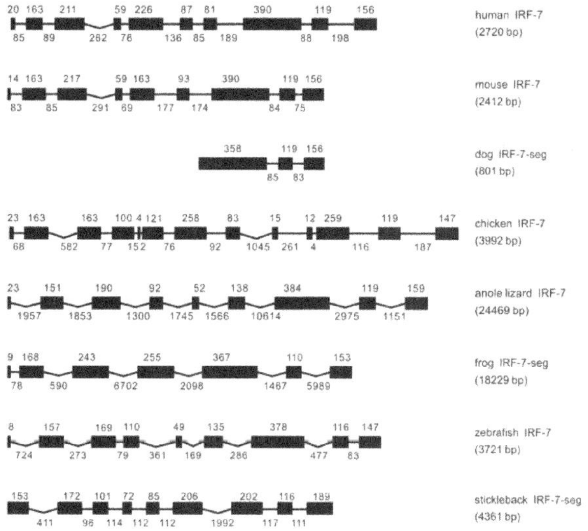

Figure 7. Genomic structure of IRF-7 in different classes of vertebrates.

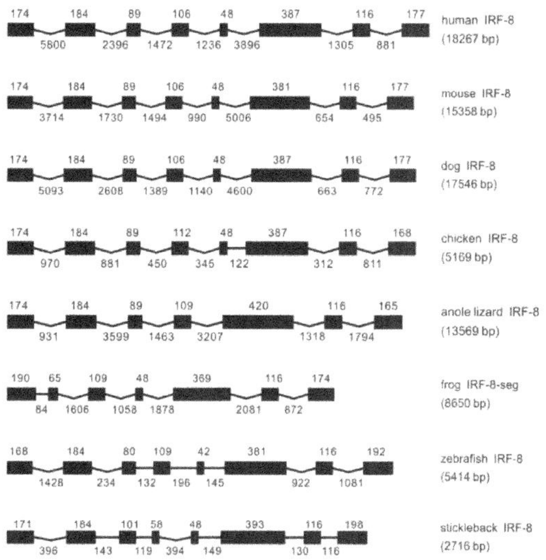

Figure 8. Genomic structure of IRF-8 in different classes of vertebrates.

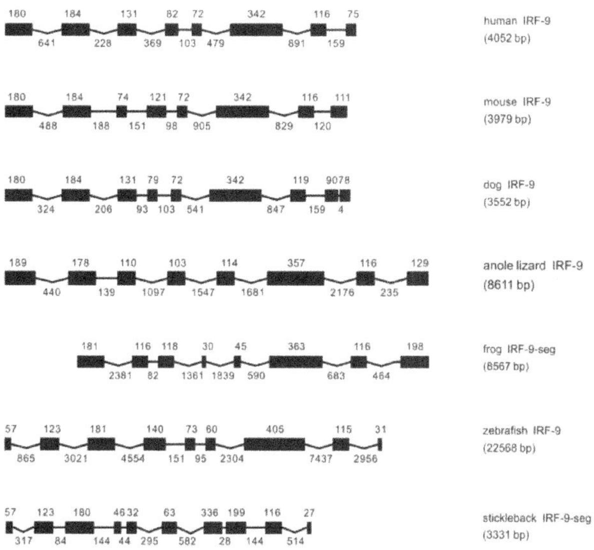

Figure 9. Genomic structure of IRF-9 in different classes of vertebrates, except that in chicken the IRF-9 was not found.

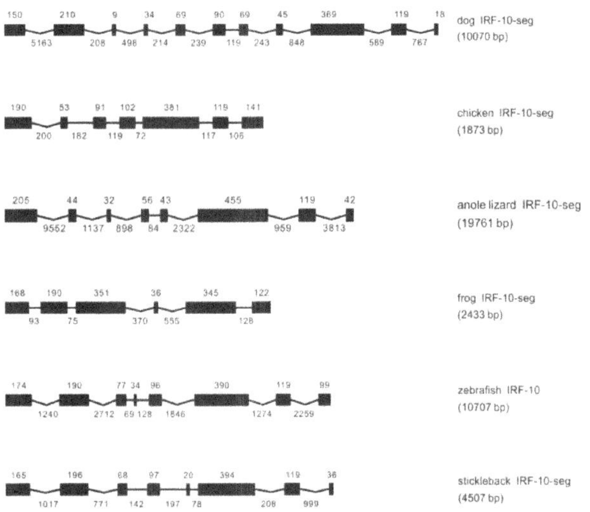

Figure 10. Genomic structure of IRF-10 in different classes of vertebrates. IRF-10 was not found in human and mouse

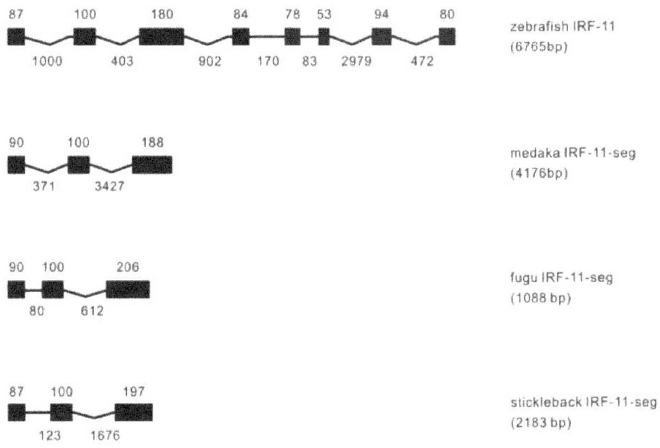

Figure 11. Genomic structure of IRF-11 in teleost fish. IRF-11 was only found in teleost, with the full length sequence available for zebrafish IRF-11. The gene name, exon-intron organization, and the fish species are expressed in a same way as in Figures 1 and 2.

The intronic architecture of the examined IRFs is listed in Table 1. All intron phases are well conserved in IRF-1 with introns 1, 4, 7 in phase 0, introns 2, 3, 5, 6, and 8 in phase 1, except that in zebrafish, the third intron in other vertebrates might have lost in zebrafish IRF-1, as also indicated in the exon-intron organization of zebrafish IRF-1 (Figure 1).

Table 1. Distribution in intron phases of IRF coding region in vertebrates and non-vertebrate deuterostomes, including human, mouse, dog, chicken, anole lizard, frog, zebrafish, stickleback, sea squirt, lancet, sea uchin and acorn worm

Gene	Start codon	Intron number												
		1	2	3	4	5	6	7	8	9	10	11	12	13
human IRF-1	*	0	1	1	0	1	1	0	1	◆				
mouse IRF-1	*	0	1	1	0	1	1	0	1	◆				
dog IRF-1	*	0	1	1	0	1	1	0	1	◆				
chicken IRF-1	*	0	1	1	0	1	1	0	1	◆				
anole lizard IRF-1	*	0	1	1	0	1	1	0	1	◆				
frog IRF-1	*	0	1	1	0	1	1	0	1	◆				
zebrafish IRF-1	*	0	1	0	1	1	0	1	◆					
stickleback IRF-1	*	0	1	1	0	1	1	0	1	◆				
human IRF-2	*	0	1	1	0	1	1	0	◆					
mouse IRF-2	*	0	1	1	0	1	1	0	◆					
dog IRF-2	*	0	1	1	0	1	1	0	◆					
chicken IRF-2	*	0	1	1	0	1	1	0	◆					
anole lizard IRF-2	*	0	1	0	1	0	1	1	0	◆				
frog IRF-2	*	0	1	1	0	1	1	0	◆					
zebrafish IRF-2-seg	*	0	1	1	0	1	1	0						
stickleback IRF-2-seg	*	0	1	1	0	1	1	1	0					
human IRF-3	*	0	1	0	1	1	0	◆						
mouse IRF-3	*	0	1	0	1	1	0	◆						
dog IRF-3-seg			1	0	1	1	0	◆						
anole lizard IRF-3	*	0	1	0	1	0	1	1	0	0	◆			
frog IRF-3	*	0	1	0	1	1	1	1	0	◆				
zebrafish IRF-3	*	0	1	1	2	2	0	0	◆					

stickleback IRF-3	*	0	1	0	1	1	1	2	1	0	◆			
human IRF-4	*	0	1	0	1	1	1	0	◆					
mouse IRF-4	*	0	1	0	1	1	1	0	◆					
dog IRF-4-seg		0	0	1	0	1	1	1	0	◆				
chicken IRF-4	*	0	1	0	1	1	1	0	◆					
anole lizard IRF-4-seg		0	1	0	1	1	1	0	◆					
frog IRF-4-seg	*	0	0	0	2	1	0	0						
zebrafish IRF-4-c2	*	0	1	0	1	1	1	0	◆					
zebrafish IRF-4-c20	*	0	0	1	0	1	1	1	0	◆				
zebrafish IRF-4-scafNA1075-seg				1	2	0	0	0	1	◆				
stickleback IRF-4-1	*	0	2	0	1	0	1	1	1	0	◆			
stickleback IRF-4-2	*	0	1	0	1	1	0	◆						
human IRF-5	*	0	1	0	1	1	1	0	◆					
mouse IRF-5	*	0	1	0	1	1	1	0	◆					
dog IRF-5-seg	*	0	1	0	1	1	1	0						
chicken IRF-5-seg					1	1	1	0						
anole lizard IRF-5-seg					1	0								
frog IRF-5	*	0	1	0	1	1	1	0	◆					
zebrafish IRF-5	*	0	0	1	0	1	1	1	0	◆				
stickleback IRF-5	*	0	1	0	2	1	1	1	0	◆				
human IRF-6	*	0	1	1	1	1	0	◆						
mouse IRF-6	*	0	1	1	1	1	0	◆						
dog IRF-6	*	0	1	1	1	1	0	◆						
chicken IRF-6	*	0	1	1	1	1	1	0	◆					
anole lizard IRF-6	*	0	1	1	1	1	0	0	◆					
frog IRF-6-1	*	0	1	1	1	1	0	◆						
frog IRF-6-2	*	0	1	1	1	1	0	◆						
zebrafish IRF-6	*	0	1	1	1	1	0	◆						
stickleback IRF-6-seg	*	0	1	1	1	1	0	0						
human IRF-7	*	2	0	1	0	1	1	1	1	0	◆			
mouse IRF-7	*	2	0	1	0	1	1	1	0	◆				
dog IRF-7-seg							1	0	0	◆				
chicken IRF-7	*	2	0	1	2	0	1	1	0	0	0	1	0	◆
anole lizard IRF-7	*	2	0	1	0	1	1	1	0	◆				
frog IRF-7-seg	*	0	0	0	0	1	0	0						
zebrafish IRF-7	*	2	0	1	0	1	1	1	0	◆				
stickleback IRF7-seg			0	1	0	0	1	0	1	0	◆			
human IRF-8	*	0	1	0	1	1	1	0	◆					
mouse IRF-8	*	0	1	0	1	1	1	0	◆					
dog IRF-8	*	0	1	0	1	1	1	0	◆					
chicken IRF-8	*	0	1	0	1	1	1	0	◆					
anole lizard IRF-8	*	0	1	0	1	1	0	◆						
frog IRF-8-seg			1	0	1	1	1	0	◆					
zebrafish IRF-8	*	0	1	0	1	1	1	0	◆					
stickleback IRF-8	*	0	1	0	1	1	1	0	◆					
human IRF-9	*	0	1	0	1	1	1	0	◆					

mouse IRF-9	*	0	1	0	1	1	1	0	◆					
dog IRF-9	*	0	1	0	1	1	1	0	0	◆				
anole lizard IRF-9	*	0	1	0	1	1	1	0	◆					
frog IRF-9-seg			1	0	1	1	1	1	0	◆				
zebrafish IRF-9	*	0	0	1	0	1	1	1	2	◆				
stickleback IRF-9-seg	*	0	0	0	1	0	0	0	1	0	0			
dog IRF-10-seg			0	0	0	1	1	1	1	1	1	0	0	
chicken IRF-10-seg			1	0	1	1	1	0	◆					
anole lizard IRF-10-seg			1	0	2	1	2	1	0	0				
frog IRF-10-seg			0	1	1	1	1	0						
zebrafish IRF-10	*	0	1	0	1	1	1	0	◆					
stickleback IRF-10-seg	*	0	1	0	1	0	1	0	0					
zebrafish IRF-11	*	0	1	1	1	1	0	1	◆					
medaka IRF-11-seg	*	0	1	0										
fugu IRF-11-seg	*	0	1	0										
stickleback IRF-11-seg			0	1	0									
sea squirt IRF-like-scaf_162	*	2	0	1	0	0	0	0	0	1	◆			
sea squirt IRF-like-3q-1	*	0	1	1	0	1	◆							
sea squirt IRF-like-3q-2	*	0	1	1	1	1	◆							
sea squirt IRF-like-3q-3	*	0	1	1	1	1	◆							
sea squirt IRF-like-12q	*	0	1	1	1	0	1	0	0	0	0	◆		
sea squirt IRF-like-14p	*	0	1	1	0	1	1	◆						
sea squirt IRF-like-seg-1	*	0	1	1	0	0	1							
sea squirt IRF-like-seg-2			1											
sea squirt IRF-like-seg-3	*	1	1											
lancet IRF-like-scaf_187-1-seg	*	0	1	1	0	1	2	2	1	0	1			
lancet IRF-like-scaf_187-2	*	2	2	1	0	1	0	◆						
lancet IRF-like-scaf_187-3	*	0	2	1	◆									
lancet IRF-like-scaf_7	*	0	0	1	1	1	1	0	◆					
lancet IRF-like-scaf_12	*	0	1	◆										
lancet IRF-like-scaf_27	*	0	1	2	1	2	2	0	1	◆				
lancet IRF-like-scaf_48	*	0	1	◆										
lancet IRF-like-scaf_136	*	0	2	0	2	2	◆							
lancet IRF-like-scaf_172	*	0	1	1	1	◆								
lancet IRF-like-scaf_196	*	0	1	1	1	◆								
sea urchin IRF-like-1	*	2	1	1	1	0	1	1	◆					
sea urchin IRF-like-2	*	2	0	1	0	1	0	1	1	1	2	◆		
acorn worm IRF-like-1	*	2	1	2	1	0	1	1	0	1	1	◆		
acorn worm IRF-like-2	*	2	0	0	1	0	1	1	1	2	◆			
acorn worm IRF-like-3-seg		1	2											

* indicating start coden; The numbers 0, 1, 2 indicating the intron phase, and phase 0 indicating that intron lies between two codons, phase 1 between the first and second positions of a codon, phase 2 between the second and third positions of a codon; ◆ indicating stop coden.

IRF-2

The genomic structure of IRF-2 has a range of 5205 to 52724 bp in length from fish to mammals, with the shortest observed in stickleback (Figure 2). The first three exons are quite conserved in size in all the vertebrates, except that the third exon in other vertebrates might have split into two exons in anole lizard. The other remaining exons were either identical in size, e.g., the two

47 bp exons and the 165 bp exons, or varied in a small range, e.g., from 112 to 118 bp in the fifth exon for all vertebrates except in fish, but in the sixth exon in anole lizard (Figure 2). However, the structure of IRF-2 in stickleback differed from other vertebrates, with the presence of nine exons. Analyses of exon sizes indicated that the sixth exon in other vertebrates might have split into two exons in stickleback (Figure 2).

Analyses of intron number and phase revealed a high degree of conservation in IRF-2 genes in vertebrates. However, the third and the sixth exons in other vertebrates were interrupted by a new phase 0 and phase 1 intron in anole lizard and stickleback, respectively (Table 1). Comparisons of the exon-intron structure and the intron phase between IRF-1 and IRF-2 in vertebrates imply that they might have been derived from a common ancestor.

IRF-3

The IRF-3 genomes varied considerably in their size, ranging from 2868 to 14802 bp and in the exon-intron organization (Figure 3). A conserved structure was observed in the size and organization for IRF-3 in human and mouse. However, it appears possible that the first exon is missing in dog IRF-3 when compared with that in human and mouse. With further searching, sequence gaps were found in the upstream of coding region in dog IRF-3; so whether the first exon is missing in dog IRF-3 requires further sequencing work. In zebrafish and frog, the IRF-3 had eight and nine exons respectively, and in both stickleback and anole lizard it has ten exons (Figure 3). Surprisingly, IRF-3 was not found in chicken.

Despite the variation in the exon-intron organization, phases for the first four introns of IRF-3 were the same in vertebrates from fish to mammals, except the difference in phases of the third and fourth introns in zebrafish IRF-3 and the sequence gap in dog IRF-3 (Table 1).

IRF-4

The genome of IRF-4 had a high degree of similarity in vertebrates from anole lizard to mammals in terms of the number of exons and their size (Figure 4). Several exons, such as the second, the fifth and sixth exons were of the same size in anole lizard, chicken, mouse and human. The size of exons was completely the same in human and mouse except the last one, and the IRF-4 in anole lizard and chicken was quite similar in the number of exons and their size. IRF-4 in dog had an additional exon, and the first two exons in dog might have been resulted from the separation of the first exon in other mammals. The genome of IRF-4 in frog and fish showed marked diversification when compared with IRF-4 in other vertebrates, with the presence of seven exons in frog.

Unexpectedly, three IRF-4-like genes were identified in zebrafish, which were named as the followings: fish species abbreviation-IRF-4-chromosome location-percent identity to human IRF-4-public protein id. For example, ddIRF-4-chr2-60.7%-EN6560, indicates zebrafish IRF-4 located on chromosome 2 with 60.7% identity to human IRF-4 and Ensembl database id ENSDARG0000006560, and other two IRF-4 were named as: ddIRF-4-chr20-1-52.0%-EN55374, ddIRF-4-scafNA1075-36.3%-EN35766. However, only two IRF-4 genes were observed in other teleost fish, such as in stickleback, gaIRF-4-groupIII-56.8%-EN16461, gaIRF-4-groupVIII-52.7%-EN4966; and in medaka, olIRF-4-chr4-56.7%-EN12712, olIRF-4-chr17-44.7%-EN17242; and in fugu, trIRF-4-scaf107-49.7%-EN2946, and trIRF-4-scaf305-52.5%-EN11568. The three IRF-4 genes in zebrafish varied in their size and genome organization, and were remarkably different from their mammalian counterparts in terms of exon sizes and intron phases (Table 1) although the whole gene structure of ddIRF-4-scafNA1075 could not be identified due to the sequence gaps in the upstream of coding region. However, the stickleback IRF-4-2 was probably comparable with the zebrafish IRF-4-C2 (Figure 2).

The intron phases were quite conserved in IRF-4 genes from anole lizard to mammals with introns 1, 3 and 7 in phase 0, and introns 2, 4, 5, 6 in phase 1 except the bias in dog which might have resulted from the insertion of the first exon in other vertebrates by a phase 0 intron. Of particular interest is the variation in intron phases in multi-copy genes of IRF-4 in teleost fish (Table 1).

IRF-5

One IRF-5 gene was found in vertebrates from fish to mammals (Figure 5). IRF-5 in mammals was comparable in genome structure and exon size. The frog IRF-5 had also eight exons as their mammalian counterparts, but the zebrafish IRF-5 had nine exons, with the size of the first two exons equivalent to the size of the first exon in frog IRF-5. The stickleback IRF-5 also had nine exons, thus comparable with the zebrafish. The IRF-5 in anole lizard was located in scaffold_7080, but was too short to characterize its gene structure. In chicken, IRF-5 was located at an unmapped region, with incomplete sequence information (Figure 5).

Comparisons of vertebrate IRF-5 exon/intron boundaries revealed a common intron position distribution from frog to mammals, with the introns 1, 3 and 7 in phase 0, and introns 2, 4, 5, 6 in phase 1 (Table 1), despite that most exons could not be characterized in the IRF-5 in anole lizard and chicken because of the incomplete genome sequences. The analyses of the genome and intron phase revealed that the first exon in other vertebrates was interrupted by a phase 0 intron in IRF-5 of zebrafish; and in addition, the IRF-5 in stickleback shared some characters in intron phases with IRF-5 in other vertebrates (Table 1).

IRF-6

The genome organization and exon sizes were quite conserved for IRF-6 in vertebrates from fish to mammals (Figure 6). In general, the gene consisted of 7 exons with 6 intervening introns, except that in chicken the IRF-6 had an eight exon structure, with the second exon in other vertebrate split into two exons (Figure 6). It is worth emphasizing that frog has two identical IRF-6 genes (Figure 6).

The intron positions of IRF-6 were strictly conserved in vertebrates from fish to mammals. The first and the last introns are in phase 0, while all other introns are in phase 1 (Table 1). It is interesting to note that the two IRF-6 genes in frog have the same size and the same position for each homologous exons and introns respectively, with the difference only in the size of introns (Table 1 and Figure 6).

IRF-7

The genome size of IRF-7 varied considerably from 2412 bp in mouse to 24469 bp in anole lizard. The size of exons and also of introns was variable. However, the IRF-7 had ten exons in human, nine in mouse, anole lizard, while it had thirteen exons in chicken. However, the chicken IRF-7 had sequence gaps in database, and IRF-7 was incomplete in dog and in frog genome databases (Figure 7).

Surprisingly, the IRF-7 differed in the phase position from other IRF members, in having a phase 2 intron, not a phase 0 intron as in other IRF members, separating the first two exons which encode the DBD region, the most conserved region in IRFs. The number of introns and intron phases varied considerably in different IRF-7 genes.

IRF-8

The IRF-8, in general, had a similar exon-intron organization in vertebrates from fish to mammal, with eight exons and seven introns (Figure 8), despite that in anole lizard and frog the IRF-8 had seven exons. The comparison of these exons may indicate that the merged-together of two exons might have occurred in anole lizard. The first exon could not be identified due to the sequence gaps in the upstream of frog IRF-8 coding region (Figure 8).

Among eight IRF-8 sequences from fish to mammal, the intron phases had exactly the same distribution. The classes of introns 1, 3, and 7 are in phase 0, leaving the others in phase 1 (Table 1), except that the fifth intron in other IRF-8 might have lost in IRF-8 in anole lizard (Table 1 and Figure 8).

IRF-9

IRF-9 was present in fish, frog, anole lizard, and mammals, but not in chicken (Figure 9). It was comparable in the whole length, and exon number and size in mammals, although the dog IRF-9 had two closely adjacent exons at last separated by only 4 bp (GCTT), which is not consistent with the GT/AG sequence at the donor and acceptor sites of RNA splicing (Figure 9). The anole lizard IRF-9 consisted of eight exons with seven introns as its counterparts in mammals but with larger variation in the size of introns. In zebrafish and stickleback, the IRF-9 had nine and ten exons, respectively; but they were quite comparable. Surprisingly, the Basic Local Alignment Search Tool (BLAST) results of predicted frog IRF-9 protein had a lack of first sixty amino acids in the IRF DBD region, which may require further study (Figure 9).

The distribution in intron phases for IRF-9 was quite comparable between mammals and anole lizard, with the introns 1 and 3 and the last one in phase 0 and introns 2, 4, 5, 6 in phase 1. However, the similar distribution in intron phases and the comparable exon sizes between IRF-8 and IRF-9 may imply, at least partially, that these two IRFs have a common origin. On the other hand, IRF-9 in lower vertebrates showed a high level of variation in intron phases (Table 1).

IRF-10

IRF-10 was not found in human and mouse, but at least an eleven-exon genome was present in dog (Figure 10). The zebrafish IRF-10 had eight exons and seven introns. However, the IRF-10 in other vertebrates was all incomplete. Nevertheless, the identified intron-exon structure in frog showed considerable variation when compared with other orthologues in other vertebrates (Figure 10). It is also quite impossible to summarize any pattern in the distribution of the intron phases (Table 1).

IRF-11

A new IRF family differing from previously-known ten members was identified only in teleost fish (Figure 11). The zebrafish IRF-11 had an eight exon structure in genome, while the IRF-11 in other teleost fish had an incomplete sequence data. Above all, these IRF-11 had the first exons similar to those in IRF-1 and IRF-2. Due to the incomplete sequences, the intron phase distribution was only analyzed for zebrafish IRF-11, with the findings of introns 1 and 6 in phase 0 and all others in phase 1 (Table 1).

IRF-like genes in other deuterostomes

The survey of several non-vertebrate deuterostome genomes may have broad implication in understanding the primitive state of vertebrate nervous, immune, and even cardiovascular systems [20, 21]. Recent studies have revealed several types of immune-related gene families in these deuterostomes, such as Toll-like receptors and Ig superfamily, providing insight into the origin of immune system in chordates and vertebrates [22, 23, 24]. Despite a recent effort to identify IRF members in invertebrates [25], any possible IRF genes were identified in available databases of deuterostomes in the present study. Two IRF-like genes were identified in sea urchin *Strongylocentrotus purpuratus* (Figure 12), three in acorn worm *Saccoglossus kowalevskii* (Figure 13), nine in sea squirt *Ciona intestinalis* (Figure 14) and ten in lancelet *Branchiostoma florida* (Figure 15). Some of these genes were just sequence segments containing IRF DBD domains (Figures 13, 14, 15). The genomic structure of these IRF-like genes was further predicted. The genome organization varied among these genes, with little similarity detected. They showed almost no similarity to any IRF members of vertebrates in genome organization (Figures 1, 2, 3, 4, 5, 6, 7, 8, 9, 10, 11, 12, 13, 14, 15) and also in sequence (Additional file 1). These IRF-like genes also had almost no similarity in intron phases, with the only exception in sea squirt that the IRF-like-3q-2 and IRF-like-3q-3 have the same intron phases and comparable gene structure, differing from sea squirt IRF-like-3q-1 (Table 1).

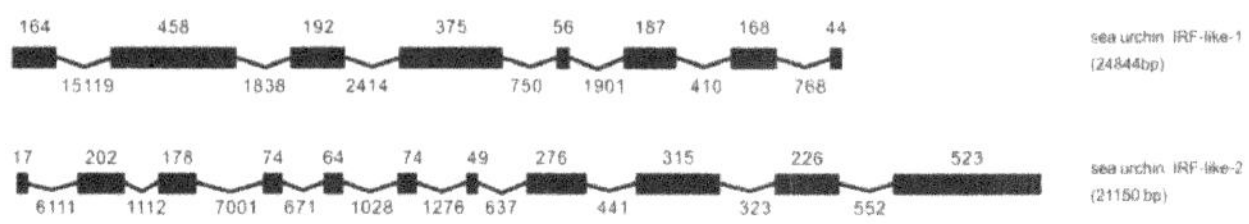

Figure 12. Predicted genomic structure of IRF-like genes in the sea urchin *Strongylocentrotus purpuratus*.

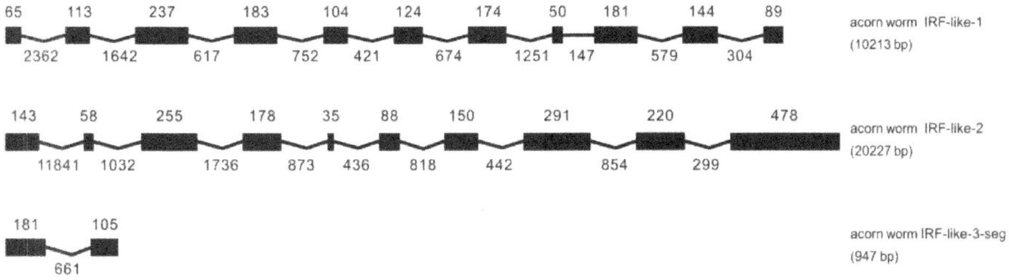

Figure 13. Predicted genomic structure of IRF-like genes and IRF-containing sequence segment in the acorn worm*Saccoglossus kowalevskii*. The sequence segment was indicated with -seg in the gene name, as also listed in Table 1 and in Additional file 2.

Figure 14. Predicted genomic structure of IRF-like genes and IRF-containing sequence segments in the sea squirt Ciona intestinalis. The sequence segment was indicated with -seg in gene names as also in Table 1 and in Additional file 2.

Figure 15. Predicted genomic structure of IRF-like genes and IRF-containing sequence segment in the lancelet*Branchiostoma floridae*. The segment was indicated with -seg, and also in Table 1 and in Additional file 2.

Synteny Analyses

IRF members were located by using BLAST algorithm to chromosomes or to scaffolds in frog and anole lizard as their genome wide sequence data is rather limited (Figures 16, 17, 18, 19, 20, 21, 22, 23, 24). The transcription orientation was well conserved in all IRF members, i.e. from IRF-1 to IRF-10 in all clearly identified genomic sequences for the vertebrates (Figures 16, 17, 18, 19, 20, 21, 22, 23, 24). The two IRF-6 genes in frog, which were connected together, were transcribed in two adverse directions (Figure 21). It is further worth noticing that IRF-3 and IRF-9 were spread on a same chromosome in zebrafish (Figure 18).

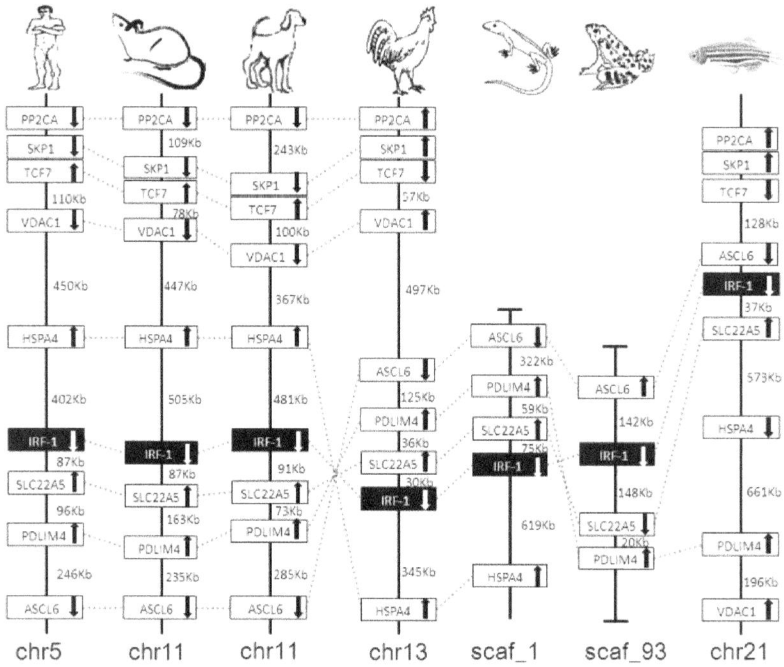

Figure 16. IRF-1 gene loci in human, mouse, dog, chicken, anole lizard, frog and zebrafish. Conserved genes are boxed, and vertical lines represent chromosomes; arrows indicate the transcription orientation of genes. Numbers on the side of the lines indicate the distance between genes. PP2CA: protein phosphatase 2 (formerly as 2A), catalytic subunit; SKP1: S-phase kinase-associated protein 1A; TCF7: transcription factor 7 (T-cell specific, HMG-box); VDAC1: voltage-dependent anion channel 1; HSPA4: heat shock 70 kDa protein 4; SLC22A5: solute carrier family 22 (organic cation transporter), member 5; PDLIM4: PDZ and LIM domain 4; ASCL6: acyl-CoA synthetase long-chain family member 6.

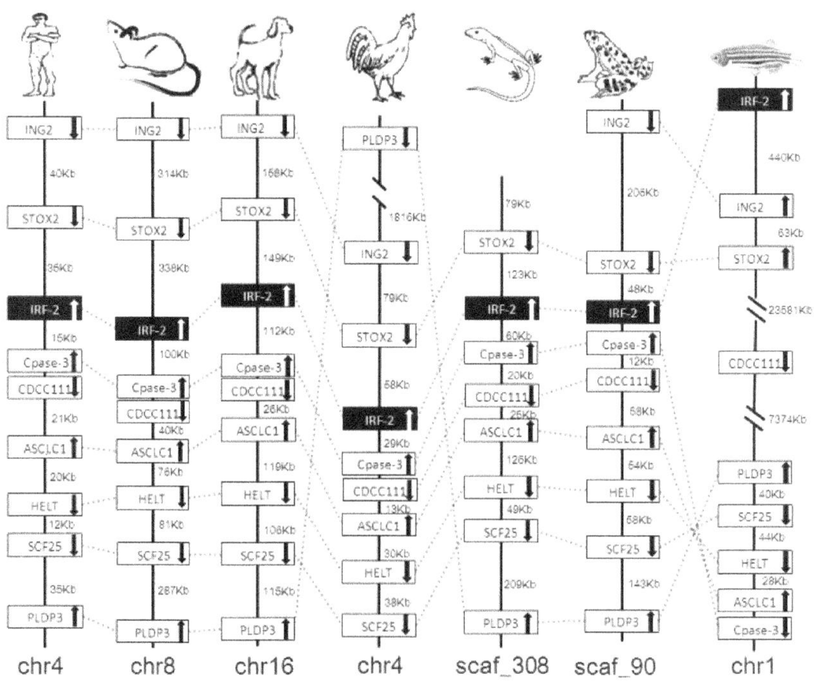

Figure 17. IRF-2 gene loci in human, mouse, dog, chicken, anole lizard, frog and zebrafish. The boxed genes, vertical lines and numbers on the side of lines in this figure and in following Figures 18 to 24 are as same as indicated in Figure 16. ING2: Inhibitor of growth protein 2; STOX2: storkhead box 2; Cpase-3: Caspase-3; CDCC111: Coiled-coil domain-containing protein 111; ASCLC1: acyl-CoA synthetase long-chain family member 1; HELT: Hey-like transcription factor; SCF25: ADP/ATP translocase 1/Solute carrier family 25 member 4; PLDP3: PDZ and LIM domain protein 3.

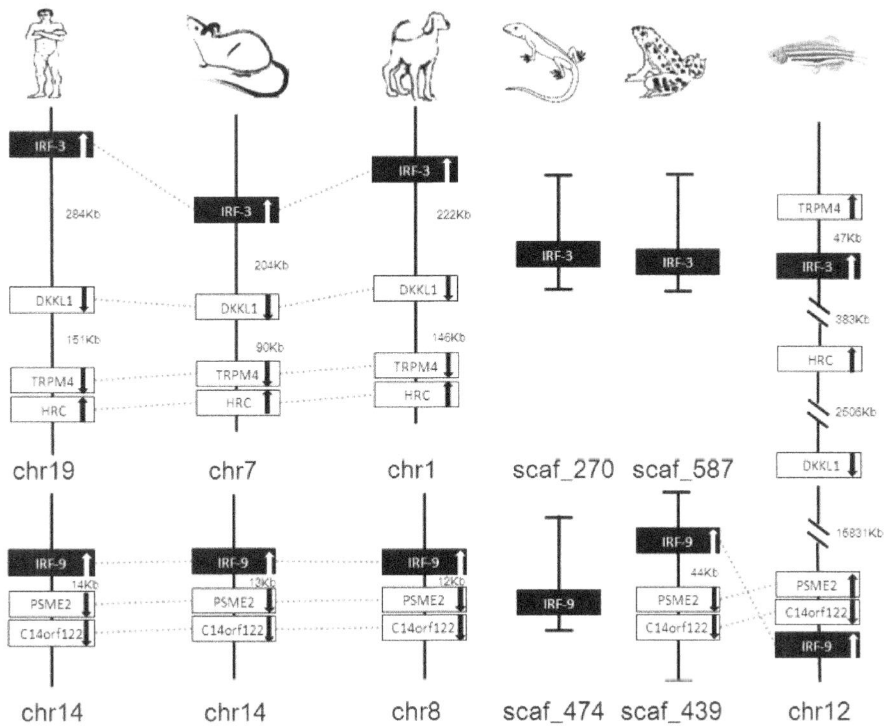

Figure 18. IRF-3 and IRF-9 gene loci in human, mouse, dog, anole lizard, frog and zebrafish. DKKL1: dickkopf-like 1; TRPM4: transient receptor potential cation channel, subfamily M, member 4; HRC: histidine rich calcium binding protein; PSME2: proteasome (prosome, macropain) activator subunit 2 (PA28 beta); C14orf122: chromosome 14 open reading frame 122.

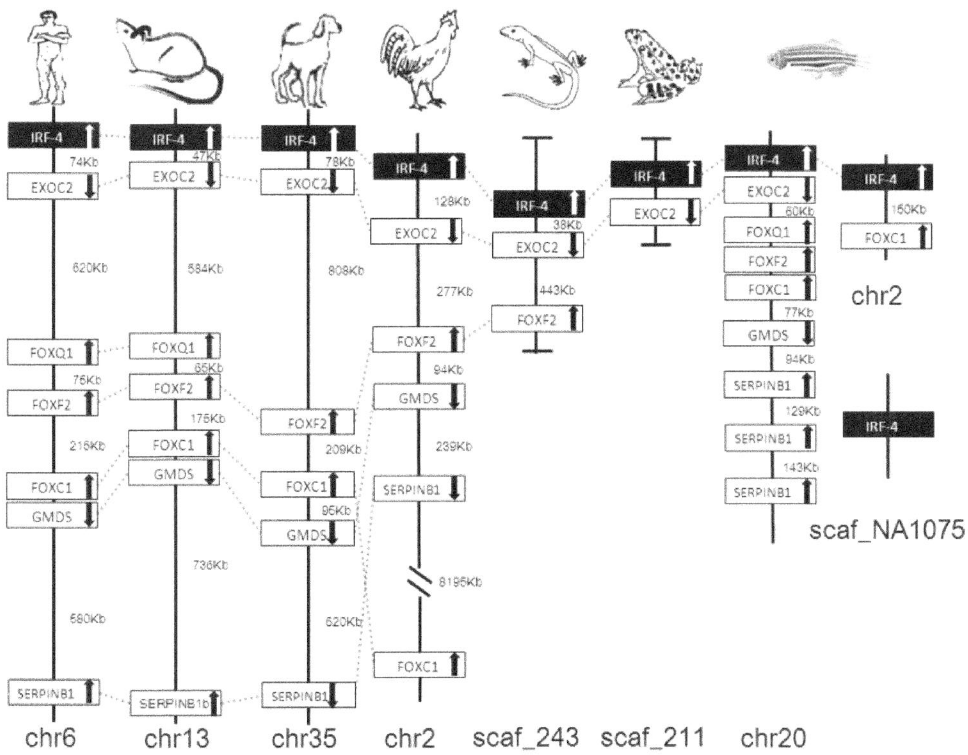

Figure 19. IRF-4 gene loci in human, mouse, dog, chicken, anole lizard, frog and zebrafish. EXOC2: exocyst complex component 2; FOXQ1: forkhead box Q1; FOXF2: forkhead box F2; FOXC1: forkhead box C1; GMDS: GDP-mannose 4,6-dehydratase; SERPINB1: serpin peptidase inhibitor, clade B (ovalbumin), member 1.

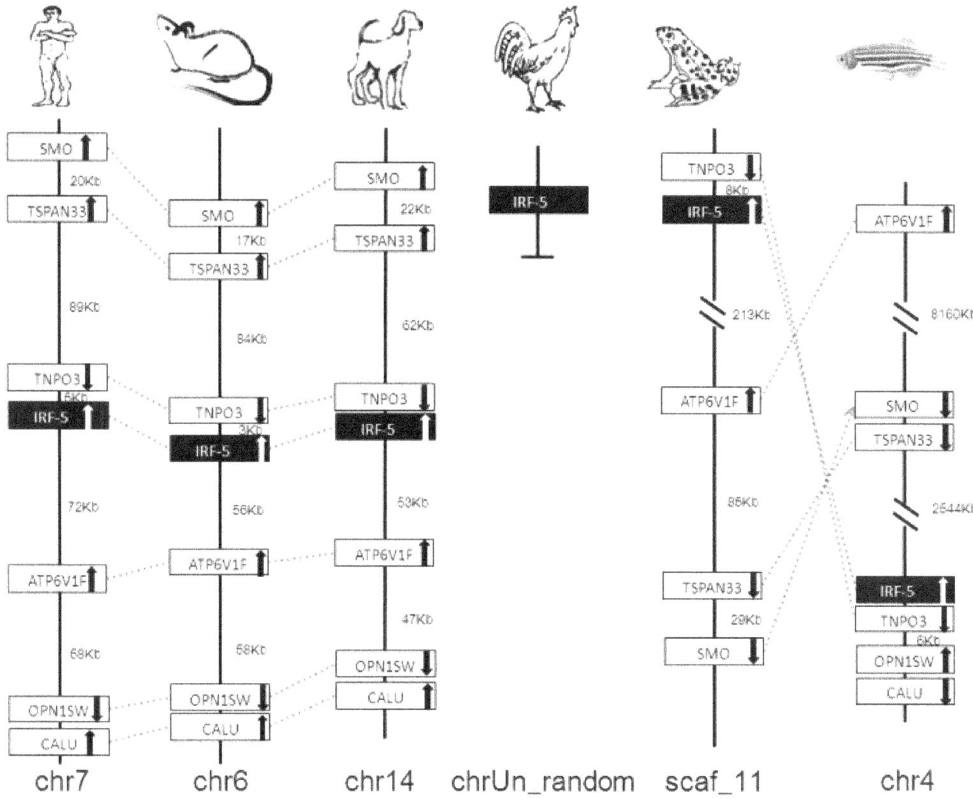

Figure 20. IRF-5 gene loci in human, mouse, dog, chicken, anole lizard, frog, and zebrafish. SMO: smoothened homologues; TSPAN33: tetraspanin 33; TNPO3: transportin 3; ATP6V1F: ATPase, H+ transporting, lysosomal 14 kDa, V1 subunit F; OPN1SW: opsin 1 (cone pigments), short-wave-sensitive (color blindness, tritan); CALU: calumenin.

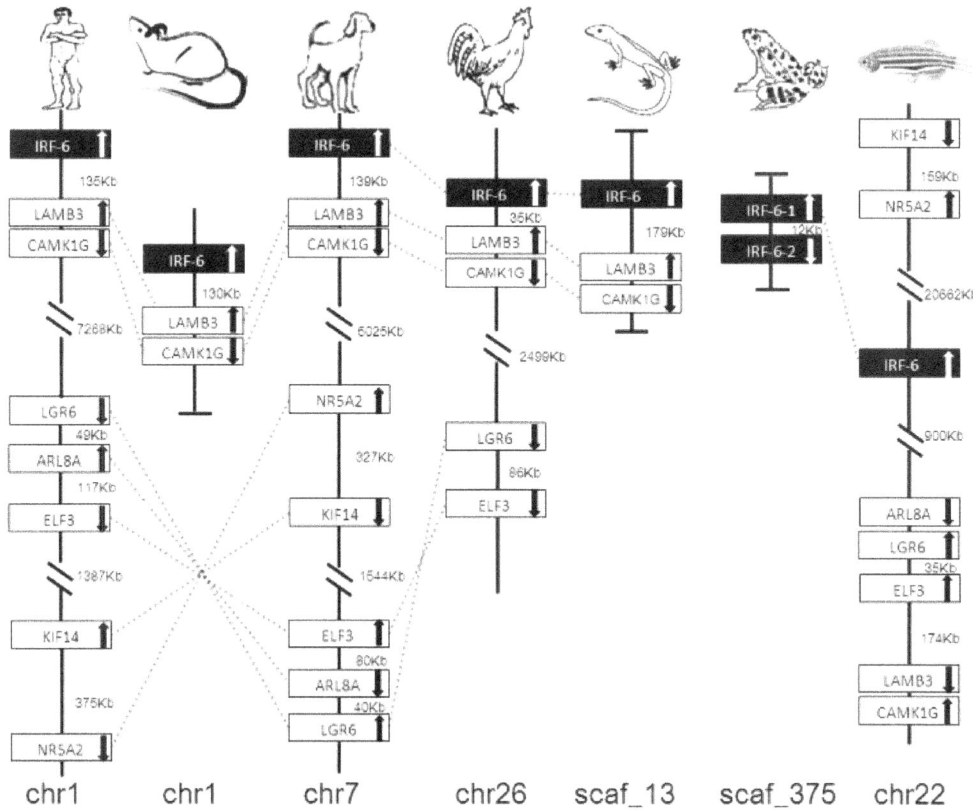

Figure 21. IRF-6 gene loci in human, mouse, dog, chicken, anole lizard, frog and zebrafish. LAMB3: laminin, beta 3; CAMK1G: calcium/calmodulin-dependent protein kinase IG; LGR6: leucine-rich repeat-containing G protein-coupled receptor 6; ARL8A: ADP-ribosylation factor-like 8A; ELF3: E74-like factor 3; KIF14: kinesin family member 14; NR5A2: nuclear receptor subfamily 5, group A, member 2.

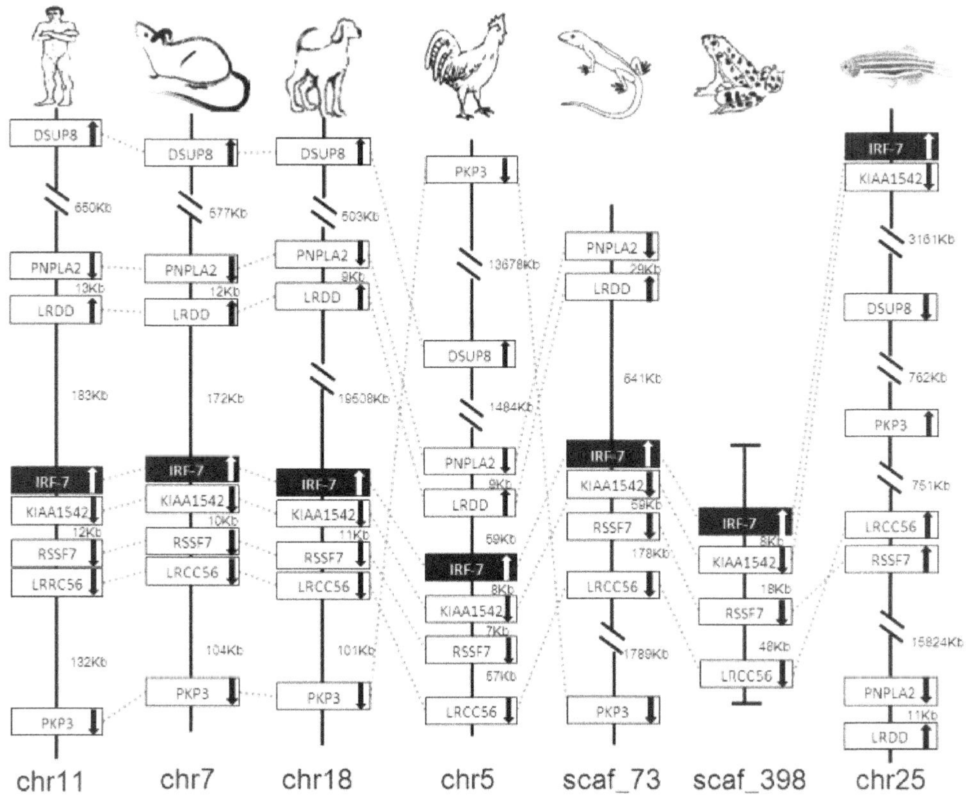

Figure 22. IRF-7 gene loci in human, mouse, dog, chicken, anole lizard, frog and zebrafish. DSUP8: dual specificity phosphatase 8; PNPLA2: patatin-like phospholipase domain containing 2; LRDD: leucine-rich repeats and death domain containing; KIAA1542: CTD-binding SR-like protein rA9, RSSF7: Ras association (RalGDS/AF-6) domain family 7, LRRC56: leucine rich repeat containing 56, PKP3: plakophilin 3

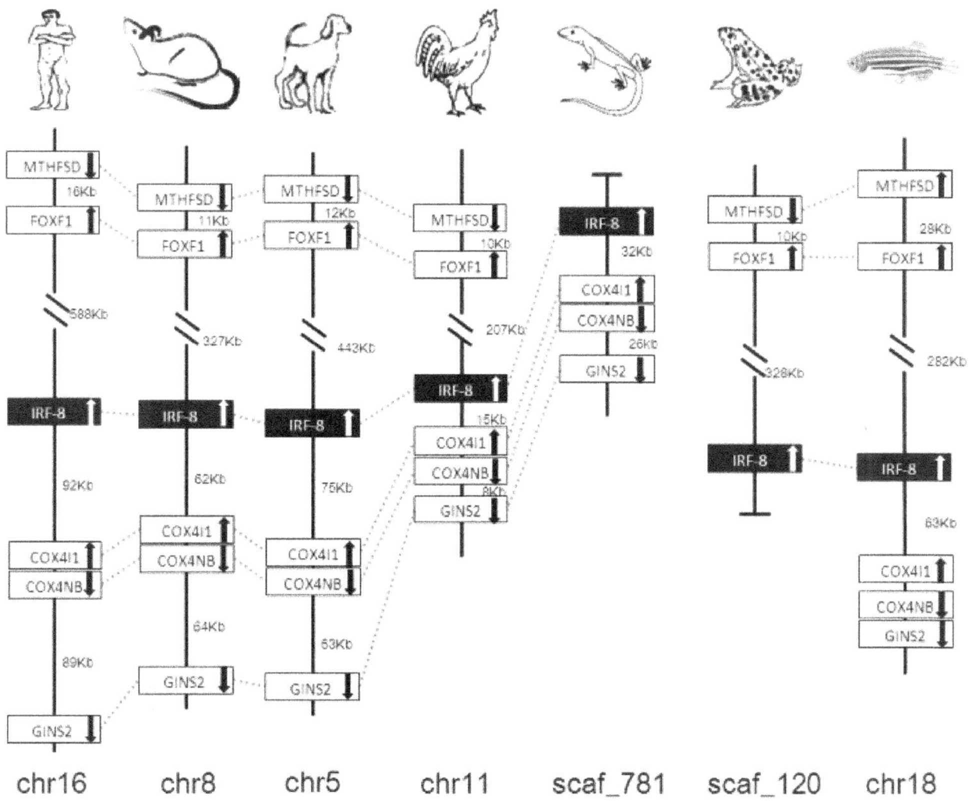

Figure 23. IRF-8 gene loci in human, mouse, dog, chicken, anole lizard, frog and zebrafish. MTHFSD: methenyltetrahydrofolate synthetase domain containing; FOXF1: forkhead box F1; COX4I1: cytochrome c oxidase subunit IV isoform 1, COX4NB: COX4 neighbor; GINS2: GINS complex subunit 2 (Psf2 homologue).

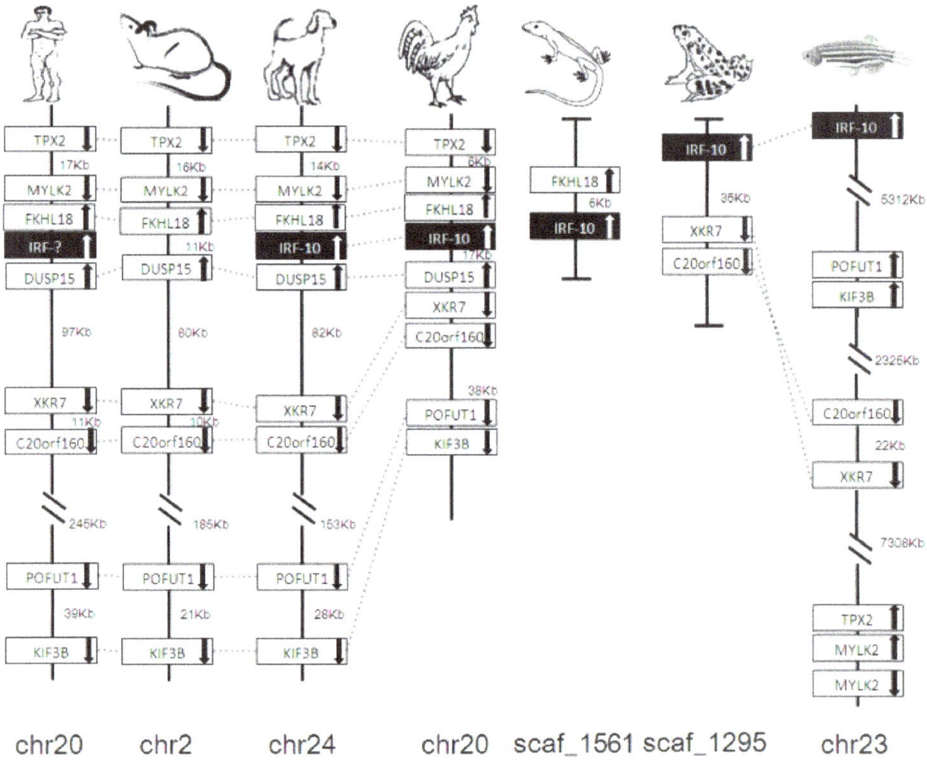

Figure 24. IRF-10 gene loci in human, mouse, dog, chicken, anole lizard, frog and zebrafish. TPX2: microtubule-associated protein homologue (*Xenopus laevis*); MYLK2: myosin, light polypeptide kinase 2, skeletal muscle; FLKHL18: forkhead-like 18 (*Drosophila*), DUSP15: dual specificity phosphatase-like 15; XKR7: X Kell blood group precursor related family member 7 homologue, C20orf160: chromosome 20 open reading frame 160, POFUT1: protein O-fucosyltransferase 1, KIF3B: kinesin family member 3B.

There is certain degree of conservation in the IRF loci in vertebrates (Figures 16, 17, 18, 19, 20, 21, 22, 23, 24). The genes adjacent to IRF genes were comparable. The arrangement or the genes close to IRF genes were conserved to a large degree in human, mouse and dog, and chicken (Figures 16, 17, 18, 19, 20, 21, 22, 23, 24). Genome sequence data in anole lizard and frog was rather limited, and thus synteny analysis was less informative in these vertebrates. Genes in IRF gene loci in zebrafish were overall comparable to their mammalian counterparts: Some loci were very similar; for example, the first gene next to IRF-4, and IRF-4-C20 in fish and to IRF-7 was the same in all vertebrates, and the IRF-8 loci were quite similar in all vertebrates. But, the conserved genes may be arranged in a much longer distance on chromosomes in fish, e.g., in IRF-2, IRF-3, IRF-5 and IRF-10 loci, and/or in different positions above or below the IRF gene, e.g., in IRF-1 loci. Despite the absence of IRF-10 gene in human and mouse, a segment encoding 210 aa was identified in an arrangement conserved in others of dog and chicken (Figure 24).

The so-called macro-synteny, i.e., comparison of gene linkage on chromosomes between amphioxus, the lancelet *Branchiostoma floridae* and human has revealed the existence of 17 linkage groups in the lancelet [24], and the locations of IRF-like genes have been identified in individual genome linkage groups, in comparison with IRF genes in different chromosomes of human [25]. But, IRF-2 could not be traced in these possible linkage groups in the present study, although Nehyba et al. [25] hypothesized its existence in linkage groups 6.

Percent Identity and Phylogeny

IRF- 1 to 9 in vertebrates were compared respectively with their counterparts in human by using Needleman-Wunsch global alignment. All identified IRFs in mouse and dog showed highest homology to their counterparts in human (Additional file 1). The percent identity appeared lower in lower vertebrates, and higher in higher vertebrates. Interestingly, some members such as IRF-2, IRF-4, IRF-6 and IRF-8, showed a higher degree of identity with their human counterparts (42.8% to 92.8%, 36.3% to 92.2%, 56.0% to 97.6, 31.8% to 91.8% respectively) (Additional file 1), implying that they may have some evolutionary conservation. Of all the IRF-10 identified in vertebrates, only avian IRF-10 showed some sequence similarity with the homologue in dog (41.3% identity, 49.9% similarity).

However, the percent identity of IRF-like genes or segments in non-vertebrate deuterostomes had less similarity to mammalian IRF members (Additional file 2). The highest similarity was below 43.6% between lancelet IRF-like-scaf187-2 and human IRF-8 (Additional file 2). In all of the four non-vertebrate deuterostomes, some identified or predicted IRF-like genes had a similarity above 30% to certain IRF members in human (Additional file 2).

The phylogenetic relationship of all IRF members including IRF-1 to IRF-11 in vertebrates and those IRF-like genes identified in all non-vertebrate deuterostomes in the preset study was shown in Figure 25. It is apparent that all well-characterized IRF members, i.e., those from IRF-1 throughout to IRF-10 were clustered into separate clades. None of IRF-like genes was clustered within the ten clades containing these already-known 10 IRF members (Figure 25). The IRF-like genes in lancelet and sea squirt were mostly clustered in each species. Interestingly, the acorn worm IRF-like-1, equivalent to SK1 in Nehyba et al. [25] (Additional file 2), was clustered with the clade containing IRF-1 and IRF-2. Furthermore, it is obvious that the zebrafish IRF-11 was clustered together with sea urchin IRF-like-1 (Figure 25).

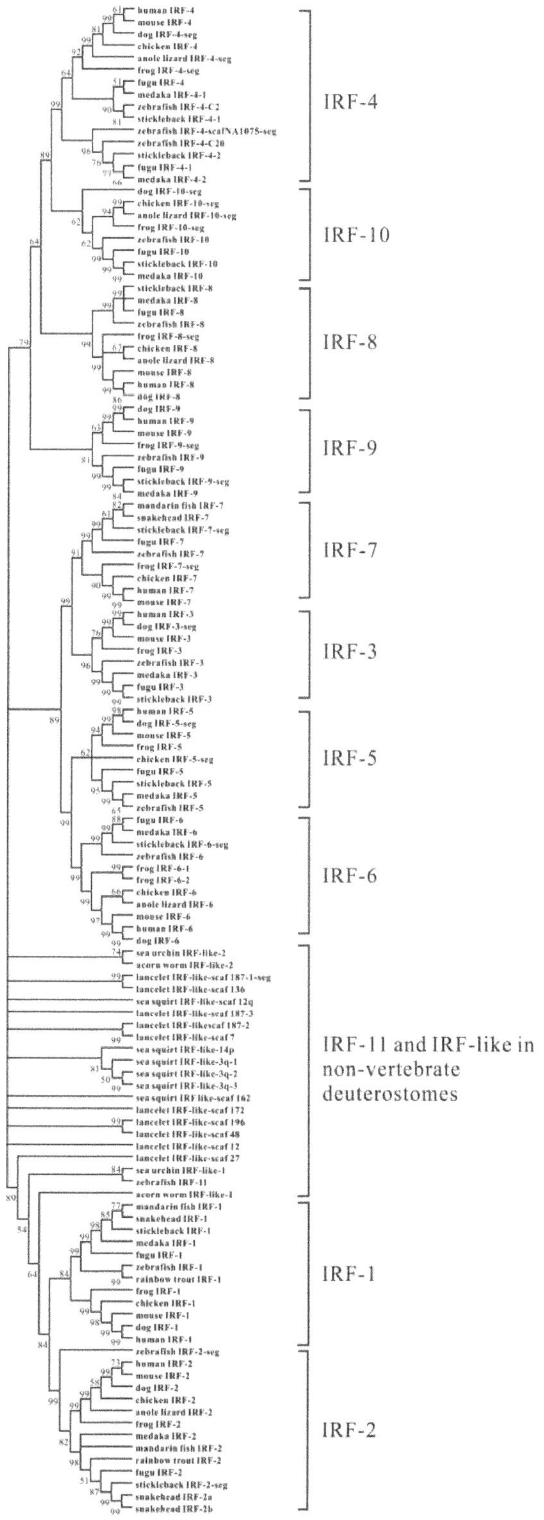

Figure 25. Phylogenetic analysis of IRF members in chordates by using neighbor-joining method within Mega4.0 program. Bootstrap values are indicated at nodes. Anole lizard IRF-5-seg, Dog IRF-7-seg and IRF-like gene segments in non-vertebrate deuterostomes were not included in the phylogenetic analysis due to the relatively short sequence lengths, and all IRF or -like genes shown in the tree are listed in Additional files 1 and 2.

DISCUSSION

All known IRF members, i.e., from IRF-1 to IRF-10 were identified in classes of vertebrates, i.e., piscine (except Chondrichthyans whose genomes are too scarce to analyze), amphibian, reptile, avian and mammal, despite that IRF-10 was not found in human and mouse, and that an additional IRF member was identified in teleost fish. The absence of IRF-3 and IRF-9 in chicken may be just a reflection of its incomplete genome sequences; alternatively these two genes may not exist in chicken at all. Overall, it is likely that vertebrates have a similar composition of IRF family members. In a recent study to investigate the evolution of IRF-like genes in invertebrates, Nehyba et al. [25] concluded that IRF family evolved distinctly in different taxonomical groups within Bilateria, as in several groups of animals the IRF family developed to a large number of family members. However, the comparable number of IRF members in classes of vertebrates may thus reflect a higher degree of conservation, and probably a similar pattern of evolution and function for these IRFs in vertebrates.

The complexity and diversity of immune systems in vertebrates increased throughout evolution, and new genes arose at various time points by gene fission and fusion [26], and retrotransposition and duplication during evolution [27]. The sequencing in whole genomes of various organisms has been providing opportunities to trace the origin of genes which comprise the immune system in vertebrates [28, 29]. It is reported that the IRF genes originated with the appearance of mulitcellular animals are present in almost all metazoan groups. In sponges and placozoans, the number of IRF family members is two, and five in cnidarians; and in several groups of animals such as in molluscan, cephalochordate, tunicate and vertebrate, the number might have evolved independently [25].

IRF members in human may be separated into IRF-1 and IRF-4 subgroups, with the former having a C-terminal associated domain (IAD2) and the later a different C-terminal associated domain (IAD1) [3, 30]. Using DBDs and IAD1 respectively, rather than the entire IRF proteins as in the present study, Nehyba et al. [25] were able to separate IRF-like genes into IRF-1 and IRF-4 groups in bilaterians, with four genes (as SK1, SP1, BF4, CI1) identified from each of the four species, acorn worm, lancelet (amphioxus), sea urchin and sea squirt being in the super group of IRF-1. In the present study, the acorn worm IRF-like-1, equivalent to SK1 (Additional file 2), has a close relationship with the big clade containing both IRF-1 and IRF-2, and the sea urchin IRF-like-1 (equivalent to SP1) is clustered with the clade containing IRF-1 and IRF-2. These two IRF-like genes may thus have a close evolutionary relationship with IRF-1 and IRF-2. The lancelet IRF-like-scaf_27 (equivalent to BF4) appeared relatively close with this clade. But, the CI1, equivalent to sea squirt IRF-like-seg-3 is just a segment (Figure 14) which was not included in the present phylogenetic study. However, all IRF members identified in vertebrates were clustered in four groups in the present phylogenetic tree, as reported by Nehyba et al. [10], i.e., IRF-1 group including IRF-1 and IRF-2, IRF-3 group including

IRF3 and IRF7, IRF-4 group including IRF-4, IRF-8, IRF-9 and IRF-10, and IRF-5 including IRF-5 and IRF-6. It can be concluded at least tentatively that the phylogenetic relationship of IRF-like genes in non-vertebrate deuterostomes with those in vertebrates remains, to a large extent, unsolved. Further sequence analysis as well as possible function analysis may be required to understand the true phylogenetic relationship of these IRF family members in chordates. If the view that IRF family evolved distinctly in different taxonomical groups by Nehyba et al. [25] were true in different groups of chordates, it might be rather difficult to trace back the origin and the true phylogeny of IRF members, unless enormous sequence data from a variety of organisms are gathered.

However, using the 17 chordate linkage groups constructed by Putnam et al. [24] for the comparison of lancelet and vertebrate genes, Nehyba et al. [25] located 10 human IRF genes, including a segment as syntenic to other vertebrate IRF-10 in the present study, to four linkage groups, each of which was related to one of the four groups of IRF family in vertebrates. But, they also considered that the distinct IRF genes in lancelet and their location in eight different linkage groups may simply represent a pattern of extensive independent evolution, and at the meantime they recognized that the lancelet IRF-like-scaf_27 gene (as BF4 in their study), which has some relationship with the IRF-1 and IRF-2 clade in the present study and was located in the sixth linkage group to which the human IRF-1 was traced, may be considered as the most likely gene linked to the predecessor of vertebrate IRF-1 gene.

It is now generally agreed that two-fold whole genome duplication might have occurred in early vertebrates after their separation from other deuterostomes [31]. Nehyba et al. [25] suggested that the generation of ten IRF genes in vertebrates might be resulted also from the duplication. On the basis of our data, a plausible model was proposed for the evolutionary history of IRF gene family (Figure 26). In this model, all IRF genes were assumed to have evolved from a common ancestor that might have undergone duplication and subsequent mutation or rearrangement prior to the divergence of urochordates, resulting in multiple IRF genes. Subsequent gene duplication of these ancestral IRF genes in some chordates and vertebrates might represent two separate evolutionary events. In vertebrates, the IRF gene might have undergone additional duplication at the time before fish diverged from their vertebrate ancestor. Polyploidization in vertebrates may have promoted such innovation while vertebrate ancestors probably only possessed single copy of gene found now in multiple copies in vertebrates [32]. Moreover, eleven to thirteen IRF genes were identified in jawed fish. Compared with other vertebrates, teleost fish have an additional member, IRF-11, which at least showed some similarity to the IRF-1 and IRF-2 in the first exons, indicating that genome, or gene and probably fragment duplication, as well as gene shuffling might have occurred in fish for the arising of this new member.

The eleventh member of IRF in fish has been reported in an early study [19], but all the IRF-11 identified previously is in fact the IRF-1 in the present study, as revealed by genome and synteny analyses in the present study. So, the IRF-11 identified in the present study is a real new member in the IRF family, and it has limited similarity with IRF-1 and IRF-2 as shown in their genome and in phylogenetic analysis.

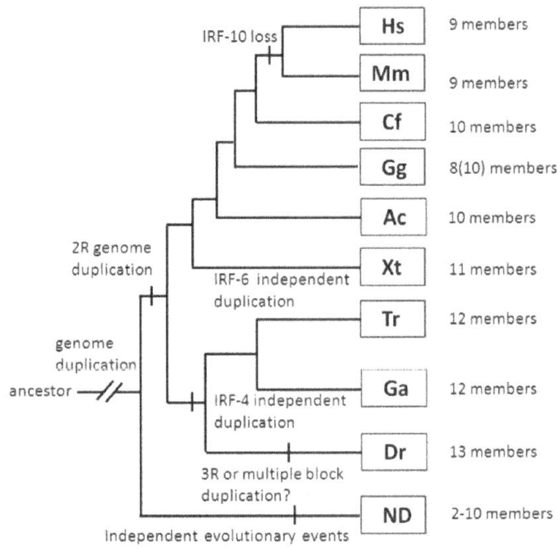

Figure 26. A cladistic model for the evolution of IRF family members. Two rounds of duplication of the whole genome are inferred to have occurred by the base of vertebrates, and tandem duplication, perhaps even an extra round of tetraploidization in zebrafish lineage. Subsequent independent gene duplication and/or loss events were marked on the tree. Hs: human *Homo sapiens*; Mm: mouse *Mus musculus*; Cf: dog *Canis familiaris*; Gg: chicken *Gallus gallus*; Ac: anole lizard *Anolis carolinensis*; Xt: frog *Xenopus tropicalis*; Tr: fugu *Takifugu rubripes* Ga: stickleback *Gasterosteus aculeatus*; Dr: zebrafish *Dario rerio*; ND: non-vertebrate deterostomes.

Interestingly, there are two IRF-4 genes present in most teleost species and three in zebrafish, but only one in higher vertebrates, suggesting that in teleost fish, especially in zebrafish tandem gene duplication or additional independent chromosome duplication might have occurred [33]. The similarity in the genome organization and intron phases in IRF-1 and IRF-2 may indicate that these two genes might have evolved from a common ancestor, by at least partial duplication or gene shuffling; and this may also be case for IRF-8 and IRF-9. The genomic organization of some IRF genes, such as IRF-5 in zebrafish (the first two exons), the middle exons of IRF-6 in chicken and dog, and the IRF-8 in frog and anole lizard might have also been resulted from gene shuffling, as also revealed by intron phase analyses. Furthermore, the copy of another IRF-6 downstream of the first IRF-6 gene in scaffold_375 in frog also provides evidence for the independent gene duplication. Regarding the IRF-10, higher vertebrate lineages might have lost the gene due to redundancy of function or compensation by other genes during evolution, given the conserved synteny of the IRF-10 locus in the genomes of vertebrates.

Nevertheless, vertebrates in different classes exhibited a high degree of conservation in the number of IRF family members, implying that the IRF family members might have evolved from common ancestors with similar evolutionary mechanisms, although multiple genes of IRF-4 and IRF-6, and even an additional IRF family member have evolved in lower vertebrates and IRF-3 and IRF-9, and IRF-10 have lost in chicken, and human and mouse, respectively. The genomic organizations for IRF-1, IRF-2, IRF-6, IRF-8 and IRF-9 and IRF-4 in higher vertebrates may be quite conservative. On the other hand, the comparison of substantial local sharing of neighbouring genes in different species may enable the understanding of phylogenetic relationships among IRF members, or even infer the genome evolution. Comparisons of IRF-4, IRF-8, IRF-9 and IRF-10 orthologues pertaining regions in vertebrates showed that these four members were co-localized with the Fox family member. These instances, therefore, were consistent with the study on Fox family. It seems likely that these four members were initially generated by an ancestral duplication [34]. The presence of conserved synteny but the lack of gene linkage in IRF orthologous regions in teleost fish when compared with other vertebrates, may suggest that inversions were more prevalent than translocations in the evolution of IRF locus, and at least some regions in teleost fish genomes might have undergone substantial intra-chromosomal rearrangements over evolution. Evidence for this conclusion comes from the comparison of IRF-2, IRF-3, IRF-5, IRF-6, IRF-7 and IRF-10 pertaining chromosome regions in vertebrates. The orders of these loci within chromosome segments in zebrafish are substantially rearranged. However, when viewed in IRF syntenic map in human, mouse, dog and chicken, most genes have obvious counterparts in the syntenic region with the same order. In these instances, a higher level of conservation in synteny suggests that similar selective forces might have been operating since these lineages diverged.

CONCLUSIONS

Ten members of IRF family, i.e., from IRF-1 to IRF-10 have been identified in vertebrates, although IRF-3 and IRF-9 were not found in chicken and IRF-10 not in human and mouse. However, lower vertebrates such as frog and fish have multiple genes of IRF-6 and IRF-4, respectively. Surprisingly, an additional number, IRF-11 was identified only in teleost fish. The genomes of some IRF family members, such as IRF-1, IRF-2, IRF-6, IRF-8 etc. have conserved structure in terms of exon-intron organization and also in the distribution of intron phases. The genes adjacent to IRF genes in vertebrates are quite comparable, especially in higher vertebrates; but it seems likely that conserved genes were spread in a much longer distance in chromosomes of zebrafish. The number of IRF family members in different classes of vertebrates might have been resulted from whole genome, or gene duplication, or even gene shuffling, and probably chromosome rearrangement especially in fish.

METHODS

Database Mining and Genome Analysis

Genomic sequences were downloaded from current assemblies within the Ensembl Database http://www.ensembl.org for human *Homo sapiens*, mouse *Mus musculus*, dog *Canis familiaris*, chicken *Gallus gallus*, anole lizard *Anolis carolinensis*, frog *Xenopus tropicalis*, zebrafish *Danio rerio*, stickleback *Gasterosteus aculeatus*, medaka *Oryzias latipes*, fugu *Takifugu rubripes*, and sea squirt *Ciona intestinalis*, the Department of Energy Joint Genome Institute database for lancelet (amphioxus) *Branchiostoma floridae* http://genome.jgi-psf.org/Brafl1/Brafl1.home.html, and National Center for Biotechnology Information http://www.ncbi.nlm.nih.gov for acorn worm *Saccoglossus kowalevskii*, sea urchin *Strongylocentrotus purpuratus* WGS sequences. Gene prediction was performed with FGENESH http://www.softberry.ru/berry.phtml and GENSCAN http://genes.mit.edu/GENSCAN.html. A quantitative sequence analysis with Hidden Markov Model (HMM) [35] was used to identify IRF genes. The DBD regions of IRF already dated in human and mouse were sorted according to their scores in HMMs trained on DBD motifs model and were used to search from the predicted gene database. All predicted IRF protein sequences were verified by BLASTP in the NCBI non-redundant protein sequence database and Ensembl database by using the BLASTP and TBLASTN programs, and any matches were used to refine prediction. The DNA region covering the query result of each predicted IRF protein coding region and the corresponding transcripts accessed by Ensemble database were extracted and the SIM4 program http://pbil.univ-lyon1.fr/sim4.php was used to reconstruct the exon-intron structure and calculate the intron phase. All IRFs and their genomic locations are listed in Additional files 1 and 2.

Pairwise Homology of all Examined IRFs

Sequence comparisons of human IRF proteins and other putative IRFs mined from the examined species were performed with Needleman-Wunsch global alignment by using the needle program from the EMBOSS package with default parameters (Gap opening penalty 10.0, Gap extension penalty 0.5) [36], and the percent identity of the top-scoring pair was obtained (Additional files 1 and 2).

Phylogenetic Analysis of IRFs

For molecular phylogenetic analyses, protein sequences were first aligned using ClustalW [37]. Phylogenetic analyses were performed using the neighbor-joining method within the Mega4.0 program [38]. Data were analyzed using Poisson correction, and gaps were removed by pairwise deletion. The bootstrap values of the branches were obtained by testing the tree 10,000 times.

Identification of Conserved Genes in IRF Syntenic Region from Human, mouse, dog, chicken, anole lizard, frog and zebrafish and Stickleback

To analyze the IRF syntenic region in vertebrates, human IRFs were used as anchor sites. Comparisons of the orthologues flanking the human anchor sites were accomplished by BLAST against the following NCBI genome assemblies: *Homo sapiens* (Build 36.3), *Mus musculus* (Build 37.1), *Canis familiaris* (Build 2.1), *Gallus gallus*(Build 2.1) on NCBI map viewer, and *Anolis carolinensis* (1.0), *Danio rerio* (Zv6) and *Xenopus tropicalis* (version 4.1) on Ensembl genome browser. Complete gene lists compiled from the syntenic region were obtained in these following regions: Human (chromosome5, 13137051-133589849; chromosome4, 184663214-186693650; chromosome9, 54346267-54860926; chromosome6, 336760-2787080; chromosome7, 128166672-128640622; chromosome1, 198263393-208046102; chromosome11, 384217-1549726; chromosome16, 84268781-85146317; chromosome14, 23678014-23705614; chromosome20, 29790565-30386475); Mouse (chromosome11, 51912326-54175041; chromosome8 46970839-48760512; chromosome7 52253030-52594343, chromosome13, 30841127-33186181; chromosome6, 29298119-29711359; chromosome1, 194979306-195196476; chromosome7, 148265677-149276175; chromosome8, 123105166-123632218; chromosome14, 56194511-56228867; chromosome2, 152673782-153157950); Dog (chromosome11, 23231763-25417744; chromosome16, 48072456-49811062; chromosome1, 109762805-110195996; chromosome35, 3765639-6182542; chromosome14, 10477159-10856486; chromosome7, 3729824-11424871; chromosome18, 28527750-48830961; chromosome5, 69307830-69996495; chromosome8, 7139778-7162946; chromosome24, 24192691-24676398); Chicken (chromosome13, 16330227-17520976; chromosome4, 40757647-41228574; chromosome2, 6766708-77050634; chromosome Un_random Contig1969; chromosome26, 316853-2993929; chromosome5, 1579936-17138835; chromosome11, 19359246-19654930; chromosome20, 9989995-10113464); Anole lizard (scaffold_1, scaffold_308, scaffold_270, scaffold_243, scaffold_7087, scaffold_13, scaffold_73, scaffold_781, scaffold_474, scaffold_1561); Frog (scaffold_93, scaffold_90, scaffold_587, scaffold_211, scaffold_11, scaffold_375, scaffold_398, scaffold_120, scaffold_439, scaffold_1295); Zebrafish (chromosome21, 44732965-48302889; chromosome1, 19605403-51285221; chromosome12, 3671869-4198128; chromosome2, 192032-182022; scaffold_NA1075, chromosome20, 19100202-31847658; chromosome4, 13813890-22022810; chromosome22, 198423-22346497; chromosome25, 5026720-25819085; chromosome18, 29472647-29955285; chromosome12, 22543362-22592525; chromosome23, 835503-15973618).

DECLARATIONS

Acknowledgments

This research was financially supported by a project (2009CB118703) of the National Basic Research Program of China (973 Program). We are very grateful to the three reviewers whose comments have significantly improved the structure and the expression of this manuscript.

REFERENCES

1. Mamane Y, Heylbroeck C, Genin P, Algarte M, Servant MJ, LePage C, DeLuca C, Kwon H, Lin R, Hiscott J: Interferon regulatory factors: the next generation. Gene. 1999, 237: 1-14. 10.1016/S0378-1119(99)00262-0.
2. Colonna M: TLR pathways and IFN-regulatory factors: to each its own. Eur J Immunol. 2007, 37: 306-309. 10.1002/eji.200637009.
3. Tamura T, Yanai H, Savitsky D, Taniguchi T: The IRF family transcription factors in immunity and oncogenesis. Annu Rev Immunol. 2008, 26: 535-584. 10.1146/annurev.immunol.26.021607.090400.
4. Taniguchi T, Ogasawara K, Takaoka A, Tanaka N: IRF family of transcription factors as regulators of host defense. Annu Rev Immunol. 2001, 19: 623-655. 10.1146/annurev.immunol.19.1.623.
5. Stellacci E, Testa U, Petrucci E, Benedetti E, Orsatti R, Feccia T, Stafsnes M, Coccia EM, Marziali G, Battistini A: Interferon regulatory factor-2 drives megakaryocytic differentiation. Biochem J. 2004, 377: 367-378. 10.1042/BJ20031166.
6. Battistini A: Interferon regulatory factors in hematopoietic cell differentiation and immune regulation. J Interferon Cytokine Res. 2009, 29: 765-780. 10.1089/jir.2009.0030.
7. Escalante CR, Yie J, Thanos D, Aggarwal AK: Structure of IRF-1 with bound DNA reveals determinants of interferon regulation. Nature. 1998, 391: 103-106. 10.1038/34224.
8. Fujita T, Sakakibara J, Sudo Y, Miyamoto M, Kimura Y, Taniguchi T: Evidence for a nuclear factor(s), IRF-1, mediating induction and silencing properties to human IFN-beta gene regulatory elements. EMBO J. 1988, 7: 3397-3405.
9. Barnes B, Lubyova B, Pitha PM: On the role of IRF in host defense. J Interferon Cytokine Res. 2002, 22: 59-71. 10.1089/107999002753452665.
10. Nehyba J, Hrdlickova R, Burnside J, Bose HR: A novel interferon regulatory factor (IRF), IRF-10, has a unique role in immune defense and is induced by the v-Rel oncoprotein. Mol Cell Biol. 2002, 22: 3942-3957. 10.1128/MCB.22.11.3942-3957.2002.
11. Klein U, Casola S, Cattoretti G, Shen Q, Lia M, Mo T, Ludwig T, Rajewsky K, Dalla FR: Transcription factor IRF4 controls plasma cell differentiation and class-switch recombination. Nat Immunol. 2006, 7: 773-782. 10.1038/ni1357.
12. Honda K, Taniguchi T: IRFs: master regulators of signalling by Toll-like receptors and cytosolic pattern-recognition receptors. Nat Rev Immunol. 2006, 6: 644-658. 10.1038/nri1900. Dougherty DC, Park HM, Sanders MM: Interferon regulatory factors (IRFs) repress transcription of the chicken ovalbumin gene. Gene. 2009, 439: 63-70. 10.1016/j.gene.2009.03.016.
13. Sun BJ, Chang MX, Chen DL, Nie P: Gene structure and transcription of IRF-2 in the mandarin fish *Siniperca chuatsi* with the finding of alternative transcripts and microsatellite in the coding region. Immunogenetics. 2006, 58: 774-784. 10.1007/s00251-006-0129-y.
14. Sun BJ, Chang MX, Song Y, Yao WJ, Nie P: Gene structure and transcription of IRF-1 and IRF-7 in the mandarin fish *Siniperca chuatsi*. Vet Immunol Immunopathol. 2007, 116: 26-36. 10.1016/j.vetimm.2007.01.001.
15. Zhang YB, Hu CY, Zhang J, Huang GP, Wei LH, Zhang QY, Gui JF: Molecular cloning and characterization of crucian carp (*Carassius auratus* L.) interferon regulatory factor 7. Fish Shellfish Immunol. 2003, 15: 453-466. 10.1016/S1050-4648(03)00025-1.
16. Yabu T, Hirose H, Hirono I, Katagiri T, Aoki T, Yamamoto E: Molecular cloning of a novel interferon regulatory factor in Japanese flounder, *Paralichthys olivaceus*. Mol Mar Biol Biotechnol. 1998, 7: 138-144.
17. Holland JW, Bird S, Williamson B, Woudstra C, Mustafa A, Wang T, Zou J, Blaney SC, Collet B, Secombes CJ: Molecular characterization of IRF3 and IRF7 in rainbow trout, *Oncorhynchus mykiss*: functional analysis and transcriptional modulation. Mol Immunol. 2008, 46: 269-285. 10.1016/j.molimm.2008.08.265.

18. Stein C, Caccamo M, Laird G, Leptin M: Conservation and divergence of gene families encoding components of innate immune response systems in zebrafish. Genome Biol. 2007, 8: R251-10.1186/gb-2007-8-11-r251.
19. Jiang Y, Doolittle RF: The evolution of vertebrate blood coagulation as viewed from a comparison of puffer fish and sea squirt genomes. Proc Natl Acad Sci USA. 2003, 100: 7527-7532. 10.1073/pnas.0932632100.
20. Cannon JP, Haire RN, Litman GW: Identification of diversified genes that contain immunoglobulin-like variable regions in a protochordate. Nat Immunol. 2002, 3: 1200-1207. 10.1038/ni849.
21. Huang S, Yuan S, Guo L, Yu Y, Li J, Wu T, Liu T, Yang M, Wu K, Liu H, et al.,: Genomic analysis of the immune gene repertoire of amphioxus reveals extraordinary innate complexity and diversity. Genome Res. 2008, 18: 1112-1126. 10.1101/gr.069674.107.
22. Putnam NH, Butts T, Ferrier DEK, Furlong RF, Hellsten U, Kawashima T, Robinson-Rechavi M, Shoguchi E, Terry A, Yu JK, et al.,: The amphioxus genome and the evolution of the chordate karyotype. Nature. 2008, 453: 1064-1071. 10.1038/nature06967.
23. Dehal P, Satou Y, Campbell RK, Chapman J, Degnan B, Tomaso DA, Davidson B, Gregorio DA, Gelpke M, Goodstein DM, et al.,: The draft genome of *Ciona intestinalis*: insights into chordate and vertebrate origins. Science. 2002, 298: 2157-2167. 10.1126/science.1080049.
24. Nehyba J, Hrdlickova R, Bose HR: Dynamic evolution of immune system regulators: the history of the interferon regulatory factor family. Mol Biol Evol. 2009, 26: 2539-2550. 10.1093/molbev/msp167.
25. Wang W, Yu H, Long M: Duplication-degeneration as a mechanism of gene fission and the origin of new genes in *Drosophila* species. Nat Genet. 2004, 36: 523-527. 10.1038/ng1338.
26. Babushok DV, Ostertag EM, Kazazian HH: Current topics in genome evolution: molecular mechanisms of new gene formation. Cell Mol Life Sci. 2007, 64: 542-554. 10.1007/s00018-006-6453-4.
27. Zou J, Chang M, Nie P, Secombes CJ: Origin and evolution of the RIG-I like RNA helicase gene family. BMC Evol Biol. 2009, 9: 85-10.1186/1471-2148-9-85.
28. Li G, Zhang J, Sun Y, Wang H, Wang Y: The evolutionarily dynamic IFN-inducible GTPase proteins play conserved immune functions in vertebrates and cephalochordates. Mol Biol Evol. 2009, 26: 1619-1630. 10.1093/molbev/msp074.
29. Qin BY, Liu C, Lam SS, Srinath H, Delston R, Correia JJ, Derynck R, Lin K: Crystal structure of IRF-3 reveals mechanism of autoinhibition and virus-induced phosphoactivation. Nat Struct Biol. 2003, 10: 913-921. 10.1038/nsb1002.
30. Kasahara M: The 2R hypothesis: an update. Curr Opin Immunol. 2007, 19: 547-552. 10.1016/j.coi.2007.07.009.
31. Yan Postlethwait YL, Gates MA, Horne S, Amores A, Brownlie A, Donovan A, Egan ES, Force A, Gong Z, et al.,: Vertebrate genome evolution and the zebrafish gene map. Nat Genet. 1998, 18: 345-349. 10.1038/ng0498-345.
32. Amores A, Force A, Yan YL, Joly L, Amemiya C, Fritz A, Ho RK, Langeland J, Prince V, Wang YL, et al.,: Zebrafish hox clusters and vertebrate genome evolution. Science. 1998, 282: 1711-1714. 10.1126/science.282.5394.1711.
33. Wotton KR, Shimeld SM: Comparative genomics of vertebrate Fox cluster loci. BMC Genomics. 2006, 7: 271-10.1186/1471-2164-7-271.
34. Eddy SR: Profile hidden Markov models. Bioinformatics (Oxford, England). 1998, 14: 755-763. 10.1093/bioinformatics/14.9.755.
35. Rice P, Longden I, Bleasby A: EMBOSS: the European Molecular Biology Open Software Suite. Trends Genet. 2000, 16: 276-277. 10.1016/S0168-9525(00)02024-2.
36. Thompson JD, Higgins DG, Gibson TJ: CLUSTAL W: improving the sensitivity of progressive multiple sequence alignment through sequence weighting, position-specific gap penalties and weight matrix choice. Nucleic Acids Res. 1994, 22: 4673-4680. 10.1093/nar/22.22.4673.
37. Tamura K, Dudley J, Nei M, Kumar S: MEGA4: Molecular Evolutionary Genetics Analysis (MEGA) software version 4.0. Mol Biol Evol. 2007, 24: 1596-1599. 10.1093/molbev/msm092.

CHAPTER 15 VERTEBRATE CARDIAC CONDUCTION SYSTEM

Neil C. Chi, Robin M. Shaw, Benno Jungblut, Jan Huisken, Tania Ferrer, Rima Arnaout, Ian Scott, Dimitris Beis, Tong Xiao, Herwig Baier, Lily Y. Jan, Martin Tristani-Firouzi, Didier Y. R. Stainier

INTRODUCTION

Vertebrate hearts have evolved into multichambered structures requiring coordinated beating of their chambers to achieve antegrade blood flow throughout the organism. Unidirectional blood flow is achieved through two specialized structures that are unique to vertebrates: cardiac valves and the specialized cardiac conduction system (CCS). In the adult heart, the initial electrical impulses are generated in the slow pacemaker sino-atrial (SA) node and then propagated across the atrium. This electrical impulse is delayed at the atrioventricular (AV) boundary through specialized slow conducting AV node cardiomyocytes. After the delay at the AV node, electrical propagation travels rapidly through the fast conduction network comprised of the His-Purkinje system, which coordinates ventricular activation to occur from the apex to the base of the heart. This apex-to-base activation allows for efficient ejection of blood from the ventricles into the outflow tracts (OFTs) at the base of the heart [1]. Despite extensive knowledge of the anatomy and physiology of the adult vertebrate CCS, the cellular and molecular events that govern the development of this specialized tissue remain unclear. Lineage tracing studies have revealed that the CCS is derived from cardiomyocyte progenitors [2,3]. Myocardial factors that regulate the specification of the CCS include Nkx2.5 and Tbx5 [2,4]. Loss of either transcriptional regulator leads to defects in the maturation and maintenance of the AV conduction system and subsequent AV heart block and bundle branch block.

Additional studies have revealed the requirement of the endocardium for cardiomyocyte specification to form the fast conduction network within the ventricle [5–7]. Secreted factors from endocardial as well as other cardiac endothelial cells, such as Endothelin 1 and Neuregulin, are able to induce cardiac conduction markers in cultured embryonic cardiomyocytes and cultured hearts [7–9]. Furthermore, hemodynamic changes regulate the secretion of Endothelin 1 from endocardial cells, thereby affecting the development of the fast conduction pathway [6]. More recently, the role of the endocardium for the development of AV conduction delay has been investigated further using the zebrafish *cloche* mutant [5], which lacks endothelial cells among other defects [10]. That study concluded that Neuregulin but not Endothelin 1 is required for the induction of AV conduction delay. Optical mapping of cardiac excitation using voltage- and calcium-sensitive dyes has allowed the spatiotemporal analysis of electrical excitation

Citation: Chi NC, Shaw RM, Jungblut B, Huisken J, Ferrer T, Arnaout R, et al. (2008) Vertebrate Cardiac Conduction System. PLoS Biol 6(5): e109. doi:10.1371/journal.pbio.0060109

wave dynamics, not only advancing our understanding of the electrical activity during cardiac arrhythmias but also allowing for further analysis of CCS development [11]. However, the use of voltage- and calcium-sensitive dyes is associated with serious shortcomings, including a lack of cellular targeting, limited live animal experimentation, the need for physical loading of these indicators into cells, and cellular toxicity. To circumvent these problems, fluorescent calcium indicator proteins have begun to replace voltage- and calcium-sensitive dyes for physiologic in vivo analysis of tissue/organ electrical activity in different animal model systems including fly and mouse [12–14]. Yet, optical mapping of mouse hearts is currently limited due to explantation for ex vivo analysis. Thus, we have taken advantage of the external fertilization and translucency of zebrafish embryos to create a cardiac-specific fluorescent calcium indicator transgenic line, *Tg*(*cmlc2*:*gCaMP*)^{s878}, to perform in vivo optical mapping analyses throughout the stages of heart development.

Here we describe a multidisciplinary approach using the zebrafish toward understanding CCS development. Utilizing the *Tg*(*cmlc2*:*gCaMP*)^{s878} optical mapping system, we identified four distinct physiologic developmental stages of the CCS that correspond to cellular and anatomical changes of the developing zebrafish heart. (1) Initially, a linear conduction travels across the heart tube from the sinus venosus to the OFT (20–24 hours postfertilization (hpf)); (2) next, a significant AV conduction delay develops during cardiac chamber formation (36–48 hpf); (3) as the heart loops and develops ventricular trabeculations (72–96 hpf), an immature fast conduction network develops within the ventricle; (4) finally, this fast conduction network fully matures to an apex-to-base activation pattern when the ventricular apex has formed.

Furthermore, to identify regulators of CCS development, we performed a diploid ethylnitrosourea (ENU) mutagenesis screen and recovered several novel as well as known cardiovascular conduction/rhythm mutants, which we have analyzed using in vivo optical imaging techniques and classified according to the affected physiologic developmental stage of the CCS. Positional cloning of *hobgoblin* (*hob*), a novel mutant with AV heart block, reveals that *tcf2*, a homeobox transcription factor gene involved in mature onset diabetes of the young, also regulates conduction between the atrium and the ventricle. Thus, these detailed electrophysiologic and genetic analyses of wild-type and mutant hearts provide further insights into the development of the vertebrate CCS and will lead to a better understanding of the pathogenesis of cardiac arrhythmias.

RESULTS

Optical Mapping of *Tg*(*cmlc2*:*gCaMP*)^{s878} Hearts Reveals Distinct Developmental Stages of the Vertebrate CCS

Previous studies have utilized calcium green, a calcium-sensitive fluorescent indicator, in zebrafish hearts to observe cardiac conduction up to 48 hpf [5,15]. However, because loading these hearts and/or embryos with calcium green is technically cumbersome and is only temporary, we created a zebrafish transgenic line *Tg*(*cmlc2*:*gCaMP*)^{s878} that specifically expresses *gCaMP*, a genetically encoded calcium reporter based on a circular permutation of green fluorescent protein (GFP) [16], at all developmental stages in the heart, using the cardiac-specific promoter *cmlc2* [17] (Figure S1).

Because cardiac contraction and blood flow begins at the linear heart tube (LHT) stage, we initiated our optical mapping studies on the LHT of 24 hpf *Tg*(*cmlc2*:*gCaMP*)^{s878} embryos. These experiments revealed that conduction travels unidirectionally in a relatively slow and linear pattern without significant pauses from the sinus venosus to the OFT, suggesting the presence of a functional SA node pacemaker activity (Figure 1A and 1B and Video S1). Additionally, an acceleration of conduction was observed within the OFT half of the heart (Figure 1B, arrow).

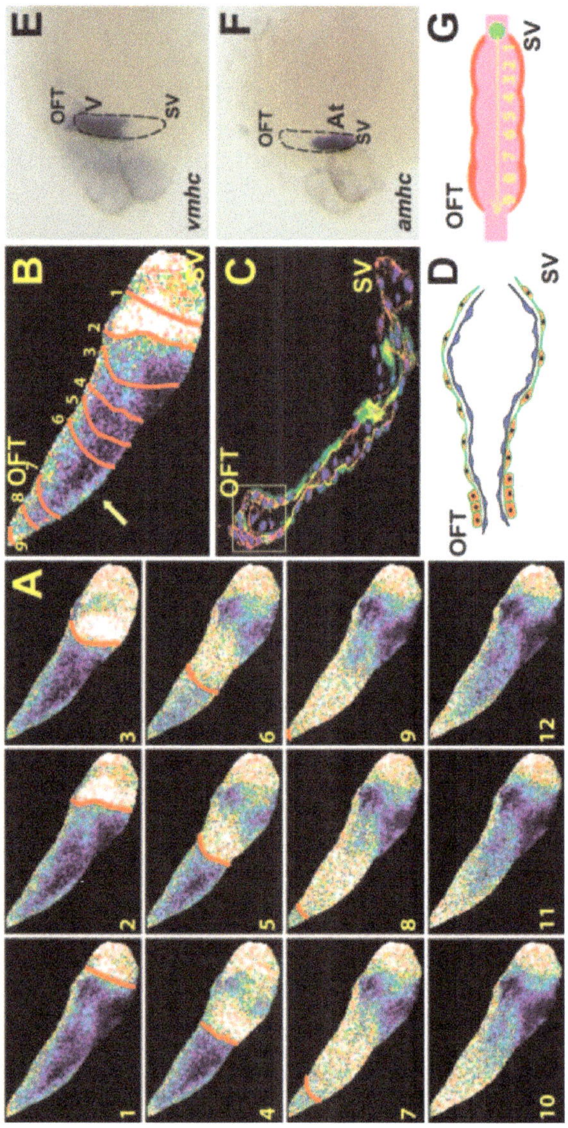

Figure 1. Cellular and In Vivo Electrophysiologic Analysis of 24 hpf Wild-Type Hearts

(A) Sequential calcium activation images of a 24 hpf wild-type *Tg*(*cmlc2*:*gCaMP*)^{s878}heart during a single cycle in a live zebrafish embryo.

(B) A 24 hpf optical map of calcium excitation represented by isochronal lines every 60 ms. Slow linear conduction with no significant delays is observed across the 24 hpf heart tube with increased conduction velocity over the presumptive ventricle (arrow).

(C) *Tg*(*cmlc2*:*eGFP-ras*)^{s883} (green) embryos at 24 hpf are stained with rhodamine phalloidin (red) and TOPRO (dark blue). The heart is a linear tube, and cardiomyocytes uniformly have a squamous shape except at the OFT (yellow box) where they have become cuboidal.

(D) Schematic representation of the heart shown in (C). Myocardium is in red. Endocardium is in blue.

(E and F) In situ analysis of 24 hpf embryos reveals molecular chamber specification within the heart tube: *atrial myosin heavy chain*, *amh*c, and *ventricular myosin heavy chain*, *vmhc*. Lateral views.

(G) Schematic representation of conduction in 24 hpf heart. Numbers indicate temporal sequence of calcium activation in the heart. Yellow arrow indicates direction of cardiac conduction. Green circle indicates slow conduction pathway/pacemaker activity.

SV, sinus venosus; V, ventricle; At, atrium.

Utilizing the *Tg(cmlc2:eGFP-ras)*^{s883} line (Jungblut B, Munson C, Huisken J, Trinh L, Stainier D, unpublished data), which outlines individual cardiomyocytes with membrane-bound GFP, confocal microscopy was performed to analyze further the cellular characteristics of the LHT. Despite displaying atrial and ventricular molecular changes [18] (Figure 1E and 1F), 24 hpf cardiomyocytes maintain a nearly uniform squamous morphology on cross-sectional analysis (Figure 1C). However, we observed a small subpopulation of cuboidal cardiomyocytes near the OFT (Figure 1C, box). These cuboidal cardiomyocytes correlate with the acceleration of conduction observed in the optical mapping of the LHT.

Cellular Changes of the Atrial, Ventricular, and AV Myocardial Cells Occur at 36–48 hpf and Correlate with AV Conduction Delay

By 36–48 hpf, the zebrafish embryonic heart has developed a distinct AV canal that separates the cardiac chambers (Figure 2C). Calcium activation travels from the sinus venosus across the atrium to the AV canal (Figure 2A and 2B). At the AV canal, a dramatic slowing of the calcium activation wave was observed as illustrated by an increased number of isochronal lines at the AV boundary between the atrial and the ventricular chambers (Figure 2B, arrowhead). Subsequently, ventricular calcium activation proceeded from the AV canal, accelerating laterally across the ventricular myocardium to the OFT where significant deceleration of conduction was observed (Figure 2B, arrow, and Video S2). These AV and OFT conduction delays may help to prevent regurgitant blood flow between the atrial and the ventricular chambers as well as between the heart and the bulbus arteriosus, respectively, thus resulting in efficient antegrade blood flow throughout the embryo.

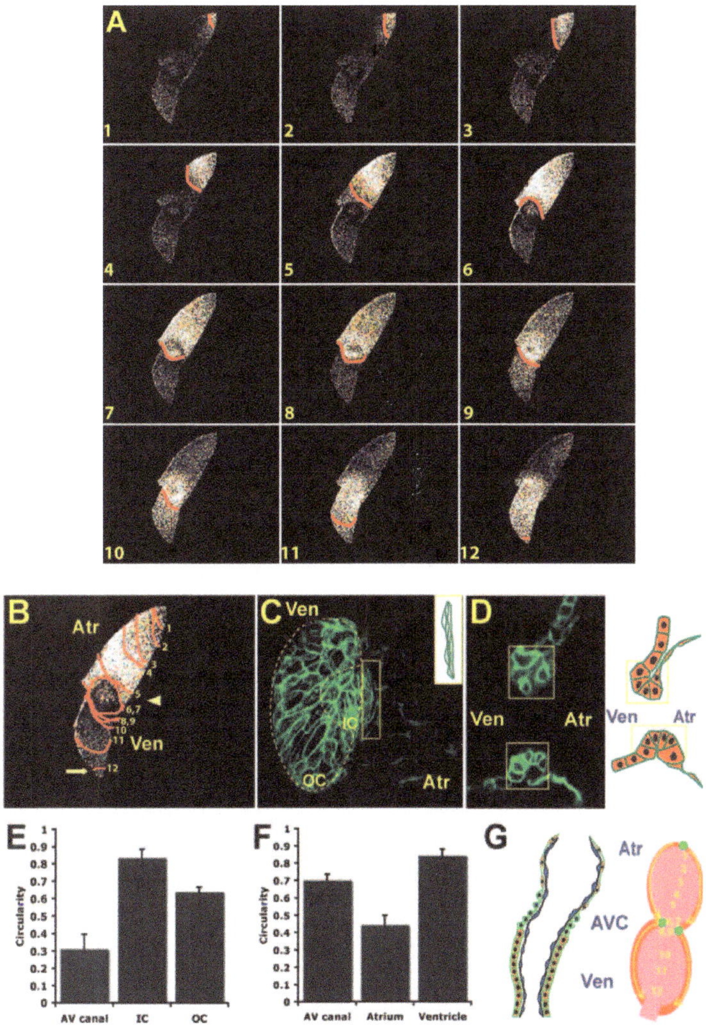

Figure 2. Development of AV Myocardium Results in AV Conduction Delay and Development of the Central CCS

(A) Sequential calcium activation images of a 48 hpf wild-type *Tg(cmlc2:gCaMP)*^{s878}heart during a single cycle in a live zebrafish embryo.

(B) The 48 hpf optical maps of calcium excitation represented by isochronal lines every 60 ms. Significant conduction delay was observed at the AV junction (arrowhead) and OFT (arrow).

(C) Surface and (D) cross-sectional analyses of *Tg(cmlc2:eGFP-ras)*^{s883} hearts at 48 hpf reveal distinct cellular morphologies and orientations among atrial, ventricular, and AV (yellow boxes) cardiomyocytes. Schematic illustration of AV canal region to the right. OC and IC are outlined on projection of ventricle in (C).

(E and F) Bar graphs represent cell morphology/circularity measurements of *Tg(cmlc2:eGFP-ras)*^{s883} hearts at 48 hpf: (E) surface of cardiomyocytes from the OC and IC of the ventricle as well as the AV myocardium and (F) atrial, ventricular, and AV canal cardiomyocytes on cross-sectional analysis.

(G) Schematic representation of a 48 hpf heart. Individual cardiomyocytes are outlined in green. Endocardium is in blue. Numbers indicate temporal sequence of calcium activation in the heart. Yellow arrows indicate direction of cardiac conduction. Green circles indicate slow conduction pathway/ pacemaker and AV conduction delay.

Atr, atrium; Ven, ventricle; AVC, AV canal.

To determine whether cell morphology or orientation may correlate with these conduction velocity differences, we analyzed and measured cardiomyocytes of 48 hpf *Tg(cmlc2:eGFP-ras)*^{s883} hearts. On cross-sectional analysis, it appeared that atrial cardiomyocytes had maintained their squamous cell morphology while ventricular cardiomyocytes had become cuboidal (Figure 2D and 2F) and cardiomyocytes at the AV boundary had initiated apical membrane constriction, resulting in cells with a distinct trapezoidal shape (Figure 2D, yellow box). Furthermore, examination of the surface of ventricular cardiomyocytes revealed that outer curvature (OC) cardiomyocytes (the greater, convex curvature of the cardiac chamber) became elongated while inner curvature (IC) cardiomyocytes (the lesser, concave curvature of the cardiac chamber) remained rounded (Figure 2C and 2E), as previously described [19]. Interestingly, cardiomyocytes at the AV boundary also became significantly elongated, forming a ring of cells around the AV canal (Figure 2C, yellow box and inset). The orientation of these AV cardiomyocytes was orthogonal to that of the OC ventricular cardiomyocytes.

To determine whether these different populations of myocardial cells exhibit distinct electrophysiologic properties, calcium transients of the atrium, ventricle, and AV canal were recorded from fluorescence of a single pixel from each region. Distinct calcium transients were recorded from each region of 48 hpf *Tg(cmlc2:gCaMP)*^{s878} hearts (Figure 3B and 3C). To further test these findings, action potentials (APs) were recorded in explanted 48 hpf wild-type hearts using patch pipettes and the current clamp technique (Figure 3D). The APs recorded from zebrafish atrium were similar in morphology to those reported in embryonic mammalian myocardium, including spontaneous diastolic depolarization. The APs recorded from zebrafish ventricle were typical of mammalian ventricles, with a flat diastolic phase and an overt systolic plateau phase. The APs recorded from the AV canal were distinguishable from atrial and ventricular APs by the presence of a slow diastolic depolarization phase in the setting of an overt systolic plateau phase. The morphology of the APs recorded from the wild-type AV canal was similar to that reported for the mammalian AV node [20], suggesting that the distinct electrophysiologic properties of AV myocardial cells contribute to the conduction delay between the cardiac chambers.

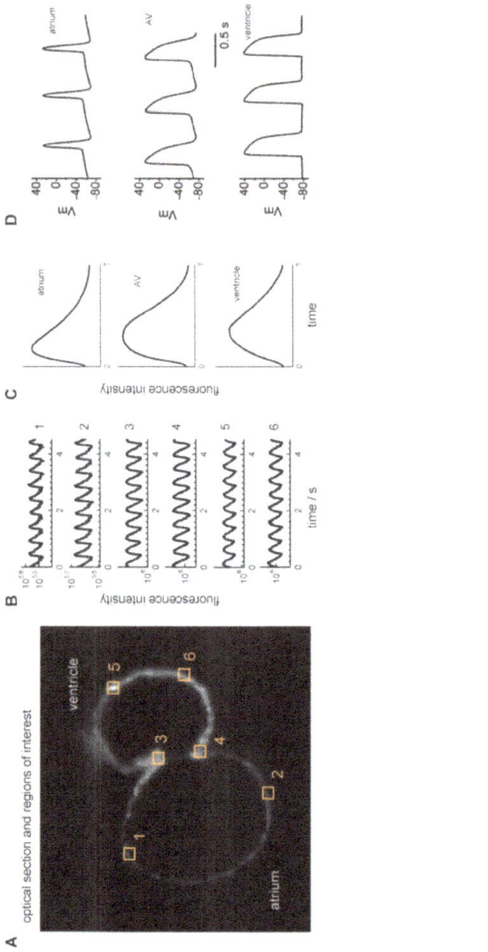

Figure 3. Calcium Transients and APs Confirm That AV Myocardium Has Distinct Electrophysiologic Properties That Distinguish It from Atrium and Ventricle

(A) Optical section of 48 hpf *Tg (cmlc2:gCaMP)*^{s878} heart. Numbers represent areas where calcium transients for the atrium, ventricle, and AV canal were recorded.

(B) Fluorescence intensity of a single pixel from each region was recorded to obtain calcium transients and plotted over time in seconds. All plots are semilogarithmic and identically scaled.

(C) Average calcium transient for each cardiac region.

(D) The APs from each of these regions were recorded in wild-type explanted 48 hpf hearts using patch pipettes and current clamp techniques. Distinct calcium transients and APs were detected in atrium, AV canal, and ventricle.

Development of the Rapid Ventricular CCS Occurs at 100hpf when Hearts Develop Ventricular Trabeculation

To achieve efficient ejection of blood from the ventricle to the arterial system, the adult vertebrate ventricle contracts from the apex to the base of the heart (i.e., where the AV canal and OFT reside). This contraction pattern is achieved through the fast cardiac conduction network (the His-Purkinje system), which passes through the ventricular septum in amniotes and allows for apex-to-base conduction across the ventricular myocardium. Previous studies have shown that the fast CCS in chick and mammalian hearts may initially develop along ventricular trabeculae after cardiac looping but prior to ventricular septation [6,21]. To understand further how the fast CCS develops, we analyzed zebrafish hearts at stages after cardiac looping.

Using the *Tg(flk1:eGFP)*^{s843} line, which marks endocardial cells with green fluorescence [22], we observed that rhodamine-phalloidin–stained 100 hpf zebrafish hearts not only have completed cardiac looping but also have initiated ventricular trabeculation (Figure 4C). By 2–3 weeks postfertilization, the ventricle has developed an apex and increased its trabeculation (Figure 4G).

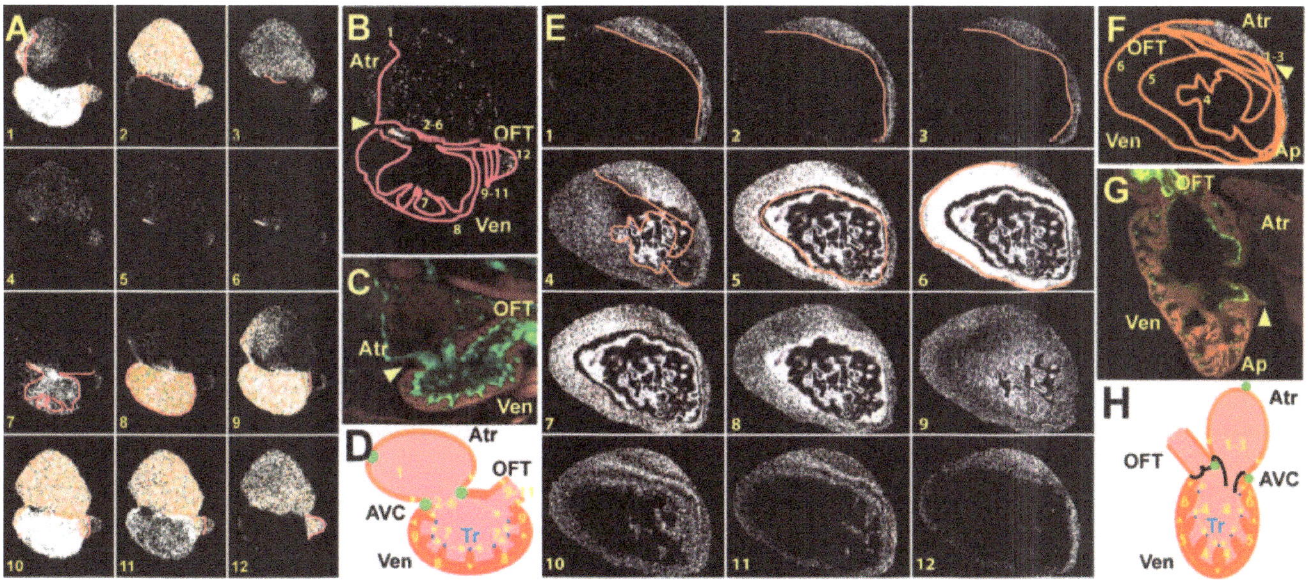

Figure 4. Development of the Fast Specialized Cardiac Conduction Tissue

Cellular and electrophysiologic analysis of wild-type *Tg(cmlc2:gCaMP)*^{s878} hearts at 100 hpf and 21 dpf.

(A and E) Sequential calcium activation images of a 100 hpf (A) *Tg(cmlc2:gCaMP)*^{s878}heart and a 21 dpf (E) *Tg(cmlc2:gCaMP)*^{s878} ventricle during a single cycle in a live zebrafish.

(B and F) The 100 hpf (B) and 21 dpf (F) optical maps of calcium excitation represented by isochronal lines every 60 ms. Ventricular calcium activation initiates along trabeculae at 100 hpf and 21 dpf and breaks through the rest of the myocardium propagating from OC to base at 100 hpf and apex to base at 21 dpf.

(C and G) Confocal images of the heart at 100 hpf and 21 dpf. (C) The 100 hpf *Tg(flk1:EGFP)*^{s843} (green) larvae stained with rhodamine phalloidin (red) reveal that the hearts have completed cardiac looping and that the ventricles have formed trabeculae.(G) At 21 dpf, the ventricle forms a distinct apex.

(D and H) Schematic representation of the hearts shown in (C) and (G). Numbers indicate temporal sequence of calcium activation in the heart. Yellow arrows indicate direction of cardiac conduction. Green circles indicate slow conduction nodes, and blue circles indicate fast conduction pathway.

Atr, atrium; Ven, ventricle; AVC, AV canal; Tr, trabeculae.

Optical mapping of 100 hpf *Tg(cmlc2:gCaMP)*^{s878} hearts revealed that ventricular conduction has transformed from a primitive linear propagation traveling from the AV canal to the OFT as observed at 48 hpf (Figure 2) to a more complex conduction pattern propagating from the OC of the ventricle to the base/IC. After the AV conduction delay, the earliest ventricular calcium activation was observed along the trabeculae (Figure 4A and 4B, activation sequence 7). Next, calcium activation rapidly proceeded from the trabeculae to the adjacent peripheral cardiomyocytes, leading to conduction from the OC to the base/IC (Figure 4A and 4B, activation sequences 8 and 9, and Video S3). Finally, similar to 48 hpf hearts, ventricular conduction terminated at the OFT where conduction velocity significantly decelerated (Figure 4A and 4B, activation sequences 9–12). Interestingly, trabeculae were not observed within the OFT (Figure 4C), further suggesting that they may be responsible for rapid ventricular conduction.

At 2–3 weeks of life, when the ventricle clearly forms an apex, we observed that the initial conduction along the trabeculae (Figures 4E and 4F, sequence activation 4) breaks through at the apex, resulting in apex-to-base/IC conduction (Figure 4E and 4F, activation sequences 5 and 6). Thus, these results indicate that the fast cardiac conduction pathway develops prior to the formation of the apex, resulting in ventricular conduction from the OC to the base/IC to allow efficient ejection of blood from the heart shortly after cardiac looping.

To understand further the development of the fast CCS, we examined the expression of gap junction proteins responsible for cardiac conduction. *Connexin40* (*Cx40*) is expressed in the atrium and the fast conduction system of mammalian hearts, and loss of this gap junction protein results in reduced cardiac conduction velocity and a predisposition to cardiac arrhythmias [23]. In contrast, *Connexin43* (*Cx43*) is expressed abundantly throughout the atrial and ventricular myocardium but in lower amounts in the fast conduction system [24,25]. Because both *cx40* and *cx43* have been suggested to be expressed in zebrafish hearts [26,27], we performed immunostaining on these hearts with antibodies against Cx40 and Cx43. Cx43 immunostaining is present throughout the heart from 24 hpf onwards (Figure 5A, 5D, and5G). However, Cx40 immunoreactivity is not observed at 24 hpf, but is detected very weakly at 48 hpf, and strongly by 100 hpf throughout the myocardium (Figure 5B, 5E, and 5H). Thus, Cx40 immunoreactivity is present during the development of the fast CCS.

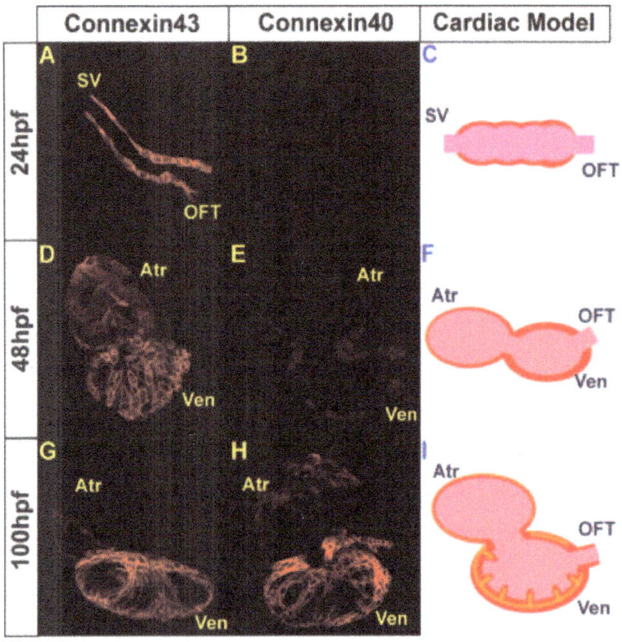

Figure 5. Immunolocalization of Cx40 and Cx43 in Wild-Type Hearts

Confocal images of the heart at 24, 48, and 100 hpf, following immunostaining with anti-Cx40 or anti-Cx43 antibodies (red).

(A, D, and G) Cx43 immunoreactivity was observed throughout the myocardium at 24, 48, and 100 hpf while (B, E and H) Cx40 immunoreactivity was observed weakly at 48 hpf and strongly at 100 hpf.

(C, F and I) Schematic representation of Connexin staining at each developmental stage. Cx43 immunoreactivity is observed throughout the myocardium at all stages and represented in red, and Cx40 immunoreactivity is observed at later cardiac developmental stages and represented in orange.

Hemodynamic Flow and Contraction Affect the Development of the CCS

Epigenetic factors, such as hemodynamic flow and cardiac contraction, previously have been suggested to influence the development of the CCS [6,9]. To determine the role of hemodynamic factors in the development of both slow/AV and fast/ventricular cardiac conduction pathways in zebrafish hearts, we performed optical mapping on the *silent heart*(*sih*^{b109}) mutant heart, which fails to contract due to a null mutation in the *cardiac troponin T*(*tnnt2*) gene [15]. Optical mapping of 48 hpf *Tg*(*cmlc2*:*gCaMP*)^{s878}; *sih* mutant hearts revealed an AV conduction delay, similar to that of 48 hpf wild-type hearts [5] (see Figure 7F). However, in contrast to the ventricular OC-to-base calcium activation pattern observed in wild-type 100 hpf hearts, 100 hpf *sih* mutant hearts displayed ventricular calcium activation laterally across the myocardium from the AV canal to the OFT as well as intermittent AV heart block (Figure 6A and 6B). Interestingly, this intermittent AV heart block occurred within the ventricle just after the AV canal (Figure 6B and 6B′). Similar conduction patterns also were observed in cardiac contractility mutants exhibiting weak contractions throughout the heart (unpublished data).

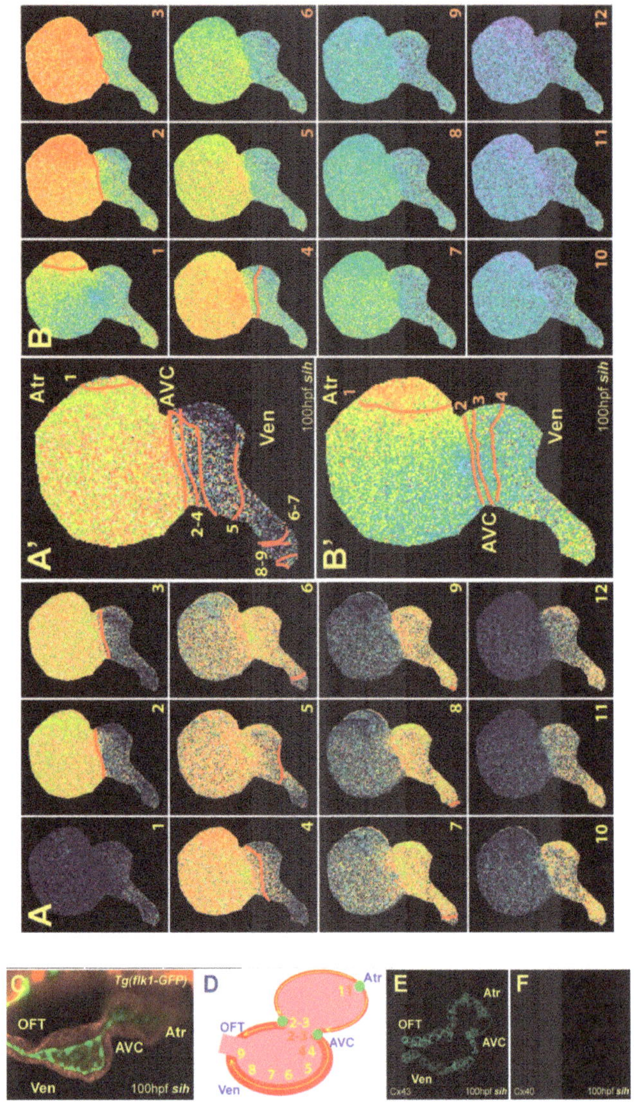

Figure 7. Optical Maps of Cardiac Conduction Mutants

Isochronal lines indicate the distance of calcium activation waves traveled every 60 ms.

(A) Ventricular conduction is absent in 48 hpf *siv* mutant hearts.

(B) Disorganized ventricular conduction was observed in 48 hpf *dcc* mutant hearts.

(C) AV conduction block is present in 48 hpf *hob* mutant hearts.

(D) AV conduction delay is absent in 48 hpf *sli* mutant hearts.

(E) Organized ventricular conduction is absent in 96 hpf *ddl* mutant hearts.

(F) AV conduction delay is present in 48 hpf *sih* mutant hearts.

Atr, atrium; AV, AV canal; Ven, ventricle.

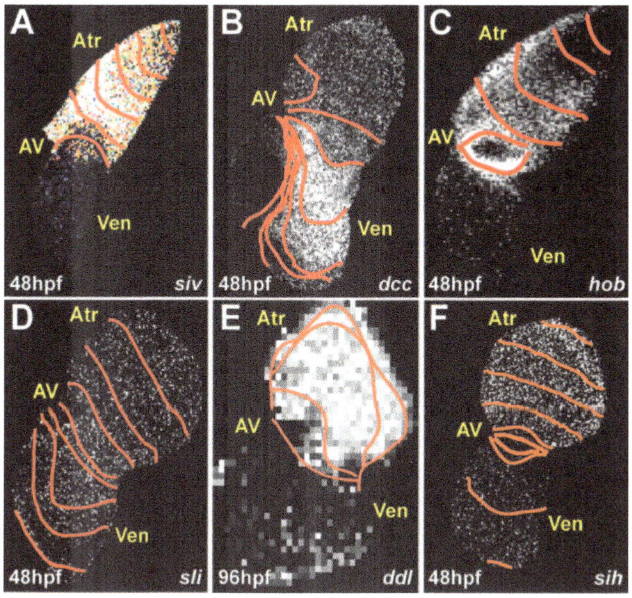

Figure 6. Hemodynamic Flow Is Required for the Development of the Fast CCS

(A and B) Sequential calcium activation images of a 100 hpf *Tg(cmlc2:gCaMP)*^{s878}; *sih*mutant heart during two different cardiac cycles in a live zebrafish embryo. (A′ and B′) Optical maps of calcium excitation of 100 hpf *sih* mutant hearts represented by isochronal lines every 60 ms.

(C) Confocal images of *sih* mutant hearts at 100 hpf. *Tg(flk1:EGFP)*^{s843}; *sih* mutant embryos were stained with rhodamine phalloidin (red). The heart has completed cardiac looping; however, the ventricle has failed to form trabeculae.

(D) Schematic representation of the heart shown in (C). Yellow and orange numbers indicate sequential calcium activation in *sih* mutant hearts from (A and B), respectively. Yellow arrows indicate direction of cardiac conduction. Green circles indicate slow conduction nodes. Note the immature lateral conduction as well as heart block across ventricular myocardium (B and B′).

(E and F) Immunolocalization of Cx40 and Cx43 in *sih* mutant hearts. Cx40 immunoreactivity was not observed in mutant hearts at 100 hpf.

Atr, atrium; Ven, ventricle; AVC, AV canal.

To determine possible etiologies for the conduction defects in *sih* mutant hearts, we analyzed rhodamine-phalloidin-stained *Tg(flk1:eGFP)*^{s843}; *sih* mutants to assess the effects of the lack of blood flow and contraction on cardiac development. At 100 hpf, *sih* mutant hearts appear to have undergone cardiac looping and possess endocardium but do not exhibit trabeculae (Figure 6C). Additionally, 100 hpf *sih* mutant hearts showed significantly diminished Cx40 immunoreactivity but unaffected Cx43 immunoreactivity compared to wild type (Figure 6E and6F). Thus, loss of hemodynamic blood flow and contraction results in the failure of the ventricle to develop trabeculae and the down-regulation of Cx40, thereby possibly affecting the development of the fast CCS.

Forward Genetic Screen for Cardiac Conduction Mutants

We performed a large-scale ENU mutagenesis screen using the *Tg(cmlc2:gCaMP)*^{s878} line as a secondary screen for physiologic analysis of cardiac conduction to identify genes that specifically regulate the development of the CCS. Intercrosses of F2 families were screened by visual inspection of live embryos for aberrant heart rates and/or coordination of atrial and ventricular contraction at each developmental stage of the CCS (24, 48, and 96 hpf). F2 carriers of putative mutations were outcrossed into the *Tg(cmlc2:gCaMP)*^{s878} background to facilitate the physiologic analysis of cardiac conduction. Recovered mutations were organized into phenotypic groups and tested by complementation analysis. We identified 17 mutations defining 14 cardiac conduction regulating loci, four of which previously had been identified (Table 1). Physiologic analyses by optical mapping revealed that the identified mutations disrupt distinct developmental stages of the CCS. Representative examples of phenotypes observed are described below.

Table 1. Cardiac Conduction/Rhythm Mutants

Class	Allele		Conduction Defect	Other Phenotypes	Reference
Class I Conduction Defects (24 hpf): Initial Organized Cardiac Conduction	*silent ventricle (siv)*	*s213/s290*	Noncontractile ventricle		[29]
	tremblor (tre)	*s645/s425*	Atrial and ventricular fibrillation	Small eyes; no circulation	[53–55]
	hiphop (hip)	*s646*	Variable AV block		[54]
	bullseye (bue)	*s885*	No heartbeat at 24 hpf; AV block at 36–48 hpf		
	kingpin (kgp)	*s886*	Atrial and ventricular fibrillation	Small eyes; no circulation	
	brady (bra)	*s887*	Slow heart rate		
Class II Conduction Defects (36–48 hpf): Development of Atrioventricular Delay	*dococ (dcc)*	*s215/s226*	Uncoordinated ventricular contraction	Small eyes	
	hobgoblin (hob)	*s634*	AV block at 48 hpf; silent ventricle at 96 hpf	Endoderm defects; kidney cysts	
	slipjig (sli)	*s644*	peristaltic contraction with no AV delay	No circulation	[54]
Class III Conduction Defects by 72–96 hpf: Development of Peripheral Cardiac Conduction	*daredevil (ddl)*	*s275/s563*	AV block; silent ventricle at 120 hpf	Swim bladder defect; retinal defect	[51]
	mobitz (mbz)	*s466*	AV block; sinus pause at 120 hpf	viable	
	elektra (elk)	*s587*	AV block	viable	
	shocker (shk)	*s633*		viable	

We recovered several noncontractile ventricle mutants including *s209*, *s264*, *s249*, *s271*, and*silent* *ventricle*[s213,s290] [28,29]. Previous work has suggested that a noncontracting ventricle phenotype may be due to contractile defects [15,19,30]. Utilizing the *Tg(cmlc2:gCaMP)*[s878], we discovered that the *silent ventricle* (*siv*) mutant heart fails to generate cardiac conduction across the ventricular myocardium (Figure 7A). Further electrophysiologic characterization of*siv* mutant hearts revealed that the ventricular membrane potential was depolarized markedly. Hyperpolarization of mutant ventricle in current clamp mode resulted in spontaneous AP generation, restoring contractility [29].

We identified several mutants that had conduction defects across the AV myocardium. The*hob* [s634] mutants develop AV heart block despite displaying a wild-type cardiac morphology and contractility (Figure 7C). However, *dococ*[s215, 226] (*dcc*) mutants display an uncoordinated contraction of the ventricle at 36–48 hpf, which was initially characterized by brightfield microscopy as ventricular fibrillation. However, optical mapping of 48 hpf*Tg(cmlc2:gCaMP)*[s878]; *dcc* mutants revealed that the AV conduction was blocked at the superior portion of the AV myocardium, resulting in heterogeneous and disorganized conduction across the ventricular myocardium (Figure 7B). In contrast, endocardial mutants including *cloche* [5,31], *santa*, and *valentine* [32] fail to display an AV conduction delay (unpublished data). Interestingly, the *slipjig*[s644] (*sli*) mutant, which has an endocardium and vasculature (unpublished data), also lacks an AV conduction delay (Figure 7D).

Finally, we identified a class of cardiac mutants that loses its ventricular conduction at 96 hpf. The mutant, *daredevil* (*ddl*), was initially characterized as a cardiovascular contractile mutant in which the ventricle became noncontractile by 96 hpf. Optical mapping of*Tg(cmlc2:gCaMP)*[s878]; *ddl* mutants revealed that these hearts developed heart block between 80–96 hpf and eventually lost all organized ventricular conduction by 96 hpf, resulting in a noncontractile ventricle (Figure 7E). Interestingly, these particular conduction mutants manifest their cardiac phenotype at the developmental stage when the fast CCS develops, suggesting that the affected genes may specifically regulate the development of the fast CCS.

Because of its resemblance to human AV heart block (Figure 8A and Video S4), we isolated the gene disrupted by the *hob* mutation. Fine mapping of 1164 diploid embryos positioned the*hob* gene to a genomic interval of ~150 kb on LG15 (Figure 8B). Sequence analysis of *tcf2*within the critical region revealed a G to A base pair change at position 552, leading to a premature stop codon after the dimerization domain thereby removing the homeobox and transactivation domains (Figure 8C). Data from morpholino (MO) knockdown, in situ analysis, and mRNA rescue experiments support the claim that *tcf2* is the gene affected by the *hob*mutation. Injection of 0.5 ng of *tcf2* MO recapitulated the *hob* mutant cardiac phenotype (Figure 8A). Whole mount in situ analyses reveal that *tcf2* is expressed at 48 hpf around the AV canal as well as the OFT portion of the ventricle (Figure 8D). Finally, injection of wild-type*tcf2* mRNA rescued the heart phenotype in ~82% of *hob* mutants, while injection of mutant*tcf2* mRNA failed to rescue (Figure 8E). Altogether, these data indicate that *tcf2* contributes to the regulation of the cardiac conduction between the atrium and the ventricle.

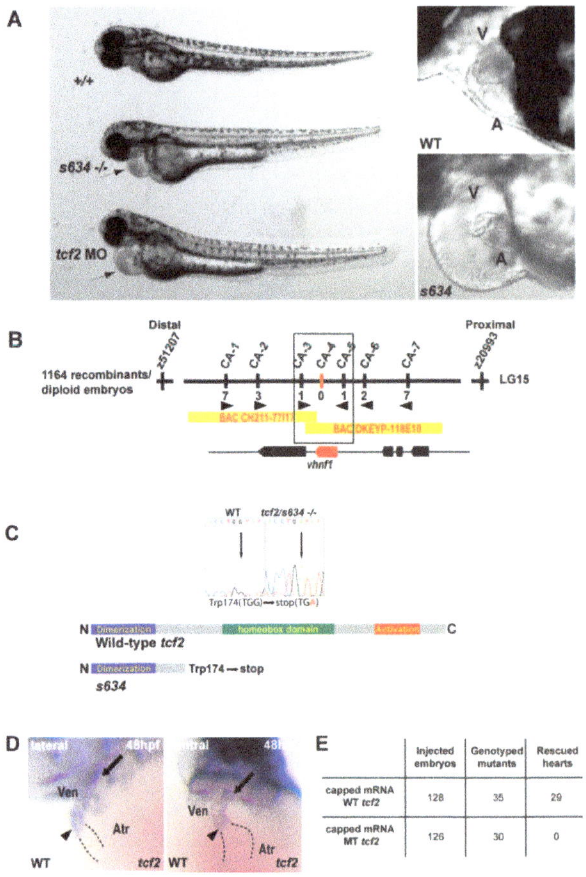

Figure 8. *s634* **Affects** *tcf2*

(A) Lateral views of brightfield images of wild-type, *s634* mutant, and 0.5 ng *tcf2* ATG MO-injected embryos at 48 hpf, anterior to the left. Despite wild-type cardiac morphology, *s634* mutants and *tcf2* MO- injected hearts exhibit an AV conduction block, resulting in reduced cardiac output and pronounced pericardial edema (arrow).

(B) Genetic map of the *s634* region. Numbers below simple sequence length polymorphism markers indicate recombination events. Two genes were identified within the critical region, which spans two bacterial artificial chromosomes.

(C) Sequencing of *tcf2*[s634] cDNA revealed a G to A change at base pair 522, resulting in a premature stop codon at amino acid 174 thereby removing the homeobox and transactivation domains.

(D) Whole mount RNA in situ hybridization at 48 hpf reveals *tcf2* expression at the AV canal (arrowhead) and the OFT (arrow) region of the ventricle.

(E) Injection of wild-type *tcf2* mRNA rescued the heart phenotype of ~82% of *s634*mutants, while injection of mutant *tcf2* mRNA failed to rescue.

A, atrium; V, ventricle.

DISCUSSION

Previous studies have utilized calcium green, a calcium-sensitive fluorescent dye, in zebrafish hearts to observe cardiac conduction up to 48 hpf [5,15]. Because of its short-lived expression, this method is inadequate to analyze later stages of CCS development. Thus, we comprehensively analyzed the development of the zebrafish CCS through cellular and physiologic studies using two new reporter lines: (1) *Tg(cmlc2:gCaMP)*[s878], a myocardial-specific line that expresses the fluorescent calcium indicator protein, gCaMP, throughout cardiac development, and (2) *Tg(cmlc2:eGFP-ras)*[s883], another myocardial-specific line that allows the morphological analysis of individual cardiomyocytes. These tools enabled us to describe four distinct physiologic cardiac conduction stages that correspond to distinct cellular and anatomical changes of the developing zebrafish heart (Figure 9). Furthermore, using this new in vivo optical mapping technique, we performed a physiology-based ENU mutagenesis screen to identify cardiac conduction phenotypes that would have escaped discovery or lacked sufficient characterization in earlier screens.

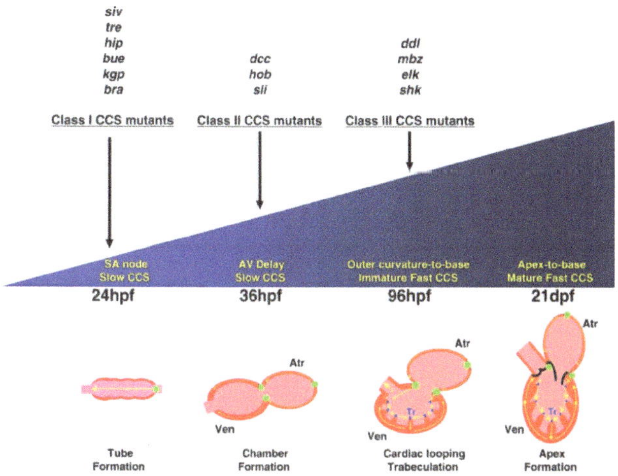

Figure 9. Development of the CCS and Classification of Mutants

The diagram illustrates the four developmental stages of the CCS. Mutations affecting each stage are listed accordingly. Yellow arrows indicate direction of cardiac conduction. Green circles indicate slow conduction pathway/pacemaker and AV conduction delay. Blue circles indicate fast conduction network.

Atr, atrium; Ven, ventricle; Tr, trabeculae.

Developmental Stages of the CCS

The vertebrate CCS can be divided into the slow conduction pathway, which regulates SA pacemaker activity and central AV conduction delay, and the fast conduction pathway, which allows apex-to-base conduction [33]. Through optical mapping of developing wild-type hearts, we observed distinct developmental stages for each of these CCS processes.

As early as the LHT stage, we observed that conduction propagates unidirectionally across the myocardium, suggesting the presence of SA node pacemaker activity [34]. Despite chamber specification, no significant conduction delay was detected at this stage. However, increased conduction velocity was observed in the ventricular half of the heart, suggesting that these cardiomyocytes have initiated cellular and molecular changes including the acquisition of relatively faster conducting properties. Interestingly, expression of *natriuretic peptide precursor a*, a marker of the OC and faster conduction velocities, recently has been observed in a subdomain of the ventricular portion of the LHT near the OFT [19].

The AV conduction delay was observed at 36–48 hpf, a time corresponding to the formation of the AV canal. Action potential and calcium transients of atrial, ventricular, and AV canal regions revealed electrophysiologic differences, suggesting that AV myocardial cells are distinct from atrial and ventricular cells. Whether these differences are due to unique channel expression versus distinct combinatorial effects of several cardiac channels remains unclear. However, weak expression of Connexin40, a marker for fast conduction, was observed at 48 hpf within the atrium and ventricle but lacking from the AV canal. In addition, the AV cardiomyocytes positioned themselves circumferentially around the AV canal. This ring-like orientation of cardiomyocytes has been suggested to contribute to the AV conduction delay [35].

Previous findings suggest that AV endocardial-specific signals, Neuregulin and Notch1b, cause the overlying AV myocardium to differentiate into slow conducting myocardium [5]. Additionally, *Tbx3*, a transcriptional repressor expressed in the developing conduction system in mouse [36], has been observed in the AV myocardium of 48 hpf zebrafish hearts [37]. Future studies exploring how these myocardial and endocardial factors may impact AV cardiomyocyte morphology, orientation, and action potential properties will be of great interest toward understanding the development of the AV conduction delay. Overall, our findings support studies suggesting that AV cardiomyocytes actively differentiate to become precursors of the AV node [5,20,38].

Despite lacking a distinct interventricular septum, adult zebrafish hearts possess a functionally fast CCS, resulting in an apex-to-base conduction across the ventricular myocardium [39]. Previous studies in mouse and chick have suggested that ventricular trabeculae and Cx40 may be responsible for an apex-to-base conduction prior to ventricular septation [6,21,40]. Consistent with these results, the rapid cardiac conduction network within the zebrafish ventricle develops as early as 96 hpf, a stage when cardiac looping is completed, trabeculae have started forming, and Cx40 expression is present. Loss of blood flow and cardiac contraction prevented the transition of ventricular conduction from a primitive linear pattern to a mature apex-to-base propagation, as previously observed in chick [6]. The absence of this transformation correlated with the lack of trabeculae and Cx40 immunoreactivity. Furthermore, we observed that *sih/tnnt2* mutants as well as weak cardiac contractility mutants also developed intermittent AV heart block below the AV canal, suggesting that this defect is within the proximal ventricular portion of the fast CCS rather than the slow AV conduction pathway. How these epigenetic factors modulate the development of the fast CCS remains unknown. However, similar to findings in chick, endocardial signals may be involved, as *cloche* mutant hearts also fail to develop trabeculae and a fast cardiac conduction network (unpublished data). Finally, a mature apex-to-base ventricular conduction was observed at 21 days postfertilization (dpf) when the hearts develop a distinct apex with increased ventricular trabeculation. Altogether, these data suggest that there is an intermediate step during the development of the mature ventricular apex-to-base conduction prior to the formation of the apex.

Overall, these cellular and electrophysiologic data provide the framework for future studies in zebrafish regarding the origin of the vertebrate CCS. Through lineage tracing and optical mapping studies, one should be able to address recent observations related to the ontogeny of the CCS from progenitors that also give rise to chambered myocardium [41].

Systematic Analysis of the Identified Mutants Will Help Elucidate Mechanisms of Cardiac Conduction Development and Cardiac Arrhythmias

Previously, a candidate gene approach in zebrafish was performed to identify AV endocardial factors, *neuregulin* and *notch1b*, as regulators of AV conduction development [5]. In contrast, we have taken an unbiased approach and performed a new physiology-based forward genetic screen to identify additional mediators of the CCS. The recovered mutants display a wide spectrum of conduction phenotypes that first appear at distinct developmental stages of the CCS (Table 1 and Figure 9). Positional cloning reveals that *hob*, an AV conduction block mutation, affects the homeobox transcription factor, *Tcf2*. Given the increased prevalence of AV block in humans with type II diabetes mellitus independent of congestive heart failure and coronary artery disease [42], *Tcf2*, a mature onset diabetes of the young/diabetes-associated gene, may also directly mediate the development and maintenance of mammalian AV conduction. Interestingly, Tcf2 is known to regulate expression of Na^+/K^+*-ATPase* [43], an ion channel gene necessary for cardiomyocyte electrical polarization. Cardiac glycosides, such as digitalis, which inhibit the Na^+/K^+-ATPase pump, also can lead to human AV conduction heart block [44]. Additionally, mouse *Tcf2* in situ hybridization and real-time PCR analyses reveal its expression in the mammalian heart (Bussen M, Martin G, unpublished data). Thus, this mutant may help to address primary cardiac risk factors for a form of hereditary diabetes from indirect effects of diabetes on heart function.

Future analysis and isolation of the genes affected by these mutations promise to uncover new molecular clues and mechanisms that underlie both genetic and acquired forms of heart disease in humans. Novel genes discovered from this screen may be potential candidates for sequencing in population genetic studies of human cardiac arrhythmias. Conversely, candidate genes identified from human genome-wide association studies for cardiac arrhythmias may be validated rapidly and characterized by employing this in vivo optical mapping technique with MO knockdown experiments. Together, these studies will help develop therapeutic options aimed at maintaining and/or improving overall cardiac conduction.

MATERIALS AND METHODS

Zebrafish husbandry and generation of the *Tg(cmlc2:gCaMP)*^{s878} line

Zebrafish were raised under standard laboratory conditions at 28 °C. We used the following transgenic lines: *Tg(flk1:EGFP)*^{s843} [22] and *Tg(cmlc2:eGFP-ras)*^{s883} (Jungblut B, Munson C, Huisken J, Trinh L, Stainier D, unpublished data). We generated the *Tg(cmlc2:gCaMP)*^{s878}construct by cloning a 900 bp fragment of the *cmlc2* promoter [17] upstream of a promoter-less*gCaMP* construct [16]. We injected 200 pg of linearized DNA into one-cell-stage embryos and selected individual transgenic carrier adults by screening for fluorescent progeny. Six*Tg(cmlc2:gCaMP)* founders were recovered with identical expression patterns and levels. Homozygous mutant embryos were obtained by incrossing *Tg(flk1:EGFP)*^{s84}/+; *sih*/+,*Tg(cmlc2:gCaMP)*^{s87}/+; *sih*/+ [15], and *Tg(cmlc2:gCaMP)*^{s87}/+; conduction mutants double heterozygotes.

Immunohistochemistry, Confocal Microscopy, cell Morphology, and in situ Analysis

Immunohistochemistry and confocal microscopy were performed as previously described [28,45,46]. The following antibodies were used at the following dilutions: rabbit polyclonal anti-Cx43 (Sigma) at 1:100 and rabbit polyclonal anti-Cx40 (Sigma) at 1:100. Cardiomyocyte surfaces and cross-sections were analyzed using ImageJ software (National Institutes of Health (NIH), http://rsb.info.hin.gov/ij/) as described previously [19]. A total of 288 cardiomyocytes were measured from 15 wild-type embryos: 165 OC, 83 IC, and 40 AV canal cardioymyocytes. Circularity measurements discriminate circular cells from elliptical cells. In situ analysis was performed as described [47].

ENU Mutagenesis and Screen

We screened approximately 9,076 F3 clutches from 2,392 ENU-mutagenized F2 families that were generated in the context of two different screens [28,48–51]. On the basis of the number of crosses per F2 family, we estimate that the screens surveyed 2,723 genomes. The specific locus test for each screen indicated a mutation rate of approximately 0.3% per gene per genome.

Positional Cloning

Utilizing a set of simple sequence length polymorphism markers, we mapped *hob*^{s634} to linkage group 5. Fine mapping with 1,164 mutant embryos narrowed the critical region to two bacterial artificial chromosomes: CHORI-211 77I17 and DanioKey pilot 118E10. This region contained two putative open reading frames, *tcf2/vhnf-1* and *gamma-synergin*. The *s634* cDNA was isolated, sequenced, and analyzed for both genes. A point mutation was discovered in*tcf2* and confirmed by sequencing *s634* genomic DNA. Approximately 0.5–1 ng of an ATG MO against *tcf2* (Gene-Tools), 5′-CTAG AGAGGGAAATGCGGTATTGTG-3′, was injected into one-cell-stage embryos. For mRNA rescue experiments, one-cell-stage embryos were injected with 50–100 pg of *tcf2* mRNA.

Optical mapping by widefield epifluorescence

Individual zebrafish between 24 hpf and 21 dpf were placed on a coverglass. Electromechanical isolation was achieved with 10 mM 2,3-butanedione monoxime (Sigma) applied 15 min before imaging. Single-plane widefield epifluorescence images of the heart were obtained with a Nikon TE-2000 inverted microscope, using 20× and 40× Plan Apo air objectives, an Xcite-120 (Exfo) widefield epifluorescent source, and standard fluorescein isothiocyanate filter set. Images were acquired with a Coolsnap HQ camera (Photometrics) using Metavue software (Molecular Devices) in stream acquisition mode at a frame rate of 30 ms/frame (512 × 512 pixels) for 48 hpf hearts and 15 ms/frame (256 × 256 pixels) for 96 hpf hearts. Image processing first consisted of manual adjustment of minor spatial shifts of the image over a temporal imaging series. Then, the fluorescence intensity of each pixel in a 2-D map was normalized to its percentage between the minimum and the maximum recorded values of the pixel over the full series. Isochronal lines at 60 ms intervals were obtained by identifying the maximal spatial gradient for a given time point. The color-coded scheme in each panel and video describes progressive activation of the heart with white/red cells and black/blue cells indicating depolarization and repolarization, respectively. Software processing was performed with Metavue software and procedures written in Matlab (Mathworks).

Isolated Calcium Transient Recordings by Selective Plane Illumination Microscopy

Videos of the cardiac conduction wave were recorded with selective plane illumination microscopy [52]. The attenuated 488-nm laser line from a diode-pumped solid-state laser (Coherent Sapphire, 30 mW) was focused to a light sheet with a thickness of 6μm. The sample was oriented such that a thin slice of atrium, AV canal, and ventricle was illuminated. The fluorescence was collected at 90 frames per second with a 20×/0.5 objective lens (Leica) and an emission filter (Chroma, HQ 525/50m) and imaged on an EM-CCD camera (Andor DV885). The microscope and camera were controlled with a Labview (National Instruments) program and analyzed with Matlab. In each sequence, several 15 × 15 pixel areas (i.e. 12 × 12μm) were selected, and the intensity in these areas was plotted over time.

Action Potential Recordings from Embryonic Heart

The 48 hpf embryos were dechorionated and anesthetized with 0.02% tricaine for 1–2 min. The heart was dissected from the thorax en bloc, and all experiments were performed at room temperature. The recording chamber was perfused with the following solution containing (in mM) 140 NaCl, 4 KCl, 1.8 $CaCl_2$, 1 $MgCl_2$, 10 glucose, and 10 HEPES, pH 7.4. Suction pipettes were made from borosilicate capillary tubes (8250 glass, A-M Systems) and fire-polished to obtain resistances of 6–9 MΩ when filled with (in mM) 120 KCl, 5 EGTA, 5 K_2ATP, 5 $MgCl_2$, and 10 HEPES, pH 7.2. Transmembrane potential was measured using an Axoclamp 2A amplifier (Molecular Devices) in the bridge mode using the disrupted patch technique. The pipette was positioned adjacent to the heart, and a seal was formed by application of minimal suction. Through the use of this technique, stable spontaneous APs were recordable for up to 2 h. Transmembrane voltage was filtered at 10 kHz and digitized at a sampling frequency of 20 kHz with a 12-bit analog-to-digital converter (Digidata 1322A Interface, Molecular Devices).

ACKNOWLEDGMENTS

We thank all of the members of the Baier and Stainier laboratory screen teams for their support during the screens; Steve Waldron, Natasha Zvenigorodsky, and Ana Ayala for expert help with the fish; Takashi Mikawa for critical comments on the manuscript; and Junichi Nakai for the*gCaMP* vector.

REFERENCES

1. Fishman GI (2005) Understanding conduction system development: a hop, skip and jump away. Circ Res 96: 809–811.
2. Pashmforoush M, Lu JT, Chen H, Amand TS, Kondo R, et al. (2004) Nkx2–5 pathways and congenital heart disease; loss of ventricular myocyte lineage specification leads to progressive cardiomyopathy and complete heart block. Cell 117: 373–386.
3. Gourdie RG, Mima T, Thompson RP, Mikawa T (1995) Terminal diversification of the myocyte lineage generates Purkinje fibers of the cardiac conduction system. Development 121: 1423–1431.
4. Moskowitz IP, Pizard A, Patel VV, Bruneau BG, Kim JB, et al. (2004) The T-box transcription factor Tbx5 is required for the patterning and maturation of the murine cardiac conduction system. Development 131: 4107–4116.
5. Milan DJ, Giokas AC, Serluca FC, Peterson RT, MacRae CA (2006) Notch1b and neuregulin are required for specification of central cardiac conduction tissue. Development 133: 1125–1132.
6. Hall CE, Hurtado R, Hewett KW, Shulimovich M, Poma CP, et al. (2004) Hemodynamic-dependent patterning of endothelin converting enzyme 1 expression and differentiation of impulse-conducting Purkinje fibers in the embryonic heart. Development 131: 581–592.
7. Rentschler S, Zander J, Meyers K, France D, Levine R, et al. (2002) Neuregulin-1 promotes formation of the murine cardiac conduction system. Proc Natl Acad Sci U S A 99: 10464–10469.
8. Patel R, Kos L (2005) Endothelin-1 and Neuregulin-1 convert embryonic cardiomyocytes into cells of the conduction system in the mouse. Dev Dyn 233: 20–28.
9. Gourdie RG, Wei Y, Kim D, Klatt SC, Mikawa T (1998) Endothelin-induced conversion of embryonic heart muscle cells into impulse-conducting Purkinje fibers. Proc Natl Acad Sci U S A 95: 6815–6818.
10. Schoenebeck JJ, Keegan BR, Yelon D (2007) Vessel and blood specification override cardiac potential in anterior mesoderm. Dev Cell 13: 254–267.
11. Myers DC, Fishman GI (2004) Toward an understanding of the genetics of murine cardiac pacemaking and conduction system development. Anat Rec A Discov Mol Cell Evol Biol 280: 1018–1021.
12. Wang JW, Wong AM, Flores J, Vosshall LB, Axel R (2003) Two-photon calcium imaging reveals an odor-evoked map of activity in the fly brain. Cell 112: 271–282.
13. Suh GS, Wong AM, Hergarden AC, Wang JW, Simon AF, et al. (2004) A single population of olfactory sensory neurons mediates an innate avoidance behaviour in Drosophila. Nature 431: 854–859.
14. Tallini YN, Ohkura M, Choi BR, Ji G, Imoto K, et al. (2006) Imaging cellular signals in the heart in vivo: Cardiac expression of the high-signal Ca^{2+} indicator GCaMP2. Proc Natl Acad Sci U S A 103: 4753–4758.
15. Sehnert AJ, Huq A, Weinstein BM, Walker C, Fishman

M, et al. (2002) Cardiac troponin T is essential in sarcomere assembly and cardiac contractility. Nat Genet 31: 106–110.

16. Nakai J, Ohkura M, Imoto K (2001) A high signal-to-noise Ca^{2+} probe composed of a single green fluorescent protein. Nat Biotechnol 19: 137–141.

17. Huang CJ, Tu CT, Hsiao CD, Hsieh FJ, Tsai HJ (2003) Germ-line transmission of a myocardium-specific GFP transgene reveals critical regulatory elements in the cardiac myosin light chain 2 promoter of zebrafish. Dev Dyn 228: 30–40.

18. Yelon D, Stainier DY (1999) Patterning during organogenesis: genetic analysis of cardiac chamber formation. Semin Cell Dev Biol 10: 93–98.

19. Auman HJ, Coleman H, Riley HE, Olale F, Tsai HJ, et al. (2007) Functional modulation of cardiac form through regionally confined cell shape changes. PLoS Biol 5: e53. .

20. McGuire MA, de Bakker JM, Vermeulen JT, Moorman AF, Loh P, et al. (1996) Atrioventricular junctional tissue. Discrepancy between histological and electrophysiological characteristics. Circulation 94: 571–577.

21. Rentschler S, Vaidya DM, Tamaddon H, Degenhardt K, Sassoon D, et al. (2001) Visualization and functional characterization of the developing murine cardiac conduction system. Development 128: 1785–1792.

22. Jin SW, Beis D, Mitchell T, Chen JN, Stainier DY (2005) Cellular and molecular analyses of vascular tube and lumen formation in zebrafish. Development 132: 5199–5209.

23. Kirchhoff S, Nelles E, Hagendorff A, Kruger O, Traub O, et al. (1998) Reduced cardiac conduction velocity and predisposition to arrhythmias in connexin40-deficient mice. Curr Biol 8: 299–302.

24. Kanter HL, Laing JG, Beyer EC, Green KG, Saffitz JE (1993) Multiple connexins colocalize in canine ventricular myocyte gap junctions. Circ Res 73: 344–350.

25. Davis LM, Kanter HL, Beyer EC, Saffitz JE (1994) Distinct gap junction protein phenotypes in cardiac tissues with disparate conduction properties. J Am Coll Cardiol 24: 1124–1132.

26. Iovine MK, Higgins EP, Hindes A, Coblitz B, Johnson SL (2005) Mutations in connexin43 (GJA1) perturb bone growth in zebrafish fins. Dev Biol 278: 208–219.

27. Christie TL, Mui R, White TW, Valdimarsson G (2004) Molecular cloning, functional analysis, and RNA expression analysis of connexin45. 6: a zebrafish cardiovascular connexin. Am J Physiol Heart Circ Physiol 286: H1623–H1632.

28. Beis D, Bartman T, Jin SW, Scott IC, D›Amico LA, et al. (2005) Genetic and cellular analyses of zebrafish atrioventricular cushion and valve development. Development 132: 4193–4204.

29. Arnaout R, Ferrer T, Huisken J, Spitzer K, Stainier DY, et al. (2007) Zebrafish model for human long QT syndrome. Proc Natl Acad Sci U S A 104: 11316–11321.

30. Berdougo E, Coleman H, Lee DH, Stainier DY, Yelon D (2003) Mutation of weak atrium/atrial myosin heavy chain disrupts atrial function and influences ventricular morphogenesis in zebrafish. Development 130: 6121–6129.

31. Stainier DY, Weinstein BM, Detrich HW III, Zon LI, Fishman MC (1995) Cloche, an early acting zebrafish gene, is required by both the endothelial and hematopoietic lineages. Development 121: 3141–3150.

32. Stainier DY, Fouquet B, Chen JN, Warren KS, Weinstein BM, et al. (1996) Mutations affecting the formation and function of the cardiovascular system in the zebrafish embryo. Development 123: 285–292.

33. Christoffels VM, Burch JB, Moorman AF (2004) Architectural plan for the heart: early patterning and delineation of the chambers and the nodes. Trends Cardiovasc Med 14: 301–307.

34. Challice CE, Viragh S (1980) Origin and early differentiation of the sinus node in the mouse embryo heart. Adv Myocardiol 1: 267–277.

35. Moorman AF, de Jong F, Denyn MM, Lamers WH (1998) Development of the cardiac conduction system. Circ Res 82: 629–644.

36. Hoogaars WM, Tessari A, Moorman AF, de Boer PA, Hagoort J, et al. (2004) The transcriptional repressor Tbx3 delineates the developing central conduction system of the heart. Cardiovasc Res 62: 489–499.

37. Ribeiro I, Kawakami Y, Buscher D, Raya A, Rodriguez-Leon J, et al. (2007) Tbx2 and Tbx3 regulate the dynamics of cell proliferation during heart remodeling. PLoS ONE 2: e398. .

38. Viragh S, Challice CE (1977) The development of the conduction system in the mouse embryo heart. I. The first embryonic A-V conduction pathway. Dev Biol 56: 382–396.

39. Sedmera D, Reckova M, deAlmeida A, Sedmerova M, Biermann M, et al. (2003) Functional and morphological evidence for a ventricular conduction system in zebrafish and Xenopus hearts. Am J Physiol Heart Circ Physiol 284: H1152–H1160.

40. Viragh S, Challice CE (1977) The development of the conduction system in the mouse embryo heart. II. Histogenesis of the atrioventricular node and bundle. Dev Biol 56: 397–411.

41. Moretti A, Caron L, Nakano A, Lam JT, Bernshausen A, et al. (2006) Multipotent embryonic isl1+ progenitor cells lead to cardiac, smooth muscle, and endothelial cell diversification. Cell 127: 1151–1165.

42. Movahed MR, Hashemzadeh M, Jamal MM (2005) Increased prevalence of third-degree atrioventricular block in patients with type II diabetes mellitus. Chest 128: 2611–2614.

43. Bagnat M, Cheung ID, Mostov KE, Stainier DY (2007) Genetic control of single lumen formation in the zebrafish gut. Nat Cell Biol 9: 954–960.

44. Zipes DP, Braunwald E (2005) Braunwald›s heart disease: a textbook of cardiovascular medicine. Philadelphia: W. B. Saunders.

45. Trinh LA, Stainier DY (2004) Fibronectin regulates epithelial organization during myocardial migration in zebrafish. Dev Cell 6: 371–382.

46. Yelon D, Horne SA, Stainier DY (1999) Restricted expression of cardiac myosin genes reveals regulated aspects of heart tube assembly in zebrafish. Dev Biol 214: 23–37.

47. Trinh LA, Yelon D, Stainier DY (2005) Hand2 regulates epithelial formation during myocardial diferentiation. Curr Biol 15: 441–446.

48. Wehman AM, Staub W, Meyers JR, Raymond PA, Baier H (2005) Genetic dissection of the zebrafish retinal stem-cell compartment. Dev Biol 281: 53–65.

49. Ober EA, Verkade H, Field HA, Stainier DY (2006) Mesodermal Wnt2b signalling positively regulates liver specification. Nature 442: 688–691.

50. Muto A, Orger MB, Wehman AM, Smear MC, Kay JN, et al. (2005) Forward genetic analysis of visual behavior in zebrafish. PLoS Genet 1: e66. .

51. Xiao T, Roeser T, Staub W, Baier H (2005) A GFP-based genetic screen reveals mutations that disrupt the architecture of the zebrafish retinotectal projection. Development 132: 2955–2967.

52. Huisken J, Swoger J, Del Bene F, Wittbrodt J, Stelzer EH (2004) Optical sectioning deep inside live embryos by selective plane illumination microscopy. Science 305: 1007–1009.

53. Ebert AM, Hume GL, Warren KS, Cook NP, Burns CG, et al. (2005) Calcium extrusion is critical for cardiac morphogenesis and rhythm in embryonic zebrafish hearts. Proc Natl Acad Sci U S A 102: 17705–17710.

54. Chen JN, Haffter P, Odenthal J, Vogelsang E, Brand M, et al. (1996) Mutations affecting the cardiovascular system and other internal organs in zebrafish. Development 123: 293–302.

55. Langenbacher AD, Dong Y, Shu X, Choi J, Nicoll DA, et al. (2005) Mutation in sodium–calcium exchanger 1 (NCX1) causes cardiac fibrillation in zebrafish. Proc Natl Acad Sci U S A 102: 17699–17704.

CHAPTER 16

MARSUPIAL GENOME SHEDS LIGHT ON THE EVOLUTION OF IMMUNITY

Emma Hill

All organisms need defense systems to ward off infection. These defenses typically mobilize in response to antigens, bits of protein fragments derived from a pathogen. Vertebrates have a superbly efficient system of adaptive immunity that uses proteins to recognize and bind billions of different antigens via a process known as antigen presentation. Ultimately, this process leads to removal of the antigen. Antigen binding proteins are encoded within one large genomic region known as the major histocompatibility complex (MHC). The high level of polymorphism, or variation, in this genomic region allows the adaptive immune system to keep pace with an ever-changing array of antigens.

Genes in the MHC region are subdivided into Class I, Class II, and Class III based on their structure and function. The organization of MHC genes in present-day species can provide insight into how immunity has evolved. To date, maps of the MHC are available from eutherian (placental) mammals and nonmammalian species, including birds, bony fish, and cartilaginous fish such as sharks. But MHC organization and complexity differ substantially between mammals and nonmammals, making it difficult to infer the region's evolutionary history, or phylogeny.

Using available marsupial opossum *(Monodelphis domestica)* genome sequences, Katherine Belov, Robert Miller, and their colleagues present a detailed map of the opossum MHC. These data from a noneutherian mammal help bridge the gap between nonmammals and eutherian mammals, to enable a more comprehensive phylogeny of the MHC.

In eutherian mammals, the MHC is large and dense, with 264 genes and pseudogenes (remnants of once functional genes) scattered over 3.6 Mb and ordered as Class I–III–II in adjacent gene blocks in humans. In nonmammals, the MHC has fewer genes (the chicken has only 19 genes over 92 kb) ordered as Class III–I–II. The authors found that the MHC in the opossum is flanked at either end by the same marker genes that border the human MHC. This noneutherian mammalian MHC spans 3.95 Mb and encodes 114 genes (identified via sequence similarity to genes from other species), sharing 87 with the human MHC. The authors have provided a Web-based genome browser (http://bioinf.wehi.edu.au/opossum) for viewing an annotated map of the opossum MHC region.

Belov et al. found that while the size and complexity of the opossum MHC is closer to eutherian mammals, its organization

Citation: Hill E (2006) Marsupial Genome Sheds Light on the Evolution of Immunity. PLoS Biol 4(3): e75. doi:10.1371/journal.pbio.0040075

is closer to fish and birds: in the opossum, Class I genes are interspersed with the Class II genes—this has not been seen in any other animals. This similar grouping of Class I and II genes implies that they may have been arranged this way in the mammalian ancestor (supported also by the presence of the Class I pseudogenes in the human Class II region).

Humans and mice have a conserved framework region, consisting of a set of genes that are interspersed among the Class I loci. The opossum also contains these genes next to the Class III region. The authors deduced that this framework region assembled in the ancestral mammalian MHC before Class I genes moved to this location in eutherians.

Only one of the opossum MHC genes, *UA1*, has the characteristics of what's known as a classical Class I gene—high polymorphism and ubiquitous expression. The authors saw transcripts of this gene in all tissues tested. Expression of a single classical Class I gene is unusual in mammals but has been seen in the chicken and the frog. Contrary to a previous hypothesis that Class I gene evolution is constrained by their proximity to antigen processing genes, the Class I genes in the opossum are significantly diverged.

The authors identified three marsupial-specific classical Class II gene families (*DA*, *DB*, and a newly discovered family designated *DC*). They also found that sequences responsible for regulating Class I genes—SXY motifs—known to be important in the eutherian MHC, have diverged in the opossum. Conversely, the Class II SXY regions are quite well conserved. This is logical given that Class I genes are more likely to evolve new functions while Class II genes are typically coexpressed.

Significantly, the authors also found data supporting the idea that there was an ancient relationship between the MHC and another critical component of the immune system, the natural killer complex (NKC), which contains natural killer (NK) cell receptor loci. Two NKC genes, *OSCAR* and *MIC*, are present in the opossum MHC. As MHC Class I molecules are ligands (activators) for NK cell receptors, it makes sense that these two gene families coevolve.

The authors conclude that vertebrates once possessed an "immune supercomplex" that performed the MHC functions. While this complex no longer exists in modern genomes, scientists can probe its remnant traces for clues to its history. Gaining a better understanding of how our mammalian immunity evolved will lead to a much broader appreciation of how organisms' internal defense systems work across a wide range of species.

Genome sequence data from the opossum fills a gap in our understanding of the evolution of the mammalian immune system.

READING THE EVOLUTIONARY HISTORY OF THE WOOLLY MAMMOTH IN ITS MITOCHONDRIAL GENOME

The woolly mammoth (Mammuthus primigenius) disappeared along with the last ice age over 10,000 years ago, yet scientists can glean insight into its evolutionary history by studying its frozen remains. Three recent independent studies applied different methods to isolate and sequence multiple DNA fragments from mammoth specimens. Two teams, reporting in December 2005, used mammoth bones extricated from the frozen permafrost of Siberia. One team, reporting in Nature, analyzed the mitochondrial DNA (mtDNA)—genetic material found in the cell's energy-producing organelles and transmitted only through the mother—of a 12,000-yearold specimen and concluded that the mammoth's closest living relative is the Asian elephant. The other team, reporting in Science, analyzed a portion of nuclear and mitochondrial DNA from a 28,000-year-old specimen, using a new method that increases the quantity of DNA fragments that can be extracted and sequenced from ancient specimens.

Ancient DNA studies have traditionally focused on mtDNA, but these investigations have relied on mostly short fragments of mtDNA. In the third and most recent study, Evgeny Rogaev, Yuri Moliaka, and their Russian colleagues sequenced the complete mitochondrial genome of a 33,000-yearold mammoth specimen. (Radiocarbon dating indicated that the mammoth lived 32,850 years ago, plus or minus 900 years.) This is the oldest sequenced mitochondrial genome and—at 16,842 base pairs—the longest stretch of sequenced DNA from an extinct Pleistocene species. This analysis, like the Nature study, points to the Asian elephant as the mammoth's closest relative. Interestingly, the Rogaev et al. study also suggests that genetic diversity among mammoths living in northeastern Siberia during the late Pleistocene was low.

The restored right back leg of a 33,000-year-old woolly mammoth found in northeastern Siberia.

Having complete mitochondrial mammoth genomes will greatly aid researchers' attempts to refine the elephant family tree and explore the genetic variation within this species. Notably, artifact mutations related to chemical modification of ancient DNA commonly contribute to errors in sequence, confounding estimates of diversity between species or individuals.

The long sequences used in the Rogaev et al. study largely avoided this problem. The ancient specimen, a right hind leg, was so well preserved that Rogaev et al. could extract DNA from mitochondria-rich muscle tissue, rather than bone. Two different labs generated multiple mtDNA products using PCR (polymerase chain reaction, the standard method for making millions of copies of DNA for sequencing) and reconstructed the mammoth genome sequence. The mtDNA sequence produced in the two labs appeared to be identical. So much DNA was retrieved that the authors needed only one round of PCR amplification, reducing the likelihood of sequence errors associated with the repeated amplification rounds needed for short DNA fragments, a longstanding problem with ancient DNA.

To determine the prehistoric pachyderm's evolutionary position relative to its Asian (Elephas maximus) and African (Loxodonta Africana) cousins, the authors also sequenced the complete mtDNA genomes of the two modern elephants from Africa and Asia. Rogaev et al. used the mammoth and elephant genomes, along with previously published mtDNA Asian and African elephant genomes, to construct an elephant evolutionary tree. They used two other placental mammals, the dugong and hyrax, to calibrate the tree, and show that the mammoth and Asian elephant are more closely related to each other than either is to the African elephant. Additional analysis using nuclear DNA and more closely related mammals could help refine the tree.

The authors also compared their sequence with other available mammoth sequences (including the recently published Nature sequence) representing populations from different locations and time periods. This analysis revealed the relatively low genetic diversity in the maternal lineage of mammoths in northeastern Siberia through the roughly 20,000-year period of the late Pleistocene. The population structure of mammoths living at that time appears less complex than that seen in the genomes of Asian and African elephants.

The mammoth and Asian elephant may have diverged around 4 million years ago, splitting soon after their lineage diverged from the African elephant. The authors point out that, given the inherent differences in mutation rates and mode of inheritance between nuclear and mitochondrial DNA, studies focusing on one or the other may produce different dates for the last common ancestor.

But these results, together with the Nature and Science studies, demonstrate the tractability of recovering large amounts of high-quality DNA from ancient extinct Pleistocene animals. Rogaev et al. found that relatively large, nondegraded DNA fragments may be extracted from animals preserved in the permafrost for many thousands of years, allowing complete genes (at least from mitochondrial genome) of extinct animals to be cloned. As scientists continue to mine the DNA-rich permafrost, they will be able to weave together the missing pieces of the evolutionary story of a long-lost, fabled creature and its modern relatives, sequence by sequence. To learn more about ancient DNA research, see the related primer (10.1371/journal.

JUMPING FOR DEAR LIFE WITH A GIANT NEURON AND A SINGLE RECEPTOR

Flies are jumpy, prone to buzzing away at the slightest moving shadow or breeze. This reflexive "jump response" is a lifesaver for the tiny insects, whose only defense against predators is a quick escape. In flies, as in many animals, evolution has forged efficient reflex circuits that transmit nerve impulses from sensory organs to muscles within milliseconds. The centerpiece of the fly's jump response circuit is the giant fiber. The large cell body of this prominent nerve cell resides in the brain and its axon extends into the fly's thorax. There, it connects with motor neurons that innervate the jump muscles in the leg and the flight muscles that power the wings. In the brain, the giant fiber receives visual inputs, caused by moving shadows, from the eyes while mechanical inputs, caused by wind or vibrations, come from the antennae.

Neurons communicate via specialized contact points called synapses, where they exchange electrical or chemical signals. The small molecule acetylcholine is a widespread chemical neurotransmitter in the fly brain that could mediate various steps in the jump response. Acetylcholine binds to a large variety of transmembrane receptors. In a new study, Amir Fayyazuddin and Hugo Bellen find that a membrane protein belonging to the nicotinic class of acetylcholine receptors, Dα7, is required within the giant fiber to link the jump response to sensory inputs.

Acetylcholine receptors clump together at neuron synapses, forming pores that open in the presence of extracellular acetylcholine. When a synapse fills with acetylcholine, the pores of the receiving neuron open and let in positively charged ions. A massive electrical impulse ensues that elicits signal release at the next synapse downstream. Fayyazuddin and his colleagues used a small piece of mobile DNA that had inserted next to the Dα7 gene in the fly genome to generate several fly strains devoid of part or most of their Dα7 gene. In the laboratory, scientists elicit the jump response in captive flies by simply turning off the light briefly. Normal flies jump while mutant flies stay put. The Dα7 mutants failed to jump in response to the light off signal, even though additional tests showed they had normal vision and locomotion.

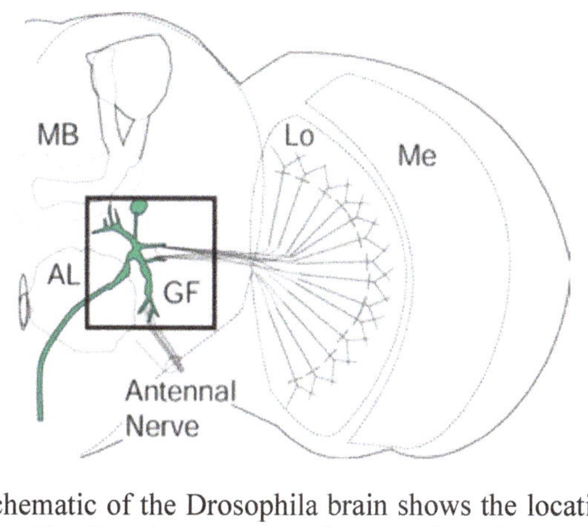

This schematic of the Drosophila brain shows the location of the giant fiber in relation to other brain regions (AL, antennal lobe; GF, giant fiber; Lo, lobula; MB, mushroom body; Me, medulla).

In immobilized and partially dissected flies, scientists can fi re up the giant fiber with an electrical stimulus and follow the nerve impulse all the way from giant fiber to twitching jump and flight muscles. Fayyazuddin and his colleagues found that in Dα7 mutants, the flight muscles failed to respond to stimulation by the giant fiber. This result was not unexpected, as earlier observations had shown that a particular neuron that connects the giant fiber to flight muscles uses acetylcholine as its messenger. By contrast, in this assay, the jump muscle behaved normally in Dα7 mutants. This was surprising, because jump precedes flight in the escape response, and the Dα7 mutants have a clear jump deficit.

The researchers reasoned that the Dα7 mutations silenced synapses farther upstream in the circuit. The Dα7 protein is abundant at brain synapses between the giant fiber and neurons that relay sensory inputs from the eyes and antennae. Acetylcholine is also found at these synapses, though its function has been unclear. Introducing an active Dα7 gene in the giant fiber of mutant flies was sufficient to restore activation of the giant fiber. This result demonstrates that the giant fiber uses its Dα7 receptor (and acetylcholine) to trigger the jump reflex in response to sensory inputs.

Nicotinic acetylcholine receptors are notoriously difficult to study, because various family members can substitute for each other and attenuate the consequences of the loss of a single representative. It is therefore remarkable that disrupting the Dα7 gene of Drosophila could have such a profound effect, especially on a behavior as vital as escape. The fly's Dα7 receptor may have evolved special properties that enhance the effectiveness of the escape refl ex, properties that now warrant further investigations.

A KEY TUMOR SUPPRESSOR PROTEIN ENSURES PROPER CHROMOSOME SEGREGATION DURING CELL DIVISION

Cell division in animals is an exquisitely exact process. Before a cell splits, it has one set of chromosomes and one centrosome, a structure just outside the nucleus that plays an important part in the process. The chromosomes replicate, the centrosome duplicates, and the resulting two centrosomes migrate to opposite sides of the nucleus. Each centrosome then sends out threads that grab onto and reel in one of the two sets of chromosomes. The result: two identical "teams" that go on to make up the nuclei of the two daughter cells. Or that's the ideal, anyway. Sometimes when cells divide they make more than two centrosomes. And sometimes cells with more than two centrosomes have daughter cells with more or fewer than the normal allotment of chromosomes. Scientists have suspected for a century that the two phenomena "supernumerary" centrosomes and the wrong number of chromosomes, called "aneuploidy," a common feature of cancer and premalignant cells—are causally linked. But, like the proverbial chicken and egg, it hasn't been clear which comes first. Do extra centrosomes sometimes cause aneuploidy? Or does aneuploidy create changes that make cells go overboard when duplicating centrosomes? The answer could help shed valuable light on the process that turns normal cells malignant.

Kimberly McDermott, Thea Tlsty, and colleagues explored these questions in a series of experiments reported in a new study. Using laboratory techniques that allowed them to observe the number and distribution of chromosomes and centrosomes at various stages of cell division, the researchers showed that the production of supernumerary centrosomes can in some cases lead to aneuploidy. They also unraveled the process by which supernumerary centrosomes develop, and showed that when supernumerary centrosomes are present, cell division produces lopsided teams of chromosomes that aren't able to play by the rules of normal cells.

To accomplish all this, the researchers focused their attention on human mammary epithelial cells (HMECs). Most HMECs can only be grown in culture for a limited amount of time, after which they stop dividing; however, a subpopulation of these cells, called vHMECs (for "variant" HMECs), can divide for longer—but eventually end up with chromosome abnormalities, including aneuploidy. The researchers discovered that if they temporarily paused chromosome replication with a reversible chemical treatment in vHMECs, some cells ended up with extra centrosomes—as though the original two centrosomes had not gotten the message that things were on hold, and duplicated again while the chromosomes sat. Knowing from other studies that a protein called p16INK4a (which normally plays a role in suppressing tumor formation) is silenced in vHMECs, they tested its possible role in the generation of extra centrosomes by blocking its expression in HMECs. As they predicted, some of the HMECs in which p16INK4a was blocked developed supernumerary centrosomes. And when they altered vHMECs to make them express p16INK4a, the tendency to produce supernumerary centrosomes was reduced.

How does the absence of p16INK4a lead to too many centrosomes? A normal centrosome contains two assemblages of microtubules, called a centriole pair that split up and make new centriole partners when a cell divides. Looking closely at the supernumerary centrosomes, the researchers discovered a significant portion with only one centriole. This, along with the fact that cells rarely end up with more than four centrosomes, led them to speculate that p16INK4a acts as a sort of centriole pair "policeman" when it's around, the centriole pair splits once as it is supposed to, but if it's disabled, the centriole pairs can split more than once to generate supernumerary centrosomes.

To clearly link production of supernumerary centrosomes resulting from loss of p16INK4a activity to aneuploidy, the researchers looked at cell division in vHMECs and in HMECs with and without blocked p16INK4a. They found that pausing replication was associated with increases in supernumerary centrosomes and aneuploidy in vHMECs and in HMECs with blocked p16INK4a, but not in normal HMECs. To prove a causal connection, they observed cell division using time-lapse microscopy. They found that vHMECs with more than two centrosomes were more likely to produce aneuploid daughter cells than were vHMECs with two centrosomes.

The researchers tested other cell types and discovered that loss of p16INK4a activity produces supernumerary centrosomes and aneuploidy in them as well. And, using human fi broblasts, they showed that p16INK4a prevents inappropriate centriole pair splitting by working with a gene called p21 that inhibits the activity of Cdk, a type of protein needed for centrosome duplication.

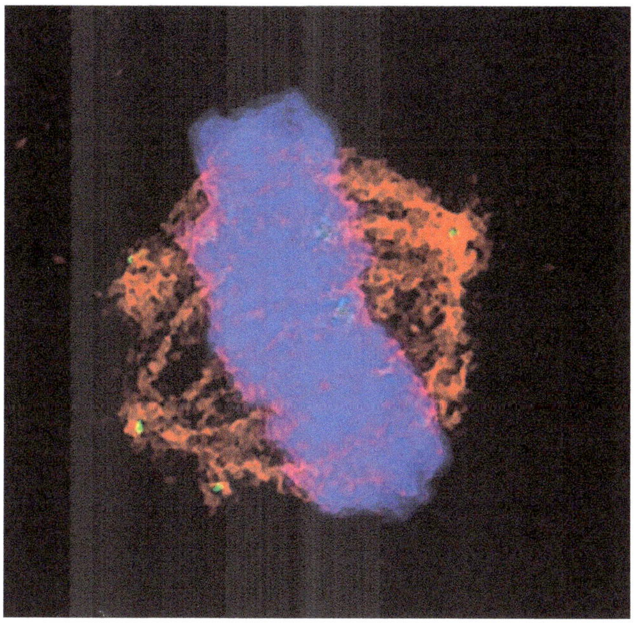

Loss of $p16^{INK4a}$ allows cells to produce too many centrosomes (the green dots) which in turn leads to genomic instability.

Cells with inactive p16INK4a are found in normal human breast tissue. This study suggests that if DNA replication gets held up—which it can if DNA needs repair due to exposure to chemotherapy—such cells would produce supernumerary centrosomes. These supernumerary centrosomes could lead to aneuploidy, loss or gain of genes regulating cell division and death, and may result in the beginning stages of a tumor.

WHEN LESS IS MORE: LOSING GENES ON THE PATH TO BECOMING HUMAN

Students of human evolution got a big boost when the draft sequence of the chimp genome was published in 2005. Now their challenge is to comb through the combined 6 billion nucleotides for clues to the evolutionary forces that made humans odd man out in the primate family tree. Chimp and human DNA nucleotides differ by just 1.23%, sufficient genetic variation for natural selection to create a bipedal, big-brained primate lineage but small enough to suggest that every mutation has an evolutionary tale to tell. Natural selection, it's typically thought, mostly favors "gain of function" mutations, which increase an organism's fitness while jettisoning "loss of function" mutations, which are considered deleterious. An alternative hypothesis proposes that gene loss— which can occur when mutations render functional genes inactive in a process called pseudogenization—can also confer selective advantage and promote evolutionary change.

In a new comparative analysis of the human and chimp genomes, Xiaoxia Wang, Wendy Grus, and Jianzhi Zhang find support for this "less is more" hypothesis. They analyzed 80 human pseudogenes—many involved in immunity and chemoreception—and show that positive selection explains how one pseudogene spread throughout human populations. Humans with the CASP12 pseudogene, which retains its immune function in other mammals, are less susceptible to severe sepsis.

Analyzing "pseudogenes" that lost their functionality since humans and chimps diverged over 6 million years ago promises to shed light on the genetic basis of human origins. (Photo: Curt Busse, Gombe National Park, Tanzania)

Several cases of gene loss have been proposed as adaptive opportunities for human evolution. To look for potentially momentous gene losses, Wang et al. turned to a human pseudogene database. Most of the 19,000 human pseudogenes in the database are "processed" pseudogenes, which derive from reverse-transcribed RNA that was reinserted into the genome, usually without the regulatory elements necessary for gene activation.

Most processed pseudo genes never functioned. And nearly half of the 1,781 "nonprocessed" pseudogenes lack the mutations that would make them clearly nonfunctional. So the authors focused on the 887 "nonprocessed" pseudogenes that had the requisite loss of-function mutations. Comparing these human pseudogenes with the chimp genome flagged 83 sequences as closest matches—assumed to be chimp orthologs (derived from an ancestral gene during speciation).

After ascertaining that the human pseudogenes have chimp orthologs and are truly nonprocessed pseudogenes, the authors whittled their list down to 67. Most of the 80 pseudogenes studied (including 13 found in previous studies) play a role in immune response or chemoreception and belong to rapidly evolving gene families with species-specific members. Interestingly, mice lacking one of the pseudogenes, Mbl1, show higher resistance to sepsis and increased survival. Though this pseudogene likely spread through the human population too long ago to detect signs of selection (about 6–7 million years ago), the authors reasoned that if mice gained an advantage from losing an immunity gene, humans might, too.

To test this possibility, they focused on another sepsis-related pseudogene, CASPASE12 (CASP12) that is still undergoing pseudogenization. In the nonfunctional (null) allele, a mutation replaces a C with a T nucleotide (called a C/T polymorphism), producing truncated, nonfunctional CASP12 proteins. The null T allele is linked to sepsis resistance in humans. All non-Africans (that is, individuals of Asian or European descent) carry two copies of the T allele, which occurs with 89% frequency in individuals of African descent.

The authors used population genetic techniques to test whether positive selection drove the high frequency of the T allele. If it did, sequences surrounding the T allele should have lower variation than those flanking the C allele. And that's just what the authors found. While the T allele existed at a much higher frequency than the C allele in the sampled populations, it had far fewer polymorphisms across nine noncoding regions adjacent to the C/T polymorphism.

Wang et al. modeled different demographic scenarios and mutation rate estimates using the polymorphism data to explain the source of the observed low diversity. Their simulation results "strongly suggest" that positive selection drove the spread of the T allele. Since non-Africans and Africans carry identical T alleles, the null mutations likely share a common ancestor, originating sometime before modern humans left Africa 40,000–60,000 years ago. Assuming that selection acted on the variant as soon as it appeared, the allele would need about 51,000–55,000 years to reach its current frequency, suggesting that the adaptive pseudogenization started shortly before the African emigration. These findings show that gene loss provides not just the opportunities but the means for selection to act. Since pseudogenization in vertebrate and other eukaryotic genomes is common, the authors speculate that adaptive gene loss may be as well. As the chimp genome sequence becomes more refined, researchers will have fresh opportunities to explore how gene loss shaped the unique traits that define human biology and formulate new hypotheses to test the breadth of a "less is more" strategy in evolutionary change.

WHEN SEEING IS MISLEADING: CLUTTER LEADS TO HIGH-CONFIDENCE ERRORS

Did you ever arrange to meet a friend at a busy street corner, and then rush up to a total stranger thinking it was your friend? Neuroscientists have a theory to explain why such potentially embarrassing mistakes occur. We're trying to detect a target (our friend) amid a noisy background filled with distracters (a street filled with strangers) that impede our visual perception. Neuroscientists probe the underlying perceptual and neural processes of visual search by studying how distracters affect performance of a visual search task under controlled conditions in the lab.

Say a person is asked to pick out a target—a line tilted clockwise—that is embedded within a set of distracters—32 vertical lines. Signal detection theory (SDT) provides a framework for making quantitative predictions about the probability that an observer will detect a target under cluttered conditions. SDT assumes the brain represents each element in a visual search display as an independent variable with its own noise. It also assumes that when the observer isn't sure which stimulus is the target, she monitors all stimuli, and performance suffers. Thus, increasing the number of distracters (trying to find your friend on a busy street or a document on a messy desk) increases the background noise of the visual system's representation while reducing the accuracy and reaction time of performing the task.

One might intuitively expect that as noise and errors increase, confidence in one's decision plummets. But in a new study, Stefano Baldassi, Nicola Megna, and David Burr show that just the opposite happens. When observers searched for a tilted target embedded in distracters, they overestimated the magnitude of the tilt—and did so with a high degree of confidence in their decision.

The probability of being sure you saw something you didn't increases in chaotic environments like this 2003 peace rally near the Campidoglio in Rome. (Image: Stefano Baldassi)

It turns out that SDT lends a logical prediction to the seemingly counterintuitive finding that observers make more high-confidence errors when confronted with clutter. The prediction flows from a "squeaky wheel gets the grease" rule about visual processing, called the "Sign Max Rule." When confronted with a set of independent, noisy responses, the visual system tends to base its decision on the largest response. Since each stimulus generates a noisy internal representation, and subjects monitor all the distracters to search for the target, as the number of distracters increases, the chance of perceiving a distracter as being more tilted than the target also increases. The authors predicted that the rule also applies to the confidence observers have in such "high magnitude" perceptual errors.

To test this prediction, the authors asked ten observers to indicate the direction and magnitude of the tilt of a target grating patch (small patches of blurred parallel black and white lines) tilted clockwise or counterclockwise. The target was briefly presented either alone or embedded in a circular array of vertical distracter patches. Though perceived magnitude is known to reflect observer confidence, the authors also got a direct measure by asking observers to report their level of confidence about each decision. Adding magnitude to the tilt discrimination task, the authors explain, helps shed light on the internal mechanisms that drive observers' decisions. As predicted by the Signed Max Model, estimates of perceived tilt increased with set size, as did the observers' confidence in their decisions. The authors conclude the visual system combines the outputs of noisy detectors and settles on the maximum signal.

These results suggest that the probability of being sure you saw something you didn't increases in chaotic environments, and could have far-reaching implications. The authors explain that while their study focused on "simple perceptual decisions about a single stimulus attribute," the same type of processes may also apply to complex cognitive tasks involving problem solving and memory. If people find themselves confronted with multiple events in a chaotic, confusing environment, they may decide about some aspect of the situation and be totally wrong even though they have full confidence in their decision. The consequences of such a phenomenon could be relatively trivial, explaining why professional athletes often end up wasting their time arguing questionable calls with an official. Or they could prove a matter of life and death, perhaps accounting for why eyewitness testimony is so unreliable—or why soldiers sometimes can't tell friend from foe in the heat of battle.

POLYGLUTAMINE LEADS TO ALTERED CHROMATIN ARCHITECTURE AND NEURODEGENERATION

In the field of neurodegenerative disease, the so-called polyglutamine diseases hold a unique place. These nine different diseases, which include Huntington disease and several forms of spinocerebellar ataxia, involve nine different genes. But each gene carries the same mutation—an increase in the number of CAG nucleotide triplets.

In the normal genes, the number of CAG units is 30 or less; in the mutated form, this grows to 35 or more. CAG codes for the amino acid glutamine, and the consequence of each mutation is to create an expanded polyglutamine tract within the protein encoded by the gene. The presence of this tract kills nerve cells, by a mechanism or mechanisms that still remain unclear.

Despite the similarity of the mutations, each disease differs in the exact set of cells it affects within the brain. This different behavior is presumed to arise from differences in the normal function of the various polyglutamine-expanded proteins, and from the cellular context in which each is expressed. But for many of these diseases, the role of the normal protein is not well understood, making it difficult to determine whether expanded polyglutamine interferes with this function. In a new study, Dominique Helmlinger, Didier Devys, and colleagues show that in spinocerebellar ataxia type 7 (SCA7), the damage done to the retina is a direct consequence of the normal protein's key role in turning on genes specifically expressed in photoreceptor cells.

The authors have previously shown that ataxin-7 (the protein made by the SCA7 gene) is a subunit in two highly related transcriptional coactivator complexes, which stimulate gene transcription by modifying the histone proteins around which DNA is wrapped. They also developed a mouse model of SCA7 in which the mutant gene is expressed only in the rod cells of the retina, leading to retinal degeneration (in humans, retinal degeneration is a prominent part of the disease). In this study, they examined the cell architecture of the degenerating rod cells, and found that the nucleus was highly swollen due to dramatically increased "decondensation" of the chromatin, or nuclear material.

Decondensation occurs when histones, which help package DNA in the nucleus, are chemically modified by acetylation (adding a two-carbon acetyl group). Acetylation opens up the chromatin, allowing the gene transcribing machinery to gain access to genes within. Rod cell degeneration was directly correlated with the degree of decondensation, suggesting that aberrant decondensation might be the primary event leading to cell death.

Decondensation usually leads to increased gene transcription, but the authors found instead that transcription of rod-specific genes was decreased, in many cases by 90% or more. This was not due to any inability of the mutant protein to become incorporated into its two transcriptional coactivator complexes, which appeared to be unchanged. Instead, the authors show that these complexes worked too well, hyperacetylating their rodspecific target genes, accounting for the increased decondensation they observed. But this hyperacetylation comes at a price—the disruption of the normal chromatin architecture of the mature rod nucleus.

The authors suggest that this disruption is the likely explanation for the loss of rod-specific gene transcription. A growing body of research has shown that within the nucleus, the arrangement of chromatin is not random, but is correlated to cell type. Some evidence suggests that highly expressed genes are preferentially located at the nuclear envelope, where they may have better access to the raw materials that sustain rapid transcription. The loss of this structure in the SCA7 mouse retina may prevent the very high levels of transcription of rod-specific genes needed to maintain normal rod function.

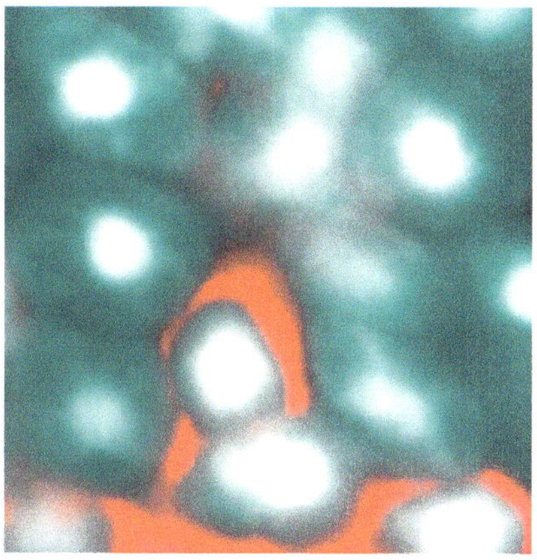

Mutations in ataxin-7 lead to retinal damage by inhibiting the expression of genes involved in chromatin regulation.

This is not likely to be the last word on disease mechanisms in the polyglutamine diseases, a field full of strong and competing hypotheses of pathogenesis. It does, though, highlight the importance of understanding the role of the normal protein, and strengthens the case for altered nuclear architecture as a key event in Neurodegeneration.

SWITCHING DRUGS FOR LIVESTOCK MAY HELP SAVE CRITICALLY ENDANGERED ASIAN VULTURES

In just ten years, tens of millions of vultures have vanished from the Indian subcontinent. Since the early 1990s, white-backed (Gyps bengalensis), long-billed (G. indicus), and slender-billed (G. tenuirostris) vulture populations have dropped by over 95%. In Europe, it was clear that human persecution eradicated bearded and griffon vultures from some countries (reintroduction and protection efforts are now restoring populations). But in India, where it is illegal to kill wildlife and the bird is valued for its ecological role, their unprecedented decline was puzzling. Dead birds found in India, Pakistan, and Nepal had extensive visceral gout (a buildup of uric acid crystals in the internal organs associated with renal failure). The birds often appeared sick and lethargic, some showed prolonged severe neck drooping, before collapsing—sometimes from their perches.

Intensive testing failed to implicate infectious disease, pesticide poisoning, starvation, and other possible causes. Then, fi nally, in 2004 a team of scientists from the United States–based Peregrine Fund made a breakthrough. In the mid-1990s, livestock farmers in India began treating their cattle and water buffaloes with the nonsteroidal antiinfl ammatory drug (NSAID) diclofenac—a known kidney toxin in mammals. Vultures, it turned out, were highly sensitive to the drug, which they ingested while feeding at carcass dumps, the traditional method of livestock disposal in South Asia. Diclofenac later came into widespread use in Pakistan and Nepal. A subsequent concentrated research effort demonstrated that diclofenac use was on a sufficient scale to fully account for the declines. As a consequence, the Indian government announced its intention to ban veterinary use of the drug in March, but progress has been frustrated in part by the lack of a safe yet effective alternative. In a new study, toxicologist Gerry Swan and a team of colleagues from South Africa, Namibia, India, and the United Kingdom show that they have found that alternative.

The consequences of the vulture collapse have already reverberated across the subcontinent. Americans and Europeans once persecuted vultures, thinking they transmitted disease, but vultures help control brucellosis, anthrax, and other livestock diseases by consuming infected carcasses. In their absence, feral dog populations have exploded, likely increasing the risk of human attacks and the spread of rabies. If rats follow suit, bubonic plague and other rodent-transmitted diseases may also increase. And because these scavengers can't match vultures' efficiency as flesh-eaters, many carcasses—including human—lay rotting. In sky burials, Zoroastrian Parsis (and Tibetans, in a slightly different ritual) leave their dead on platforms for vultures to devour, to avoid defiling earth, water, or fi re with an unholy corpse. Where corpses once attracted 300 vultures—which could pick a body clean in half an hour—today so few remain that many Parsis must find new ways to send off their dead.

Tens of millions of Asian vultures including the oriental white backed vulture (Gyps bengalensis), above—have vanished from the Indian subcontinent since the early 1990s. (Image: Goran Ekstrom)

To find an anti-inflammatory that could treat livestock without killing vultures, Swan et al. collected records on NSAID use and effects on captive Gyps vultures from veterinarians at zoos and bird of prey collections around the world. They settled on meloxicam—the only NSAID that had been used extensively on vultures with no evidence of kidney damage—as a promising candidate. They first tested the drug's safety on a species that faces no risk of extinction, but suffers the same diclofenac toxicity as its endangered brethren: the African white-backed vulture (G. africanus).

The six-phase safety trial was designed to minimize experimental birds' suffering and risk of death. In the fi rst three phases, five vultures orally received meloxicam through a tube; three controls received water by the same method. After ensuring the health of all the birds by analyzing blood levels of uric acid and other markers, the authors increased the dose for the next phase. By the third phase, the dose just exceeded the estimated maximum likely exposure for wild vultures.

Though the difference in mortality risk between meloxicam and diclofenac was statistically significant—all vultures in Phases I–III survived meloxicam treatment while both G. africanus vultures treated with diclofenac in a previous study died—the small sample size can't preclude all risk. But strong corroborating evidence of safety comes from comparing blood samples of treated and control birds from both studies: diclofenac-treated vultures had a marked and dose-dependent elevation of uric acid levels compared to controls; meloxicamtreated vultures showed no such differences.

The authors next expanded the number of birds receiving the highest dose of meloxicam, treating 11 captive and 21 wild G. africanus vultures (plus captive and wild birds as controls). The wild birds were captured on a special expedition to Namibia, held in temporary facilities, and released after the experiment. All the vultures survived and showed no changes in blood uric acid levels. In the fifth phase, captive G. africanus vultures ate liver and muscle tissue from cattle treated with above-standard doses of meloxicam, to mimic the natural route of exposure and account for the possibility that treated cattle might produce toxic metabolites. Again, all survived without elevated uric acid levels or ill effects. As a final test, the authors treated ten endangered Asian vultures of two species with meloxicam; five received the maximum likely exposure. All ten were alive and healthy four months after the treatment. These results make a strong case that the recovery of the Asian vulture depends on immediate action to replace diclofenac with meloxicam. The authors hope other researchers use this approach to evaluate the safety of veterinary drugs on vultures and other scavengers—preferably before the drugs reach the market. For more information, go to Darwin Initiative, http://www.darwin.gov.uk/projects/details/ 10013.html; Vulture Rescue, http://www.vulturedeclines.org; and BirdLife International, http://www.birdlife.org/action/ science/species/ asia_vulture_crisis/index.html.

CLASSIC ILLUSION SHEDS NEW LIGHT ON THE NEURAL SITE OF TACTILE PERCEPTION

Imagine a race of people with tiny torsos, arms, and legs, but gigantic fingers, lips, and tongues. That's what we would look like if each of our body parts were sized proportionally to the brain area that processes their sensory inputs. Each of our body parts is represented in the somatosensory cortex on a map (called the somatotopic map), which expands the representation of the more sensitive body parts. Sensory inputs travel from touch receptors in the skin to neurons in the appropriate sectors of the somatosensory cortex.

As much as neuroscientists know about these neural projections for touch, surprisingly little is understood about the neural correlates of conscious tactile perception. In a new study, Felix Blankenburg, Jon Driver, and their colleagues turn to a classic somatosensory illusion—called the cutaneous rabbit—that is perfectly suited to decoupling real and illusory touch. In the illusion, a rapid succession of taps is delivered first to the wrist and then to the elbow, which creates the sensation of intervening taps hopping up the arm (hence the illusion's name), even when no physical stimulus is applied at intervening sites on the arm. Blankenburg et al. took advantage of this somatosensory illusion to investigate which brain regions play a role in illusory tactile perceptions.

Previous studies had implicated the somatosensory cortex in the rabbit illusion, but did not directly test this possibility. To do this, the authors used state-of-the-art functional magnetic resonance imaging technology (called 3T fMRI) to scan the brains of people experiencing the illusion. With the enhanced image quality and resolution of this scanner (deriving from the stronger magnetic field plus a specially customized imaging sequence), the authors show that the same brain sector is activated whether the tactile sensation is illusory or real.

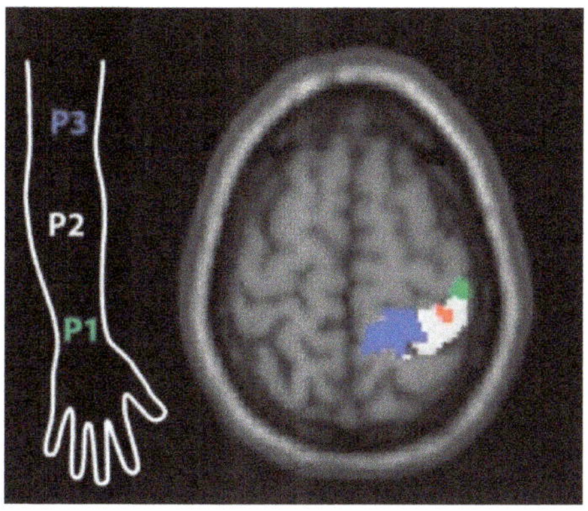

The cutaneous rabbit illusion engages the same sector of the brain that would respond if that body site (P2) had actually been touched.

To identify brain-related activity associated with real and illusory perceptions, the researchers taped three electrodes to the inner side of participants' left forearms, one just above the wrist, the others spaced equidistantly toward the elbow. Electrical stimulation could be applied to these points (P1, P2, and P3) while participants lay in the scanner. For the genuine rabbit experience, each point received three pulses in P1-P2- P3 succession. For the illusion, six pulses were applied to P1 (the last three substituting for pulses at the intervening P2 site), followed by three pulses to P3; this resulted in the same P1-P2-P3 tactile experience as when P2 was actually stimulated. For the control condition, P1 received three pulses, followed by three pulses to P3 and then to P1 (a sequence that does not produce the illusion of being stimulated at P2). After each sequence, participants indicated whether or not they felt any stimulation at P2. Blankenburg et al. looked for brain regions that showed similar increases in neuronal activity during the real and illusory rabbit conditions, compared with the controls, and also looked for any regions that differed between the two conditions. Only one area showed similar and heightened activity during the genuine and illusory rabbit sequences, compared with controls: the precentral gyrus, where the first cortical area to represent touch is located (called S1). The increased activity within S1 fell in the exact sector corresponding to the P2 position on the forearm (even though it was not actually stimulated during the illusion). The researchers confirmed this correspondence by separate somatotopic mapping of the skin sites' representation in each participant's brain when each site was stimulated (with no illusion produced).

Altogether, these results suggest that the illusion of being touched at a particular place on the body engages exactly the same sector of the brain that would respond if that body part had actually been touched. This connection between conscious perception and somatotopic cortical processing for illusory percepts may shed light on conditions such as phantom limb pain following amputation, and other perceptual illusions associated with disease. The authors point out that recent fMRI studies have shown somewhat analogous effects in the visual system, with the primary visual cortex involved in some conscious visual illusions. It's still unclear if this phenomenon will hold for all other perceptual systems as well, but future studies can now explore how the brain bridges the gap between actual stimulation and conscious experience.

Clues to Our Past: Mining the Human Genome for Signs of Recent Selection

Within the past 100,000 years, Homo sapiens left Africa in search of new opportunities, likely crossing paths with H. erectus in Asia and H. neanderthalensis in Europe. These early pioneers encountered unfamiliar climates, habitats, and food sources (not to mention alien human species). Then, after adjusting to a major climate change following the last ice age, they underwent a dramatic lifestyle switch, from hunting and gathering to agriculture—a change that brought crowded living conditions and new infections. All these radical changes likely precipitated significant genetic adaptations, with selection favoring genotypes most suited to the novel conditions. Indeed, recent studies have found evidence of strong selection on new gene variants reflecting adaptations to disease (conferring resistance to malaria) and dietary changes (lactose tolerance).

In a new study, Benjamin Voight, Sridhar Kudaravalli, Jonathan Pritchard, and their colleagues take a global approach to search for such signals across the genome and characterize the types of gene variants, or alleles, targeted by selection. Analyzing variants called single nucleotide polymorphisms (SNPs) in three populations—one sampled from Africa; one, a combined sample of Japanese and Chinese individuals from Asia; and one from Europe— they found widespread signals of recent positive selection in all three populations. These signals highlight the types of biological processes that have been targeted by selection during the evolution of modern humans.

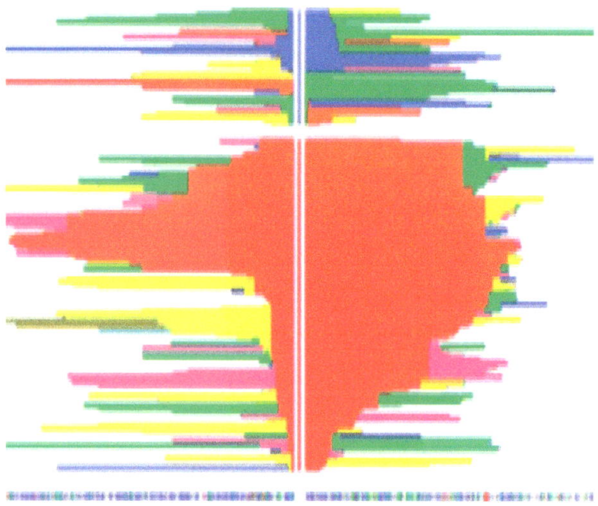

In this graphical representation, a new selected allele (red, center) is sweeping to fixation, replacing the ancestral allele (blue).

Voight et al. based their analysis on data recently collected by the HapMap Project, a repository of the SNPs that researchers used to identify genetic variants involved in human health and disease risk. Humans differ by just 0.1% at the nucleotide level, but these polymorphisms and the linked loci they are inherited with (called haplotypes) contain a wealth of clues to our evolutionary history.

The authors analyzed about 800,000 SNPs from 309 individuals, looking for genomic regions where strong selection has pushed new alleles to intermediate frequency. While these polymorphisms could become fixed (widespread in the population) or remain as SNPs, they likely reflect recent adaptations. To systematically detect this type of signal across the genome, they developed a new statistical scoring method (called integrated haplotype score, or iHS) that builds on an existing test for positive selection.

To detect selection, both tests rely on the relationship between an allele's frequency and the distance it maintains with homozygous, or invariable, loci that occur on either side of it along the chromosome (referred to as extended haplotype homozygosity). Old alleles with high population frequency will have only short-range correlations with adjacent loci; variants that quickly increase in frequency due to selection tend to sit within longer haplotypes with lower nucleotide diversity than predicted under models of neutral evolution. By scoring the strength of selection on an SNP, the new test identifies SNPs that stand out against the rest of the genome. Since strong selection is likely to sweep large blocks of adjacent loci along the path to fixation, the authors focused on SNP clusters with extreme iHS scores rather than on individual SNPs.

The authors saw extreme values clustered within distinct genomic regions, including a region containing the lactase gene and other known targets of strong selection. Additionally, when they looked for the strongest signs of selection, they found many regions classified as selection candidates by the HapMap Project in their results.

Most of the detected signals occurred recently, based on the lengths of the haplotypes. And since most of the selection signals are found in different populations, they likely occurred after the populations segregated, during the agricultural transition. Many of the selected genes play a role in fertility and reproduction (relating to sperm motility and egg fertilization) and morphology (relating to skin pigmentation and skeletal development). Others are involved in food metabolism, reflecting regional changes in diet.

With this genome-wide map of selected targets, future studies can explore what types of selective pressures these genes likely responded to and fill in the gaps in our recent evolutionary history. Although the nature and target of these selection signals are not yet clear, the authors argue that many might have medical implications and that such "selection maps" should become an integral part of annotating the human genome sequence. To facilitate this application, they have created fi les of SNPs that can be used to tag the strongest selection signals and a Web tool (http://pritch.bsd. uchicago. edu/data.html) for querying any HapMap SNP for evidence of selection. And by providing a more sensitive method of detecting selection, the authors' new test statistic should prove a valuable tool for uncovering the significant events in our history that lay hidden in the sequence of our DNA.

HOW FEMALES KEEP MALE HOVERFLY VISUAL NEURONS FROM DISTRACTION

Legendary athletes like running back Gale Sayers and center fi elder Willie Mays earned fame and fortune by pursuing a target traveling at high speeds against a moving, chaotic background. But most animals rely on this evolutionarily honed skill for more fundamental needs. Peregrine falcons snag prey while flying at speeds approaching 100 miles per hour. Males from several fly species chase females during dazzling courtship dances with impressive aerial maneuvers. How the brain accomplishes visual pursuit has been a longstanding question, which becomes even more fascinating when one considers how the tiny nervous system of the male fly executes the visual precision and flight control necessary during courtship chase behavior.

The visual system of the male fly lends an attractive model for investigating this question, partly because females don't chase males during courtship, allowing scientists to link the male-specific neurons to behaviorally relevant functions. In a new study, Karin Nordström, Paul Barnett, and David O'Carroll present surprising insight into the neural basis of visual pursuit in male hoverflies (Eristalis tenax), and identify a novel class of neurons associated with the task.

While hoverflies responded to pixel-sized dots moving against a cluttered, moving background, the authors measured neural activity in the hoverfly visual system (called the lobula complex). Even though the distracting background stimulus sometimes moved at the same velocity as the target, one class of neurons showed highly specialized tuning properties for the targets. The authors refer to these neurons as "small target motion detectors" (STMD). For a sense of the neurons' performance, imagine driving a golf ball high up into a partly cloudy sky and trying to keep your eye on it while it soars through the air. The hoverfly neurons might be able to track the golf ball even if the sky was moving at the same speed as the ball.

A novel class of visual neurons helps hoverflies stay focused on their target as they travel in complex visual environments. (Photo: Roman Prokhorov)

To get this intriguing data, Nordström et al. secured 74 wild-caught male hoverflies in front of a monitor and presented the flies with a range of target and background stimuli that varied in contrast, movement, direction, and speed. The authors recorded intracellular electrical activity from several hundred individual neurons. Of these, 206 neurons shared the striking sensitivity to small moving targets, and the authors classified them as STMD neurons. To further characterize the STMD neurons' response properties, the authors varied the speed of the background motion. They expected the moving background to suppress target response based on the "feedback hypothesis," which predicts that target neuron response would be dampened by the surrounding inhibitory circuitry. Unexpectedly, most STMD neurons maintained their sensitivity to the targets despite the distracting background motion. The authors also empirically confirmed the absence of inhibition in one class of STMD neurons.

To further understand the basis of the tuning properties of these STMD neurons, Nordström et al. explored the contribution of relative contrast between the target and background on the neurons' response profile. This class of STMD neurons, they found, displayed a striking sensitivity to contrast. The neurons even responded when the luminance of the background and target was similar, though the response was attenuated. This result suggests that even a slight difference in contrast is enough to evoke a response, and that when a female hoverfly moves against foliage with similar luminance properties as her body; the male's STMD neurons might be able to track her.

From the perspective of object recognition, the robust rejection of background clutter in favor of the target suggests that STMD neurons may be true feature detectors faithfully tuned to their preferred stimulus. Although predicted to exist, neurons that display such remarkable selectivity have not been particularly forthcoming. A compelling future direction of this work will be to uncover how the neural circuits of the insect compound eye and visual system mediate such exquisite feature detection. Eventually, the authors hope to link their physiological studies of visual neurons to the complex social behaviors of hoverflies.

DIATOMS RELY ON SOPHISTICATED SIGNALING SYSTEMS FOR POPULATION CONTROL

When you're a single-celled organism at the bottom of the food chain, it pays to be resourceful. Diatoms, highly successful photosynthetic plankton responsible for 40% of the net primary production in the oceans, undergo seasonal population explosions called phytoplankton blooms that attract billions of krill, copepods, and other grazing predators. As a defense, wounded diatoms release aldehyde compounds that minimize future diatom casualties by compromising the hatching success of grazers. But these diatom-derived aldehydes can also kill diatoms.

In a new study, Assaf Vardi, Chris Bowler, and their colleagues investigated the possibility that the contrasting effects of aldehydes reflect their role as "infochemicals" that trigger different responses attuned to changing conditions in the diatoms' habitat. The authors found that different concentrations of aldehydes produce different diatom responses.

At low doses, aldehydes induce resistance to the compound's toxic effects. High aldehyde concentrations, on the other hand, trigger cell death, which may lead to termination of a bloom. Thus, diatom-derived aldehydes regulate the population dynamics of both diatoms and their predators.

To investigate aldehyde effects on diatom cell fate and population dynamics, the authors studied how two cultured diatom species responded to a highly reactive aldehyde called decadienal. Thalassiosira weissflogii is a ubiquitous, cosmopolitan species; Phaeodactylum tricornutum is a standard model for understanding diatom biology.

Zooming into the bloom: cell-to-cell signaling in the sea. (Illustration: Nivi Alroy)

Reactive compounds like decadienal are likely to generate a variety of potentially harmful molecules called reactive oxygen species (ROS). But the authors detected increased levels of just one ROS, nitric oxide, an unstable compound involved in a wide range of physiological processes. Monitoring nitric oxide levels with a nitric oxide– sensitive fluorescent dye and time-lapse imaging revealed that both diatom species experienced similar bursts of nitric oxide production about five minutes after decadienal treatment.

Treated cells succumbed to decadienal in a time- and dosedependent manner, with significant increases in fatalities above a specific threshold. Below this threshold, cells survived, but underwent cell cycle arrest. To clarify nitric oxide's role in cell death, the authors stimulated nitric oxide production without using decadienal by using molecules called nitric oxide donors. Next, they pretreated cells with a nitric oxide inhibitor before exposing them to decadienal. The number of dying cells increased along with the levels of nitric oxide in the first experiments, and incidence of decadienal-related cell death decreased with the inhibitor. These results clearly implicate nitric oxide in cell death.

How does the cell stimulate nitric oxide production? Since plant and animal cells use calcium to perceive a wide range of environmental signals, Vardi et al. reasoned that diatoms might, too. Using P. tricornutum cells that express a calcium-sensitive bioluminescent protein, they tracked changes in intracellular calcium levels in response to aldehydes. As predicted, intracellular calcium levels spiked following decadienal exposure. None of the nitric oxide donors stimulated intracellular calcium production, suggesting that nitric oxide functions downstream of calcium. And, indeed, the nitric oxide synthase that produces nitric oxide was shown to be calciumactivated: after perceiving ambient aldehyde levels, the cell undergoes transient calcium increases that result in nitric oxide production.

Interestingly, nitric oxide production levels varied among the diatoms. Some cells showed rapid increases in nitric oxide production while their neighbors showed delayed responses, suggesting that the signal to produce nitric oxide was propagating through the diatom population. Healthy cells sensed the level of stressed cells in their midst by detecting the wounded cells' aldehydegenerated signal. Cells pretreated with a lower dose of decadienal before receiving a higher dose had far better survival and growth rates than cells treated with only a single high dose. These results suggest that lower decadienal doses may immunize cells, stimulating resistance to normally lethal aldehyde concentrations. This induced resistance may provide diatoms who escape grazing predators with a better chance of surviving the toxic aldehydes released by the dying diatoms.

Altogether, these results suggest that decadienal-like aldehydes not only affect the reproductive capacity of grazers but also act as info chemicals that monitor stress levels in diatom populations. During phytoplankton blooms, this stress surveillance system can induce resistance or death. The authors propose that this differential response, regulated by the sophisticated use of intracellular calcium and nitric oxide signals, may determine the fitness and succession of phytoplankton communities.

The finding that diatoms use chemical signaling for cell–cell communication provides new insights into the cellular mechanisms mediating biotic interactions at the population level and challenges traditional concepts of phytoplankton bloom dynamics. With thousands of different diatom species potentially producing a diverse array of aldehydes, these reactive compounds may prove to be even more versatile than shown here—and may be a key factor contributing to the ecological success of these organisms. Now researchers can begin to unravel the mechanisms that endow one group of molecules with the means to mediate such diverse responses.

INDEX

Q

R

S

T

V